Condensed Encyclopedia of Polymer Engineering Terms

Condensed Encyclopedia of Polymer Engineering Terms

Nicholas P. Cheremisinoff, Ph.D.

Boston Oxford Auckland Johannesburg Melbourne New Delhi

Library of Congress Cataloging-in-Publication Data
Cheremisinoff, Nicholas P.
  Condensed encyclopedia of polymer engineering terms / Nicholas P. Cheremisinoff.
    p. cm.
  Includes index.
  ISBN 0-7506-7210-2 (alk. paper)
    1. Polymers—Encyclopedias. 2. Plastics—Encyclopedias. 3.
  Polymerization—Encylopedias. I. Title.

TP1087 .C477 2001
668.9'03—dc21

                                                          00-068904

British Library Cataloguing-in-Publication Data
A catalogue record for this book is available from the British Library.

The publisher offers special discounts on bulk orders of this book.
For information, please contact:

Manager of Special Sales
Butterworth–Heinemann
225 Wildwood Avenue
Woburn, MA 01801-2041
Tel: 781-904-2500
Fax: 781-904-2620

For information on all Butterworth–Heinemann publications available, contact our World Wide Web home page at: http://www.bh.com

10 9 8 7 6 5 4 3 2 1

Printed in the United States of America

PREFACE

This volume has been prepared as a concise reference to assist practitioners who work with polymers. It is intended as a general resource for engineers, chemists, technologists, and technicians, as well as for students. Working definitions and explanations are given for nearly four hundred terms encountered in industry. The intent of the volume is to provide the user with sufficient background on key terminology. Most terms are cross-referenced to other terminology, so that the reader can obtain a general background and working definition for the subject entry covered. In a number of cases references are cited so that the user can obtain more in-depth information and engineering data and formulas.

The user is encouraged to heavily access the Internet when referring to this volume. The World Wide Web can be used as a primary source for detailed information and supplier specific information on the properties of many of the elastomers and plastics discussed in the volume. The author has compiled an extensive list of suppliers and trade names for commercial polymers, which is included in the volume.

The volume is organized in a straight alphabetical listing. As such, no table of contents is included. The user can either scan the subject entries by letter category, or review the subject index at the end of the volume to find terms that are included. Immediately before the first subject entry, the user will find an extensive table of abbreviations for polymer and plastic products. The user should refer to this table when looking up information on specific polymers. Polymer property data for neat (i.e., unvulcanized) plastics and elastomers are average values reported by several suppliers of these materials. Properties data should not be misconstrued as grade-specific, but rather general.

In addition to specific polymers, subject entries on equipment, processing and compounding techniques, synthesis, and quality control test methods are included. Rubber and plastic compounders, polymer processing engineers, product development specialists, as well as students will likely find this volume useful. The author has tried to use industry descriptions of terms for the most part, as opposed to oligarchical definitions. References to specific trade names and polymer producers is not an endorsement of products on the part of the author or the publisher. Likewise, the omission of specific product and company name references should not be viewed as a lack of endorsement. This volume is intended as a general aid to industry and academia and not an exhaustive review of the industrial press and scientific literature.

The volume is by no means intended to be definitive. Indeed, many important terms have been omitted and not all subject entries are treated in balance. It is planned that this volume will be added to over the years to enhance its service to industry and students.

A special thanks is extended to Butterworth-Heinemann for their fine production of this volume.

Nicholas P. Cheremisinoff, Ph.D.

ABOUT THE AUTHOR

Nicholas P. Cheremisinoff is a technical and business development consultant to both the private sector and government agencies, specializing in pollution prevention and environmental strategies. He has more than 20 years experience in the petrochemicals industry, which includes industry experience in polymer product development, processing troubleshooting, product failure analysis and redesign, and market development. Among his many clients are Sunykong Corp. of South Korea, Avtovaz Auto Making Corp. of Russia, OHIS Chemical Corp (PVC and Chlor-Alkali divisions) of Macedonia, the World Bank Organization, the U.S. Trade and Development Agency, the U.S. Department of Energy, Bateman Engineering of Israel, the U.S. Export Import Bank, and others. Dr. Cheremisinoff has contributed extensively to the industrial press, having authored, co-authored, or edited more than 150 technical books. He received his B.S., M.S. and Ph.D. degrees in chemical engineering from Clarkson College of Technology. The author can be contacted by email at ncheremisi@aol.com.

ABBREVIATIONS OF POLYMERS

ABR	Acrylate-butadiene rubber
ABS	Acrylonitrile-butadiene-styrene rubber
ACM	Acrylate rubber
AES	Acrylonitrile-ethylene-propylene-styrene quater-polymer
AMMA	Acrylonitrile-methyl methacrylate copolymer
ANM	Acrylonitrile-acrylate rubber
APP	Atactic polypropylene
ASA	Acrylonitrile-styrene-acrylate terpolymer
BUR	Brominated isobutene-isoprene, (butyl) rubber
BR	Cis-1,4-butadiene rubber (cis-1,4-polybutadiene)
BS	Butadiene-styrene copolymer (see also SB)
CA	Cellulose acetate
CAB	Cellulose acetate-butyrate
CAP	Cellulose acetate-propionate
CF	Cresol-formaldehyde resin
CHC	Epichlorohydrin-ethylene oxide rubber
CHR	Epichlorohydrin rubber (see also CO)
CMC	Carboxymethyl cellulose
CN	Cellulose nitrate (see also NC)
CNR	Carboxynitroso, rubber; (tetrafluoroethylene-tri-fluoronitrosomethane-unsat. monomer terpolymer)
CO	Poly[(chloromethyl)oxirane]; epichlorohydrin rubber (see also CHR)
CP	Cellulose propionate
CPE	Chlorinated polyethylene
CR	Chloroprene rubber
CS	Casein
CSM	Chlorosulfonated polyethylene
CTA	Cellulose triacetate
CTFE	Poly(chlorotnfluoroethylene); (see also PCTFE)
EAA	Ethylene-acrylic acid copolymer
EVA	Ethylene-vinyl acetate copolymer
EC	Ethyl cellulose
ECB	Ethylene copolymer blends with bitumen
ECTFE	Ethylene-chlorotrifluoroethylene copolymer

EEA	Ethylene-ethyl acrylate copolymer
EMA	Ethylene-methacrylic acid copolymer or ethylene-maleic anhydride copolymer
EP	Epoxy resin
E/P	Ethylene-propylene copolymer (see also EPM, EPR)
EPDM	Ethylene-propylene-nonconjugated diene terpolymer (see also EPT)
EPE	Epoxy resin ester
EPM	Ethylene-propylene rubber (see also E/P, EPR)
EPR	Ethylene-propylene rubber (see also E/P, EPM)
EPS	Expanded polystyrene; polystyrene foam (see also XPS)
EPT	Ethylene-propylene-diene terpolymer (see also EPDM)
ETFE	Ethylene-tetrafluoroethylene copolymer
EVA, E/VAC	Ethylene-vinyl acetate copolymer
EVE	Ethylene-vinyl ether copolymer
FE	Fluorine-containing elastomer
FEP	Tetrafluoroethylene-hexafluoropropylene rubber; see PFEP
FF	Furan-formaldehyde resins
FPM	Vinylidene fluoride-hexafluoropropylene rubber
FSI	Fluorinated silicone rubber
GR-I	Butyl rubber (former US acronym) (see also IIR, PIBI)
GR-N	Nitrile rubber (former US acronym) (see also NBR)
GR-S	Styrene-butadiene rubber (former US acronym; see PBS, SBR)
HDPE	High-density polyethylene
HEC	Hydroxyethylcellulose
HIPS	High-impact polystyrene
HMWPE	High molecular weight polyethylene
IIR	Isobutene-isoprene rubber; butyl rubber (see also GR-I, PIBI)
IPN	Interpenetrating polymer network
IR	Synthetic cis- I .4-oolyisoprene
LDPE	Low-density polyethylene
LLDPE	Linear low density polyethylene
MABS	Methyl methacrylate-acrylonitrile-butadiene-styrene
MBS	Methyl methacrylate-butadiene-styrene terpolymer
MC	Methyl cellulose
MDPE	Medium-density polyethylene (ca. 0.93-0.94 g/cm^3)
MF	Melamine-formaldehyde resin
MPF	Melamine-phenol-formaldehyde resin
NBR	Acrylonitrile-butadiene rubber; nitrile rubber; GR-I

NC	Nitrocellulose; cellulose nitrate (see also CN)
NCR	Acrylonitrile-chloroprene rubber
NIR	Acrylonitrile-isoprene rubber
NR	Natural rubber (cis- 1,4-polyisoprene)
OER	Oil extended rubber
OPR	Propylene oxide rubber
PA	Polyamide (e.g., PA 6,6 = polyamide 6,6 nylon 6,6 in U.S. literature)
PAA	Poly(acrylic acid)
PAI	Polyamide-imide
PAMS	Poly(alpha-methylstyrene)
PAN	Polyacrylonitrile (fiber)
PARA	Poly(arylamide)
PB	Poly(I-butene)
PBI	Poly(benzimidazoles)
PBMA	Poly(n-butyl methacrylate)
PBR	Butadiene-vinyl pyridine copolymer
PBS	Butadiene-styrene copolymer (see also GR-S, SBR)
PBT,PBTP	Poly(butylene terephthalate)
PC, PCO	Polycarbonate
PCD	Poly(carbodiimide)
PCTFE	Poly(chlorotrifluoroethylene)
PDAP	Poly(diallyl phthalate)
PDMS	Poly(dimethylsiloxane)
PE	Polyethylene
PEA	Polv(ethyl acrylate)
PEC	Chlorinated polyethylene (see also CPE)
PEEK	Poly(arylether ketone)
PEI	Poly(ether imide)
PEO,PEOX	Poly(ethylene oxide)
PEP	Ethylene-propylene polymer (see also E/P, EPR)
PEPA	Polyether-polyamide block copolymer
PES	Polyethersulfone
PET,PETP	Poly(ethylene terephthalate)
PF	Phenol-formaldehyde resin
PFA	Perfluoroalkoxy resins
PFEP	Tetrafluoroethylene-hexafluoropropylene copolymer; FEP
PI	Polyimide
PIB	Polyisobutylene,
PIBI	Isobutene-isoprene copolymer; butyl rubber; GR-I
PIBO	Poly(isobutylene oxide)
PIP	Synthetic poly-cis-1,4-polyisoprene; (also CPl, IR)
PIR	Polyisocyanurate
PMA	Poly(methyl acrylate)

PMI	Polymethacrylimide
PMMA	Poly(methyl methacrylate)
PMMI	Polypyromellitimide
PMP	Poly(4-methyl- I-pentene)
PO	Poly(propylene oxide); or polyolefins; or phenoxy resins
POM	Polyoxymethylene, polyformaldehyde
POP	Poly(phenylene oxide) (also PPO/PPE)
PP	Polypropylene
PPC	Chlorinated polypropylene
PPE	Poly(phenylene ether)
PPMS	Poly(para-methylstyrene)
PPO	Poly(phenylene oxide) (also PPO/PPE)
PPOX	Poly(propylene oxide)
PPS	Poly(phenylene sulfide)
PPSU	Poly(phenylene sulfone)
PPT	Poly(propylene terephthalate)
PS	Polystyrene
PSB	Styrene-butadiene rubber (see GR-S, SBR)
PSF, PSO	Polysulfone
PSU	Poly(phenylene sulfone)
PTFE	Poly(tetrafluoroethylene)
P3FE	Poly(trifluoroethylene)
PTMT	Poly(tetramethylene terephthalate) = poly(butylene terephthalate) (see also PBTP)
PUR	Polyurethane
PVA,PVAC	Poly(vinyl acetate)
PVAL	Poly(vinyl alcohol) (also PVOH)
PVB	Poly(vinyl butyral)
PVC	Poly(vinyl chloride)
PVCA	Vinyl chloride-vinyl acetate copolymer (also PVCAC)
PVCC	Chlorinated poly(vinyl chloride)
PVDC	Poly(vinylidene chloride)
PVDF	Poly(vinylidene fluoride)
PVF	Poly(vinyl fluoride)
PVFM	Poly(vinyl formal) (also PVFO)
PVI	Poly(vinyl isobutyl ether)
PVK	Poly(N-vinylcarbazole)
PVP	Poly(N-vinylpyrrolidone)
RF	Resorcinol-formaldehyde resin
SAN	Styrene-acrylonitnle copolymer

SB	Styrene-butadiene copolymer
SBR	Styrene-butadiene rubber (see also GR-S)
SCR	Styrene-chloroprene rubber
S-EPDM	Sulfonated ethylene-propylene-diene terpolymers
SHIPS	Superhigh-impact polystyrene
Sl	Silicone resins; poly(dimethylsiloxane)
SIR	Styrene-isoprene rubber
SMA	Styrene-maleic anhydride copolymer
SMS	Styrene-alpha-methylstyrene copolymer
TPE	Thermoplastic elastomer
TPR	1,5-trans-Poly(pentenamer)
TPU	Thermoplastic polyurethane
TPX	Poly(methyl pentene)
UF	Urea-formaldehyde resins
UHMW-PE	Ultrahigh molecular weight poly(ethylene) (also UHMPE) (molecular mass over 3.1×10^6 g/mol)
UP	Unsaturated polyester
VC/E	Vinyl chloride-ethylene copolymer
VC/E/VA	Vinyl chloride-ethylene-vinyl acetate copolymer
VC/MA	Vinyl chloride-methyl acrylate copolymer
VC/MMA	Vinyl chloride-methyl methacrylate copolymer
VC/OA	Vinyl chloride-octyl acrylate
VC/VAC	Vinyl chloride-vinyl acetate copolymer
VC/VDC	Vinyl chloride-vinylidene chloride
VF	Vulcan fiber
XLPE	Cross-linked polyethylene
XPS	Expandable or expanded polystyrene; (see also EPS)

A

ABS (ACRYLONITRILE BUTADIENE STYRENE)

ABS is a terpolymer of acrylonitrile, butadiene and styrene. Unusual compositions are about 50 % styrene with the balance divided between butadiene and acrylonitrile. There are a considerable number of variations possible leading to a large number of commercially available grades. Many blends with other materials such as polyvinyl chloride, polycarbonates and polysulfones have been developed and are among the most common class of plastics used in electroplated metal coatings for decorative hardware. The primary advantages of this plastic are good impact resistance with toughness and rigidity; metal coatings have excellent adhesion to ABS; the plastic can be formed into articles using conventional thermoplastic methods; and it is a lightweight plastic. The disadvantages of this material are poor solvent resistance; low dielectric strength; grades are only available with low elongation properties; and it has a low continuous service temperature.

As noted, ABS is produced by a combination of the three monomers: acrylonitrile, butadiene, and styrene. Each of the monomers impart different properties: hardness, chemical and heat resistance from acrylonitrile; processability, gloss and strength from styrene; and toughness and impact resistance from butadiane. Morphologically, ABS is an amorphous resin.

The polymerization of the three monomers produces a terpolymer, which has two phases: a continuous phase of styrene-acrylonitrile (SAN) and a dispersed phase of polybutadiene rubber. The properties of ABS are affected by the ratios of the monomers and the molecular structure of the two phases. This allows a good deal of flexibility in product design and consequently, as already noted, there are literally hundreds of commercial grades. Commercially available grades offer different characteristics such as medium to high impact, low to high surface gloss, and high heat distortion.

Applications

The most common applications for this plastic are as follows:

Automotive Hardware: instrument and interior trim panels, glove compartment doors, wheel covers, mirror housings, etc.

Appliance Cases: refrigerators, small appliance housings and power tools applications such as hair dryers, blenders, food processors, lawnmowers, etc.

Miscellaneous: pipe, telephone housings, typewriter housings, typewriter keys and various plated items.

Properties

Table 1 provides some of the general properties of this polymer based on average properties reported for different grades; note the use of the following legend: A = amorphous; Cr = crystalline; C = clear; E = excellent; G = good; P = poor - O = opaque; T = translucent; R = Rockwell; S = Shore.

Table 1. General Properties of ABS Polymer

Structure: A	Fabrication		
	Bonding	**Ultrasonic**	**Machining**
	E	E	G
Specific density: 1.05	**Deflection temperature (° F)**		
	@ 66 psi		**@ 264 psi**
	206		193
Water absorbtion rate (%): 0.27	**Utilization temperature (° F)** min: -40 max: 194		
Elongation (%): 20	**Melting point (° F)**: 221		
Tensile strength (psi): 4300	**Coefficient of expansion**: 0.000053		
Compression strength (psi): 9000	**Arc resistance**: 80		
Flexural strength (psi): 9200	**Dielectric strength (kV/mm)**: 16		
Flexural modulus (psi): 300000	**Transparency**: T		
Impact (Izod ft. lbs/in): 6.6	**Uv resistance**: P		
Hardness: R110	**Chemical resistance**		
	Acids	**Alkalis**	**Solvents**
	G	E	P

ACCELERATORS

Accelerators and activators are chemical additives used extensively in rubber and plastic vulcanization. First, it should be noted that the use of rubber or elastomers as a material for finished products dates back to the 1830s. Unvulcanized natural rubber was originally used, but this suffered from problems of softening during the summer, hardening during the winter, and noxious odors after use during the summer. Prompted by this, Charles Goodyear in the U.S. and Thomas Hancock in England mixed sulfur with natural rubber and then heated the sample. This improved the strength of the rubber and limited the hardening and softening of the product with changes in temperatures. Sulfur alone was used as the vulcanizing agent up to the discovery of organic accelerators in the early part of the 20th century. It was quickly realized that the use of accelerators gave improved properties and significantly reduced the required cure times.

Table 1.

Common Accelerators Used for Accelerated Sulfur Vulcanization

Compound	Abbreviation	Structure
Benzothiazoles		
2-mercaptobenzothiazole	MBT	
2,2′-dithiobisbenzothiazole	MBTS	
Benzothiazolesulfenamides		
N-cyclohexylbenzothiazole-2-sulfenamide	CBS	
N-t-butylbenzothiazole-2-sulfenamide	TBBS	
2-morpholinothiobenzothiazole	MBS	
N-dicyclohexylbenzothiazole-2-sulfenamide	DCBS	
Dithiocarbamates		
tetramethylthiuram monosulfide	TMTM	
tetramethylthiuram disulfide	TMTD	
zinc diethyldithiocarbamate	ZDEC	
Amines		
diphenylguanidine	DPG	
di-o-tolylguanidine	DOTG	

The first accelerators were amine-based compounds, with other classes of accelerators following quickly. The commercially used accelerators can generally be categorized as belonging to one of three main categories: amine-based, alkyl thiocarbarnates, and benzothiazole-based (MBT and derivatives). Examples of compounds for accelerated sulfur vulcanization are summarized in Table 1.

Thiocarbamate accelerators are well known for their superior accelerating capabilities, but suffer from very short (if any) induction periods, which limits processing safety. Another way of stating this is that thiocarbamate-based systems are sensitive to premature vulcanization. On the other hand, sulfenamides, amide derivatives of mercaptobenzothiazole (MBT) enjoy good processing safety due to their lengthy induction period, but do not have the accelerating power of thiocarbarnates. Several other accelerators consist of mixed thiocarbarnates and benzothiazole moieties, and, thus, incorporate the higher accelerating power of the thiocarbamates with the lengthy induction period of the sulfenamides.

Another option in vulcanization formulations is the use of binary accelerator systems. Binary accelerator formulations involve the use of two different accelerators in the system, often leading to improved properties. In many cases, a synergistic behavior is observed, leading to better properties in the combined system than would be expected by a rule of mixtures for the individual accelerator systems. Examples of commercially available accelerators and activators, along with their trade names are listed in Table 2.

Table 2.

Common Accelerators and Activators along with Trade Names

Chemical Name/Synonyms	Chemical Family	Trade Names
Calcium oxide, lime, quicklime	Alkaline earth	Desical P
Magnesium aluminum hydroxy carbonate	Alkaline earth	Hysafe 50
Talc, hydrous magnesium silicate	Silicates	Mistron Compound
Azodicarbonamide-azobisformamide	Azo amide	Kempore
Diethylene glycol	Dihydroxyethyl ether	Cumate
Zinc dietheldithio-carbamate	Dithiocarbamate	Ethyl Zimate

Table 2 continued

Chemical Name/Synonyms	Chemical Family	Trade Names
Nickel dibutyldithio-carbamate	Dithiocarbamate	NBC
Bismuth dimethyldithio-carbamate	Dithiocarbamate	Bismate
Zinc dibutyldithio-carbamate	Dithiocarbamate	Butyl Zimate
Zinc dimethyldithio-carbamate	Dithiocarbamate	Methyl Zimate
Selenium dimethyldithio-carbamate	Dithiocarbamate	Methyl Selenac
Lead dimethyldithio-carbamate	Dithiocarbamate	Methyl Ledate
Tellurium diethyldithio-carbamate	Dithiocarbamate	Ethyl Tellurac
Cadmium diethyldithio-carbamate	Dithiocarbamate	Ethyl Cadmate
Zinc salt of stearic acid	Metal soap	U.S.P. S-1271, DLG-10, DLG-20, Zinc Stearate
Di-ortho-tolylguanidine	Guanidine	DOTG Accelerator
Copper mercaptobenzo-thiazole	Thiazole	Cupsac Accelerator
Thiram	Thiuram	MOTS No. 1 Accelerator
Tetramethylthiuram	Thiuram	Monex
2-Benzothiazolethiol, zinc salt	Organic acid salt	ZMBT

Table 2 continued

Chemical Name/Synonyms	Chemical Family	Trade Names
Bis(Dimethylthiocarba myl) disulfide	Organic sulfide	TETD
1,3-Diphenyl-2-thiourea	Rubber accelerator	A-1 Thiocarbanilide
Carbamide	Activator	Arazate
1,3-Diphenyl guanidine	Guanidine	DPG

Refer to *Activators* and *Sulfur Vulcanization*. (Source: Elastomer Technology Handbook, N. P. Cheremisinoff - editor, CRC Press, Boca Raton, FL, 1991).

ACRYLIC RESIN

The term refers to any of a group of thermoplastic resins formed from the polymerization of acrylic acid, methacrylic acid, esters of these acids, or acrylonitrile. It is used in the manufacture of lightweight, weather-resistant, exceptionally clear plastics. Acrylic fibers are produced from acrylonitrile, which is derived from elements taken from natural gas, air, water and petroleum. The acrylonitrile is usually combined with small amounts of other chemicals to improve the ability of the fiber to absorb dyes. Some acrylic fibers are dry spun and others are wet spun. Acrylic fibers are used in staple or tow form. Acrylic fibers are made with many modifications to give special properties best suited for different types of constructions, to blend harmoniously with various other fibers or to meet requirements of an end-use product. Acrylic fibers are unique among manufactured fibers because they have an uneven surface, even when extruded from a round hole spinneret. Man-made fibers, such as acrylic, begin with a polymer. The polymer, with an addition of a solvent, is forced through a spinnerete from which come the individual filament fibers. The uncrimped tow is then crimped and put through an annealer which via extreme high pressure and temperature enables the fiber to shrink, stabilize and become more receptive to dyeing later in the yarn or fabric process. This creates tow silver, which is more compact. In the next stage, the tow is cut into staple. At this point, the acrylic fiber can be dyed during the production process rather than later in the yarn of fabric process. This is called "producer dyed" acrylic. The staple is then shipped to the mills who create the carded silver and then yarn. The first U.S. commercial production acrylic resin was in 1950 by E.I. du Pont de Nemours & Company, Inc. As manufactured fiber, the fiber-forming substance is any long chain synthetic polymer composed of at least 85% by weight of acrylonitrile units. Refer to *Acrylic (Polyacrylate)*.

ACRYLIC (POLYACRYLATE)

These are special purpose polymers developed for high temperature use. General properties include moderate mechanical properties; poor cold resistance; excellent resistance to oils, including oils with sulphur additives such as those commonly used in cars; good ozone, oxidation, and weathering resistance. Approximate working temperature range -20° C to +120° C. Typical uses include high temperature gaskets and seals.

ACTIVATORS

Other compounds commonly used in vulcanization, in addition to sulfur and accelerators, are zinc oxide and saturated fatty acids such as stearic or lauric acid. These materials are termed activators (as opposed to accelerators). Zinc oxide serves as an activator, and fatty acids are used to solubilize the zinc into the system. Rubber formulations can also include fillers such as fumed silica and carbon black, and compounds such as stabilizers and antioxidants. Further complicating the situation is the engineering practice of blending various elastomers to obtain the desired properties.

ADDITIVE

The term refers to any chemical substance added to a plastic or elastomer compound to impart or improve certain end-use properties. Common additives are: foam agents, colorants, curatives like sulfur or peroxide, accelerators, activators, tackiness agent, stabilizer packages, carbon black, clay. Refer to *Carbon Black, Antizonant, and Antioxidants*.

ADHESIVES

There are three general classes of adhesives. These are hot melt adhesives (HMA), pressure sensitive adhesives (PSA), and hot melt pressure sensitive adhesives (HMPSA). A HMA is a 100 % solids, thermoplastic composition which is compounded and applied molten at elevated temperatures and whose strength is obtained solely by the removal of heat. A PSA is a viscoelastic material which is permanently tacky at room temperature, such that a low force contact to a surface will cause the material to adhere instantaneously. A HMPSA is a viscoelastic material which is permanently tacky at room temperature and which is applied as a 100 % solids, thermoplastic composition in the molten state. Solvent based systems are the oldest technology. Water based systems have the advantage that they avoid flammability and generally do not give off obnoxious odors. Basic adhesive components within each of the three classes are:

1. HMA - thermoplastic polymers, denoted TP (e.g., EVA, APP, PE), tackifier resin, wax, and stabilizers.

2. PSA (solvent-based or water-based) - elastomer (e.g., NR, SBR/CSBR, Butyl/Polyisobutylene, acrylic), tackifier resin, plasticizers, fillers, stabilizers.

3. HMPSA - TP elastomers such as styrene block copolymers, tackifier resin, plasticizers, fillers, stabilizers.

Typical constructions of a PSA system are illustrated in Figure 1. For solvent based adhesive blending, the equipment that is generally used for processing is high speed, rotating saw-blade type mixers or a high speed, single turbine blade mixer. For higher viscosity compounds, the equipment includes double-arm planetary mixers and sigma blade kneaders. The general blending techniques for a solution mixing process involves four basic steps that are performed in the standard mixing equipment. The steps involved are outlined below.

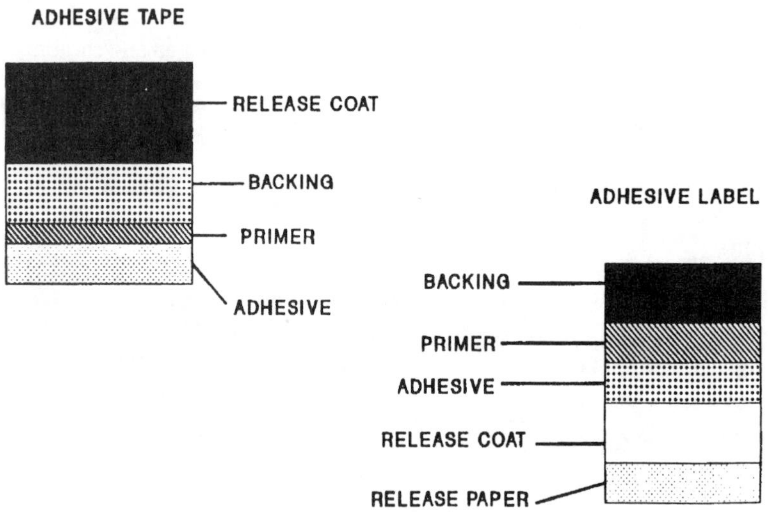

Figure 1. Pressure sensitive adhesive system constructions.

1. Adhesives can be blended by charging all the ingredients to the mixer at once (i.e., the elastomer, resin, etc.)

2. The desired amount of solvent is added next.

3. The solution is then blended until homogeneous. Natural rubber is usually premasticated. Solvation is facilitated by increasing the surface area of the elastomer (i.e., chopping or shredding the polymer). Some polymers such as Butyl or EP (ethylene propylene) rubber may be added to the mixer as a solvent solution. Where needed the blend can be sequentially processed on a paint mill or ball mill to attain sufficient dispersion/"grind" of the filler.

4. Alternatively, elastomers that are difficult to solvate (e.g., the higher

molecular weight butyls and polyisobutylenes) may be added in chopped form gradually to the agitated solvent. Other ingredients are then added to the elastomer solution.

For an HMA, the blending techniques for an ethylene vinyl acetate based system are usually as follows:

1. Wax and stabilizer are mixed until melted.

2. EVA is added in small amounts, allowing time for homogenization.

3. Resin is added and the adhesive is mixed until homogeneous.

For styrene block copolymer systems:

1. Mixing is done in low shear mixers. Resins, plasticizers and stabilizers are added first and melted, followed by adding SBC in incremental amounts.

2. Filler materials are added last.

3. An inert gas blanket is applied in order to prevent degradation during blending.

For water based adhesive blending, low shear/low speed mixers are used to minimize shearing, which can coagulate the emulsion. This tends to minimize foam formation. The blending procedure for a simple mixing scheme is as follows:

1. Polymer and resin emulsions are typically combined and then other ingredients are added.

2. The order of addition of ingredients can be important to minimizing shock and resultant stability problems.

3. Simple mixing is done until the solution is homogenous.

4. The adhesive is monitored for a period of time after mixing to check for instability (i.e., separation and flocculation).

The coating requirements that an adhesive system must fulfill include:

1. Stability and Appropriate Rheology

2. Release of Vehicle

3. Safety

4. Predictability

5. Clean Up Ease

6. Continuity

HMA applications methods include the use of the following: roll coaters; calendering; extrusion coating; slot die coating; multibead applications; air extrusion; spray; foam applications. Examples of slot die coating and multibead coating equipment for HMAs are illustrated in Figures 2 and 3.

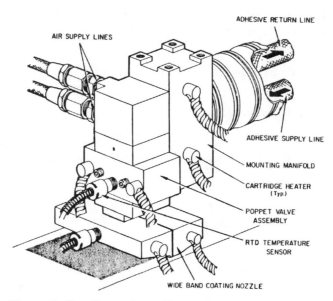

Figure 2. Slot die coating of hot melt adhesives.

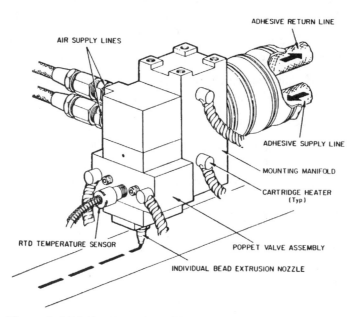

Figure 3. Multibead coating of hot melt adhesives.

AIR COOLED HEAT EXCHANGER

Air cooled heat exchangers are used to transfer heat from a process fluid to ambient air. The process fluid is contained within heat conducting tubes.

Atmospheric air, which serves as the coolant, is caused to flow perpendicularly across the tubes in order to remove heat. In a typical air cooled heat exchanger, the ambient air is either forced or induced by a fan or fans to flow vertically across a horizontal section of tubes. For condensing applications, the bundle may be sloped or vertical. Similarly, for relatively small air cooled heat exchangers, the air flow may be horizontal across vertical tube bundles. In order to improve the heat transfer characteristics of air cooled exchangers, the tubes are provided with external fins. These fins can result in a substantial increase in heat transfer surface. Parameters such as bundle length, width and number of tube rows vary with the particular application as well as the particular finned tube design. The choice of whether air cooled exchangers should be used is essentially a question of economics including first costs or capital costs, operating and maintenance expenses, space requirements, and environmental considerations; and involves a decision weighing the advantages and disadvantages of cooling with air. The advantages of cooling with air may be seen by comparing air cooling with the alternative of cooling with water. These issues should be examined on a case by case basis to assess whether air cooled systems are economical and practical for the intended application. The major components of air cooled heat exchangers include the finned tube, the tube bundle, the fan and drive assembly, an air plenum chamber, and the overall structural assembly.

ALCOHOL

Alcohols refer to those compounds that are formed when a hydroxyl group (-OH) is substituted for a hydrogen. They have the general formula R-OH. The hydroxyl radical looks exactly like the hydroxide ion, but it is not an ion. Where the hydroxide ion fits the definition of a complex ion - a chemical combination of two or more atoms that have collectively lost or (as in this case) gained one or more electrons - the hydroxide radical is a molecular fragment produced by separating the -OH from another compound, and it has no electrical charge. It does have an unpaired electron waiting to pair up with another particle having its own unpaired electron. The alcohols, as a group, are flammable liquids in the short-chain range, combustible liquids as the chain grows longer, and finally solids that will burn if exposed to high temperatures, as the chain continues to become longer. As in the case of the halogenated hydrocarbons, the most useful alcohol compounds are of the short-carbon-chain variety. Just as in the case of the halogenated hydrocarbons, the simplest alcohol is made from the simplest hydrocarbon, methane. Its name is methyl alcohol and its molecular formula is CH_3OH.

Nature produces a tremendous amount of methyl alcohol, simply by the fermentation of wood, grass, and other materials made to some degree of cellulose. In fact, methyl alcohol is known as wood alcohol, along with names such as wood spirits and methanol (its proper name; the proper names of all alcohols end in -ol). Methyl alcohol is a colorless liquid with a characteristic alcohol odor. It has a flash point of 54° F, and is highly toxic. It has too many commercial uses to list here, but among them are as a denaturant for ethyl alcohol (the addition of the toxic chemical methyl alcohol to ethyl alcohol in order to form denatured alcohol), antifreezes, gasoline additives, and solvents. No further substitution of hydroxyl radicals is performed on methyl alcohol.

The most widely known alcohol is ethyl alcohol, simply because it is the alcohol in alcoholic drinks. It is also known as grain alcohol, or by its proper name, ethanol. Ethyl alcohol is a colorless, volatile liquid with a characteristic odor and a pungent taste. It has a flash point of 55° F, is classified as a depressant drug, and is toxic when ingested in large quantities. Its molecular formula is C_2H_5OH. In addition to its presence in alcoholic beverages, ethyl alcohol has many industrial and medical uses, such as a solvent in many manufacturing processes, as antifreeze, antiseptics, and cosmetics. The substitution of one hydroxyl radical for a hydrogen atom in propane produces propyl alcohol, or propanol, which has several uses. Its molecular formula is C_3H_7OH. Propyl alcohol has a flash point of 77° F and, like all the alcohols, burns with a pale blue flame. More commonly known is the isomer of propyl alcohol, isopropyl alcohol. Since it is an isomer, it has the same molecular formula as propyl alcohol but a different structural formula. Isopropyl alcohol has a flash point of 53° F. Its ignition temperature is 850° F, while propyl alcohol's ignition temperature is 700° F, another effect of the different structure. Isopropyl alcohol, or 2-propanol (its proper name) is used in the manufacture of many different chemicals, but is best known as rubbing alcohol.

The above-mentioned alcohols are by far the most common. Butyl alcohol is not as commonly used as the first four in the series, but it is used. Secondary butyl alcohol and tertiary butyl alcohol, so named because of the type of carbon atom in the molecule to which the hydroxyl radical is attached, must be mentioned because they are flammable liquids, while isobutyl alcohol has a flash point of 100° F. All of the alcohols of the first four carbon atoms in the alkanes, therefore, are extremely hazardous because of their combustion characteristics.

Whenever a hydrocarbon backbone has two hydroxyl radicals attached to it, it becomes a special type of alcohol known as a glycol. The simplest of the glycols, and the most important, is ethylene glycol, whose molecular formula is $C_2H_4(OH)_2$. The molecular formula can also be written CH_2OHCH_2OH and may be printed as such on some labels. Ethylene glycol is a colorless, thick liquid with a sweet taste, is toxic by ingestion and by inhalation, and among its many uses is a permanent antifreeze and coolant for automobiles. It is a combustible liquid with a flash point of 240° F.

The only other glycol that is fairly common is propylene glycol which has a molecular formula of $C_3H_6(OH)_2$. It is a combustible liquid with a flash point of 210° F, and its major use is in organic synthesis, particularly of polyester resins and cellophane.

The last group of substituted hydrocarbons produced by adding hydroxyl radicals to the hydrocarbon backbone are the compounds made when three hydroxyl radicals are substituted; these are known as glycerols. The name of the simplest of this type of compound is just glycerol. Its molecular formula is $C_3H_5(OH)_3$. Glycerol is a colorless, thick, syrupy liquid with a sweet taste, and has a flash point of 320° F, and is used to make such diverse products as candy and explosives, plus many more. Other glycerols are made, but most of them are not classified as hazardous materials. (Source: N. P. Cheremisinoff, *Handbook of Industrial Toxicology and Hazardous Materials*, Marcel Dekker, Inc., New York, 1999).

ALKALI

A hydroxide or carbonate of an alkali metal (e.g., lithium, sodium, potassium, etc), the aqueous solution of which is characteristically basic in chemical reactions. The term may be extended to apply to hydroxides and carbonates of barium, calcium, magnesium, and the ammonium ion. The term alkali should be viewed in relation to the terms corrosive, bases and acids. The EPA defines corrosivity in terms of pH (i.e., wastes with pH <2 or ≥ 2.5) or in terms of ability to corrode steel (SAE 20) at a rate of >6.35 mm (0.250 in.) per year at a temperature of 55° C (13° F). This discussion will address corrosivity as it applies to acids and caustics (i.e., alkali materials). Acids are compounds that yield H^+ ions (actually H_3O^+ ions) when dissolved in water. Common industrial acids include acetic, nitric, hydrochloric, and sulfuric acids. The terms *concentrated* and *dilute* refer to the concentrations in solution. Mixing a concentrated acid with enough water will produce a dilute acid. For example, a bottle of concentrated HCl direct from the manufacturer is approximately 12N in HCl, while a solution of HCl used in a titration may be only 0.5N. The latter is a dilute acid solution.

Strong and *weak* acids are classified by how completely they ionize in solution. For example, HCl is classified as a strong acid because it is completely ionized to H^+ and Cl^- ions. Acetic acid is classified as a weak acid because it does not totally ionize in solution. Weak acids such as acetic acid have higher pK_as. The pK_a for acetic acid is 4.75. The negative antilog of this value (1.76×10^{-5}) can be used to calculate the concentrations at equilibrium of the acetate and hydrogen ions. Strong acids include perchloric, hydrochloric, sulfuric, nitric, and hydriodic acids. Examples of weak acids include boric, hydrocyanic, carbonic, and acetic acids. Thus, the terminology "strong versus weak acid" may bear little relationship to the nature or extent of potential hazard, while the terms "concentrated versus dilute" most often do.

The acidic nature of a given solution is characterized by its pH, where pH is the negative logarithm of the molar H^+ concentration ($-\log (H^+)$). A solution with pH <7 is acid, a solution with pH 7 is neutral, and a solution with pH >7 is basic. For example, the pH of lemon juice is 2, while the pH of lye is about 14.

Acids may be inorganic, such as H_2SO_4, and are then known as mineral acids, or they may be organic, like acetic acid. Mineral acids may be weak or strong, but organic acids tend to be uniformly weak. Table 1 gives a list of commonly occurring acids along with their relative strengths. It should be noted that salts of several metals (e.g., Al^{3+}, Fe^{3+}, and Zn^{4+}) dissolve in water to produce acid solutions. Acids include a variety of compounds, many of which have other significant properties that contribute to their reactivity. Typical reactions of acids are: neutralization of bases (strong and weak) and oxidation of substances. Characteristics of common acids are presented in Table 2. Examples of neutralization of bases are the following reactions:

$H^+ + OH^- \rightarrow H_2O$

$HCl + NaOH \rightarrow H_2O + NaCl$

$CaCO_3 + 2HCl \rightarrow CaCl_2 + H2O + CO_2$

Examples of oxidation reactions are as follows:

$Zn^\circ + 2HCl \rightarrow Zn^{2+} + 2Cl^- + H^2\uparrow$

$2NaI + 2H_2SO_4 \rightarrow I_2 + SO_2 + 2H_2O + Na_2SO_4$

A base is any material that produces hydroxide ions when it is dissolved in water. The words alkaline, basic, and caustic are often used synonymously. Common bases include sodium hydroxide (lye), potassium hydroxide (potash lye), and calcium hydroxide (slaked lime). The concepts of strong versus weak bases, and concentrated versus dilute bases are exactly analogous to those for acids. Strong bases such as sodium hydroxide dissociate completely while weak bases such as the amines dissociate only partially. As with acids, bases can be either inorganic or organic. Typical reactions of bases include neutralization of acids, reaction with metals, and reaction with salts.

Table 1. Relative Strengths of Acids in Water

Perchloric acid	$HClO_4$	↑
Sulfuric acid	H_2SO_4	↑
Hydrochloric acid	HCl	↑
Nitric acid	HNO_3	↑
Phosphoric acid	H_3PO_4	Increasing
Hydrofluoric acid	HF	Acid
Acetic acid	CH_3COOH	Strength
Carbonic acid	H_2CO_3	↑
Hydrocyanic acid	HCN	↑
Boric acid	H_3BO_3	↑

Table 2. Properties of Some Common Acids and Bases

Acids—Sulfuric, Nitric, Hydrochloric, Acetic

a. *These acids are highly soluble in water.*

b. *Concentrated solutions are highly corrosive and will attack materials and tissue.*

c. *If spilled on skin, flush with lots of water.*

d. *Sulfuric and nitric acids are strong oxidizers and should not be stored or mixed with any organic material.*

e. *Sulfuric, nitric, and hydrochloric acids will attack metals upon contact and generate hydrogen gas which is explosive.*

f. *Acetic acid (glacial) is extremely flammable. Its vapors form explosive mixtures in the air. It is dangerous when stored with any oxidizing material, such as nitric and sulfuric acids, peroxides, sodium hypochlorite, etc.*

g. *Breathing the concentrated vapors of any of these acids can be extremely harmful. Wear appropriate equipment.*

h. *When mixing with water, always add acids to water, never water to acids.*

Bases (Caustics)—Sodium Hydroxide, Ammonium Hydroxide, Calcium Hydroxide (Slaked Lime), Calcium Oxide (Quick Lime)

a. *These bases are highly soluble in water.*

b. *Concentrated solutions are highly corrosive. They are worse than most acids because they penetrate the skin (Saponification reactions).*

c. *If spilled on skin, flush immediately with lots of water.*

d. *When mixed with water, they generate a significant amount of heat-- especially sodium hydroxide and calcium oxide.*

e. *Unless unavoidable, do not store or mix concentrated acids and bases, as this gives off much heat--dilute, then mix.*

f. *Do not store or mix ammonium hydroxide with other strong bases. It can release ammonia gas which is extremely toxic.*

g. *Do not store or mix ammonium hydroxide with chlorine compounds (i.e., sodium hypochlorite). It can release chlorine gas which is extremely toxic.*

An example of a reaction with a metal is:

$$2Al + 6NaOH \rightarrow 2Na_3AlO_3 + 3H_2 \uparrow$$

(reaction goes slowly).

An example of a reaction involving salt is:

$$Pb(NO_3)_2 + 2NaOH \rightarrow Pb(OH)_2 + 2NaNO_3$$

(Source: N. P. Cheremisinoff, *Handbook of Industrial Toxicology and Hazardous Materials*, Marcel Dekker, Inc., New York, 1999).

ALKYL

Any of a series of monovalent radicals having the general formula C_nH_{2n+1}, derived from aliphatic hydrocarbons by the removal of a hydrogen atom: for example, CH_3. (methyl radical, from methane). Refer to *Halogenated Hydrocarbons*.

AMBIENT

Pertaining to any localized conditions, such as temperature, humidity, or atmospheric pressure, that may affect the operating characteristics of equipment or the performance of a resin; e.g., a high ambient temperature may cause solvents or plastisols to evaporate during compounding operations.

AMERICAN NATIONAL STANDARDS INSTITUTE (ANSI)

ANSI is a federation of standards provided from commerce and industry, professional, trade, consumer, and labor organizations and government. ANSI helps to perform the following:

1. identifies the needs for standards and sets priorities for their completion;

2. assigns development work to competent and willing organizations;

3. sees to it that public interests, including those of the consumer, are protected and represented;

4. supplies standards writing organizations with effective procedures and management services to ensure efficient use of their manpower and financial resources and timely development of standards; and

5. follows up to assure that needed standards are developed on time.

Another role is to approve standards as American National Standards when they meet consensus requirements. It approves a standard only when it has verified evidence presented by a standards developer that those affected by the standard have reached substantial agreement on its provisions.

ANSI's other major roles are to represent U.S. interest in nongovernmental international standards work, to make national and international standards available, and to inform the public.

AMERICAN SOCIETY FOR TESTING AND MATERIALS

ASTM is a scientific and technical organization formed for "the development of standards on characteristics and performance of materials, products, systems and services and the promotion of related knowledge." ASTM is the world's largest source of voluntary consensus standards.

The society operates through more than 135 main technical committees with 1550 subcommittees. These committees function in prescribed fields under regulations that ensure balanced representation among producers, users, and general interest participants. The society currently has 28,000 active members, of whom approximately 17,000 serve as technical experts on committees, representing 76,200 units of participation.

Membership in the society is open to all concerned with the fields in which ASTM is active. An ASTM standard represents a common viewpoint of those parties concerned with its provisions, namely, producers, users, and general interest groups. It is intended to aid industry, government agencies, and the general public.

The use of an ASTM standard is voluntary. It is recognized that for certain work, ASTM specifications may be either more or less restrictive than needed. The existence of an ASTM standard does not preclude anyone from manufacturing, marketing, or purchasing products or using products, processes, or procedures not conforming to the standard.

Because ASTM standards are subject to periodic reviews and revision, it is recommended that all serious users obtain the latest revision. A new edition of the Book of Standards is issued annually. On the average about 30% of each part is new or revised. Table 1 provides a list of pertinent ASTM test methods for plastics. Some of these test methods are described under various subject entries throughout the volume. Every polymer testing and quality assurance laboratory should invest in the ASTM reference volumes as a matter of due diligence in meeting ISO 9000.

Table 1.

List of Pertinent ASTM Testing Methods for Polymeric Materials

Property	Units of Measure	ASTM Test Ref.
Permanence		
Mold Shrinkage	in/in	D-955
Water Absorption	%	D-570
Mechanical Properties		
Impact Strength, Izod	ft-lb/in	D-256
Tensile Strength	psi	D-638
Tensile Elongation	%	D-638
Tensile Modulus	$psi \times 10^6$	D-638
Flexural Strength	psi	D-790

Table 1 continued

Property	Units of Measure	ASTM Test Ref.
Flexural Modulus	psi $\times 10^6$	D-790
Compressive Strength	psi	D-695
Harness, Rockwell	Rockwell R	D-785
Electrical Properties		
Dielectric Strength	V/mil	D-149
Dielectric Constant	-	D-150
Dissipation Factor	-	D-150
Arc Resistance	sec	D-495
Volume Resistivity	Ω-cm	D-257
Thermal Properties		
Deflection Temperature	° F	D-648
Flammability	-	UL-94
Linear Thermal Expansion	in/in/°F $\times 10^{-5}$	D-696
Thermal Conductivity	Btu/Hr/ft^2/°F/in	C-177
Wear		
Wear Factor	in^3/min/ft/Lb/Hr	D-3702
Coefficient of Friction		D-3702

AMORPHOUS

Polymers exhibit two types of morphology in the solid state: amorphous and semi-crystalline. In an amorphous polymer the molecules are oriented randomly and are intertwined, much like cooked spaghetti, and the polymer has a glasslike, transparent appearance. In semi-crystalline polymers, the molecules pack together in ordered regions called crystallites. Linear polymers, having a very regular structure, are more likely to be semi-crystalline. Semi-crystalline polymers tend to form very tough plastics because of the strong intermolecular forces associated with close chain packing in the crystallites. Also, because the crystallites scatter light, they are more opaque.

Crystallinity may be induced by stretching polymers in order to align the molecules--a process called drawing. In the plastics industry, polymer films are commonly drawn to increase the film strength. At low temperatures the molecules of an amorphous or semi-crystalline polymer vibrate at low energy, so that they are essentially frozen into a solid condition known as the glassy state. As the polymer is heated, however, the molecules vibrate more energetically, until a transition occurs from the glassy state to a rubbery state. The onset of the rubbery state is indicated by a marked increase in volume, caused by the increased molecular motion. The point at which this occurs is called the glass transition temperature.

In the rubbery state above Tg, polymers demonstrate elasticity, and some can even be molded into permanent shapes. One major difference between plastics and rubbers, or elastomers, is that the glass transition temperatures of rubbers lie below room temperature--hence their well-known elasticity at normal temperatures. Plastics, on the other hand, must be heated to the glass transition temperature or above before they can be molded.

Crystallinity is very important to product form. With rubbery products for example, the degree of crystallinity will play a dominant role in what the form of the neat rubber can be made as. For example, elastomers that have a high degree of crytallinity can be pelletized, or made into crumb (both product forms that are suitable to supply to customers in bags); or friable and semi-friable bales that can be easily mixed by compounders, as opposed to dense bales of rubber which are characteristic of amorphous elastomers.

ANSI

ANSI stands for the American National Standards Institute. This is an organization of industrial firms, trade associations, technical societies, consumer organizations, and government agencies, intended to establish definitions, terminologies, and symbols; improve methods of rating, testing, and analysis; coordinate national safety, engineering, and industrial standards; and represent U.S. interests in international standards work. ANSI standards are heavily referenced by the ASTM standards and are important to ISO 9000.

ANTIOXIDANTS

Antioxidants can be divided into two basic chemical types: amines and phenolics. In most rubber systems, amines are more effective in preventing long-term oxidative degradation. However, amine antioxidants usually discolor with aging and may not be the system of choice for light or brightly colored rubber articles where color retention is important.

Phenolic antioxidants, in contrast to amine antioxidants, do not discolor on aging but are generally less effective in preventing long-term oxidative degradation. Thus compromises may be necessary in formulating light or brightly colored materials with the use of the generally less effective phenolic antioxidants. If the rubber compound is black, then one can formulate using the more effective amine antioxidants. The amine antioxidants can be further subdivided into several categories by common chemical types:

1. Secondary diaryl amines: phenyl naphthylamines, substituted diphenylamines, and para-phenylenediamines;

2. Ketone-amine condensates;

3. Aldehyde-amine condensates;

4. Alkyl aryl secondary amines; and

5. Primary aryl amines.

Representative chemical structures of the various amine antioxidant types are shown in Figure 1. Similarly, phenolic antioxidants can also be subdivided by basic chemical types:

1. Hindered phenols;

2. Hindered bisphenols;

3. Hindered thiobisphenols; and

4. Polyhydroxy phenols. Representative chemical structures of the various phenolic antioxidants are shown in Figure 2.

In addition, many of the organosulfur compounds and organometallic compounds used in rubber compounds as vulcanization accelerators are known to have antioxidant activity. Mercaptobenzimidazole and its zinc salt have been shown to have antioxidant activity. In addition, mercaptobenzimidazole, when combined with other known antioxidants, has been shown to have a synergistic effect in oxidation prevention when metal ion catalyzed oxidation is prevalent. Also, metal and amine salts of dialkyldithio-carbarnates have been shown to have antioxidant action.

Selecting the correct combination of antioxidants is specific to the elastomer polymer type as well as the compound formulation and the end use application. It is important to note that many of these materials may be considered toxic or hazardous in nature. Chemical specific Material Safety Data Sheets (MSDS) should be consulted for safe handling practices. Particular attention should be given to the proper selection and use of personal protective equipment, including proper ventilation and/or the use of respiratory protection. The MSDS will also provide information on how to handle spills and proper disposal procedures. Disposal methods should not be overlooked since these chemicals are all regulated, and hence waste disposal must conform to EPA and local disposal regulations. Refer to *Rubber Oxidation.* (Source: *Handbook of Polymer Science and Technology: Volume 2 - Performance Properties of Plastics and Elastomers*, N. P. Cheremisinoff - editor, Marcel Dekker Inc., New York, 1989).

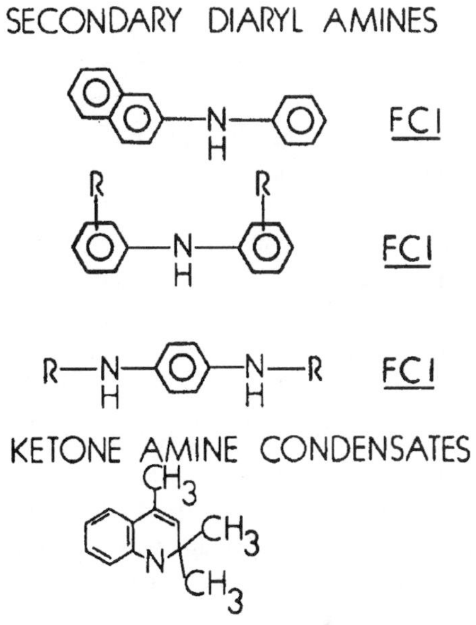
SECONDARY DIARYL AMINES
FCI
FCI
FCI
KETONE AMINE CONDENSATES

Figure 1. Structures of amine antioxidants.

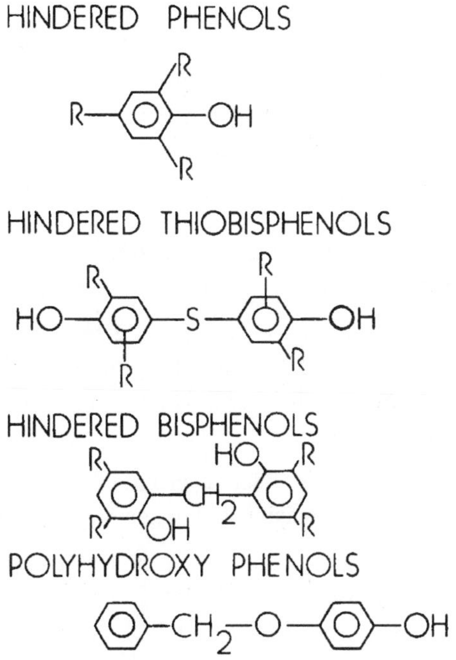
HINDERED PHENOLS
HINDERED THIOBISPHENOLS
HINDERED BISPHENOLS
POLYHYDROXY PHENOLS

Figure 2. Structures of phenolic antioxidants.

ANTIOZONANT

The term *antiozonant* denotes any additive that protects rubber against ozone deterioration. Most frequently, the protective effect results from a reaction with ozone, in which case the term used is *chemical antiozonant*. Ozone is generated naturally by electrical discharge and also by solar radiation in the stratosphere. These sources produce ground-level ozone concentrations of 1-5 parts per hundred million (pphm). In urban environments, however, ozone reaches much higher levels, up to 25 (pphm) due to the ultraviolet photolysis of pollutants. Only a few parts per hundred million of ozone in air can cause rubber cracking, which may destroy the usefulness of elastomer products. Some desirable properties of an antiozonant additive are as follows:

1. A *physical* antiozonant must provide an effective barrier against the penetration of ozone at the rubber surface. In this regard, the barrier should be continuous at the surface, unreactive and impenetrable to ozone, and capable of renewing itself if damaged (as by abrasion). Flexibility (extensibility) under dynamic stress conditions is also desirable. A *chemical* antiozonant, on the other hand, must first be extremely reactive with ozone. A compound that is not reactive enough will not protect adequately. An antiozonant must not be too reactive with oxygen, however, or it will not persist long enough in the rubber to afford long-term ozone protection.

2. The antiozonant should possess adequate *solubility* and *diffusivity* characteristics. Since ozone attack is a surface phenomenon, the antiozonant must migrate to the surface of the rubber to provide protection. Poor solubility in rubber may result in a problem with excessive bloom, or else the loading level obtainable may be insufficient for long-term protection. The diffusion rate to the exposed surface must be high enough to meet the incoming ozone flux but low enough to ensure a reservoir in the rubber bulk during the useful lifetime of the article. Because antiozonants must diffuse and replenish themselves on the surface, it is not possible to have an effective bound antiozonant.

3. The antiozonant should have no adverse effects on the rubber processing characteristics (mixing, fabrication, vulcanization, physical properties). In general purpose rubbers, this implies that the antiozonant must be compatible with sulfur curing systems.

4. The antiozonant should be effective under both *static* and *dynamic* conditions over a wide range of extension and temperature conditions.

5. The antiozonant should persist in the rubber over its entire life cycle. It should be resistant to loss via oxidation, vaporization, or extraction by water or other solvents.

6. For non-carbon black-filled rubbers, it must be *nondiscoloring* and *nonstaining*.

7.   The antiozonant should have a *low toxicity* and should be nonmutagenic.

8.   The antiozonant should be acceptable *economically*. That is, it should have a low manufacturing cost and be usable at low bulk concentration levels.

Rubber is protected against ozone attack by the addition of physical and/or chemical antiozonants. Hydrocarbon waxes are the most common type of physical antiozonant, and p-phenylenediamine derivatives are the prevalent chemical antiozonants. Refer to *Hydrocarbon Waxes* and *Chemical Antiozonants*. (Source: *Handbook of Polymer Science and Technology, Volume 2: Performance Properties of Plastics and Elastomers*, N. P. Cheremisinoff - editor, Macrel Dekker Publishers, New York, 1989).

APPARENT VISCOSITY

Viscosity of a fluid that holds only for the shear rate (and temperature) at which the viscosity is determined. This is usually a solution viscosity. Refer to *Brookfield Viscosity*.

AROMATIC

Unsaturated hydrocarbon identified by one or more benzene rings or by chemical behavior similar to benzene. The benzene ring is characterized by three double bonds alternating with single bonds between carbon atoms (compared with olefins). Because of these multiple bonds, aromatics are usually more reactive and have higher solvency than paraffins and naphthenes. Aromatics readily undergo electrophylic substitution; that is, they react to add other active molecular groups, such as nitrates, sulfonates, etc. Aromatics are used extensively as petrochemical building blocks in the manufacture of pharmaceuticals, dyes, plastics, and many other chemicals. The unsaturated hydrocarbons are the alkenes with one double bond and the alkynes with one triple bond. There are other straight-chain hydrocarbons that are unsaturated containing more than one multiple bond, some with more than one double bond, and some with a mixture of double bonds and triple bonds. The combinations and permutations are endless, but there are only a few of the highly unstable materials.

From a commercial standpoint, there is a large body of hydrocarbons that is very important and hence these are of relevance to first responders to a hazardous-materials incidents. These hydrocarbons are different in that they are not straight-chain hydrocarbons but have a structural formula that can only be called cyclical. The most common and most important hydrocarbon in this group is benzene. It is the first and simplest of the six-carbon cyclical hydrocarbons referred to as aromatic hydrocarbons.

Benzene's molecular formula is C_6H_6, but it does not behave like hexane, hexene, or any of their isomers. One would expect it to be similar to these other six-carbon hydrocarbons in its properties. Table 1 provides a comparison between benzene, hexane and 1-hexene. The table shows that there are major differences between benzene and the straight-chain hydrocarbons of the same carbon content. Hexene's ignition temperature is very near to hexane's. The flash point difference is not great, however, there are significant differences in melting points. The explanation for these differences is structure, which in the case of benzene is a cyclical form with alternating double bonds.

Table 1.

Comparison Between Benzene and of Straight-Chain Hydrocarbons

Compound	Formula	Melting Point (°F)	Boiling Point (°F)	Flash Point (°F)	Ignition Temp. (°F)	Molecular Weight
Hexane	C_6H_{14}	-139.5	156.0	-7	500	86
1-Hexene	C_6H_{12}	-219.6	146.4	< -20	487	84
Benzene	C_6H_6	41.9	176.2	12	1,044	78

The alternating double bonds are illustrated in Figure 1A. Initially, it was believed that the alternating double bonds impart very different properties to benzene, however, and the fact is that they do not. The only possible way for the benzene molecule to exist is illustrated in Figure 1B, in which a circle is drawn within the hexagonal structural to show that the electrons that should form a series of alternating double bonds are really spread among all six carbon atoms. It is the only structure possible that would explain the unique properties of benzene. This

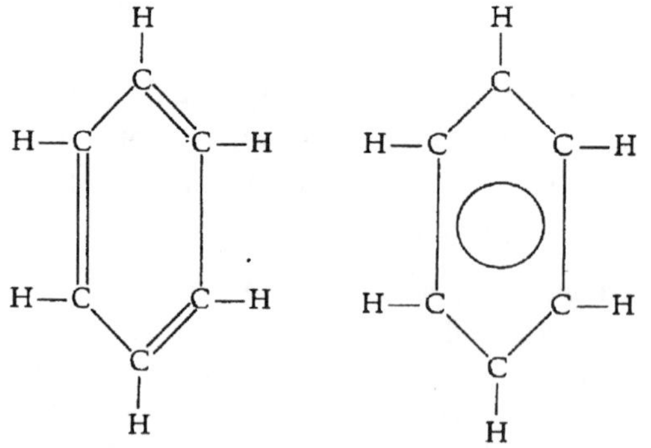

(A)   **(B)**

Figure 1. Illustrates the structure of benzene: (A) conventional illustration of double bonds, (B) illustration implying resonance.

structural formula suggests resonance; that is, the possibility that the electrons represented by the circle are alternating back and forth between and among the six carbon atoms. This particular hexagonal structure is found throughout nature in many forms, almost always in a more complicated way, usually connected to many other "benzene rings" to form many exotic compounds. Of importance to the immediate discussions are benzene and a few of its derivatives. Benzene's derivatives include toluene and xylene, whose structural formulas are illustrated in Figure 2 along with that of benzene.

Some typical properties are given in Table 2, which illustrates the differences caused by molecular weight and structural formulas. There are other cyclical hydrocarbons, but they do not have the structural formulas of the aromatics, unless they are benzene-based. These cyclical hydrocarbons may have three, four, five, or seven carbons in the cyclical structure, in addition to the six-carbon ring of the aromatics. None of them has the stability or the chemical properties of the aromatics.

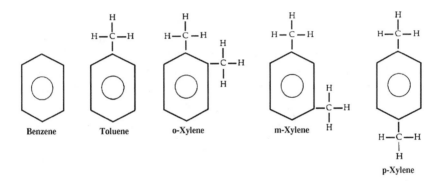

Figure 2. Structures of benzene and common derivatives.

Table 2.

A Comparison of Benzene and Some of its Derivatives

Compound	Formula	Melting Point (°F)	Boiling Point (°F)	Flash Point (°F)	Ignition Temperature (°F)	Molecular Weight
Benzene	C_6H_6	41.9	176.2	12	1,044	78
Toluene	C_7H_8	-138.1	231.3	40	997	92
o-xylene	C_8H_{10}	-13.0	291.2	90	867	106
m-xylene	C_8H_{10}	-53.3	281.9	81	982	106
p-xylene	C_8H_{10}	-55.8	281.3	81	984	106

The aromatic hydrocarbons are used mainly as solvents and as feedstock chemicals for chemical processes that produce other valuable chemicals. With regard to cyclical hydrocarbons, the aromatic hydrocarbons are the only compounds discussed. These compounds all have the six-carbon benzene ring as a base, but there are also three-, four-, five-, and seven-carbon rings. After the alkanes, the aromatics are the next most common chemicals shipped and used in commerce. The short-chain olefins (alkenes) such as ethylene and propylene may be shipped in larger quantities because of their use as monomers, but for sheer numbers of different compounds, the aromatics will surpass even the alkanes in number, although not in volume. (Source: N. P. Cheremisinoff, *Handbook of Industrial Toxicology and Hazardous Materials*, Marcel Dekker, Inc., New York, 1999).

ATOMIC ABSORPTION SPECTROSCOPY

In atomic absorption spectrometry (AA) the sample is vaporized and the element of interest atomized at high temperatures. The element concentration is determined based on the attenuation or absorption by the analyte atoms, of a characteristic wavelength emitted from a light source. The light source is typically a hollow cathode lamp containing the element to be measured. Separate lamps are needed for each element. The detector is usually a photomultiplier tube. A monochromator is used to separate the element line and the light source is modulated to reduce the amount of unwanted radiation reaching the detector. Conventional AA instruments use a flame atomization system for liquid sample vaporization. An air-acetylene flame (2300° C) is used for most elements. A higher temperature nitrous oxide-acetylene flame (2900° C) is used for more refractory oxide forming elements. Electrothermal atomization techniques such as a graphite furnace can be used for the direct analysis of solid samples.

Atomic absorption is used for the determination of ppm levels of metals. It is not normally used for the analysis of the light elements such as H, C, N, 0, P and S, halogens, and noble gases. Higher concentrations can be determined by prior dilution of the sample. AA is not recommended if a large number of elements are to be measured in a single sample. Although AA is a very capable technique and is widely used worldwide, its use in recent years has declined in favor of ICP and XRF methods of analysis for polymers. The most common application of AA is for the determination of boron and magnesium in oils. Conventional AA instruments will analyze liquid samples only. Dilute acid and xylene solutions are common. The volume of solution needed is dependent on the number of elements to be determined. AA offers excellent sensitivity for most elements with limited interferences. For some elements sensitivity can be extended into the sub-ppb range using flameless methods. The AA instruments are easy to operate with cookbook methods available for most elements. The determination of several elements per sample is slow and requires larger volumes of solution due to the sequential nature of the method. Chemical and ionization interferences must be corrected by modification of the sample solution. Chemical interferences arise from the formation of thermally stable compounds such as oxides in the flame. The use of electrothermal atomization, a hotter nitrous oxide-acetylene flame or the addition of a releasing agent such as lanthanum can help reduce the interference. Refer to *Flame Ionization.* (Source: Cheremisinoff, N.P. *Polymer Characterization: Laboratory Techniques and Analysis*, Noyes Publishers, New Jersey, 1996).

AUTO-IGNITION TEMPERATURE

Lowest temperature at which a flammable gas or vaporized liquid will ignite in the absence of a spark or flame, as determined by test method ASTM D 2155; not to be confused with flash *point* or fire *point,* which is typically lower. Auto-ignition temperature is a critical factor in heat transfer oils and transformer oils, and in solvents used in high temperature applications. The auto-ignition temperature should be viewed in relation to the entire phenomenon of combustion.

Fire, or combustion is a chemical reaction, and specifically it is an oxidation reaction. Oxidation is defined as the chemical combination of oxygen with any substance. In other words, whenever oxygen (and some other materials) combines chemically with a substance, that substance is said to have been oxidized. Rust is an example of oxidized iron. In this case, the chemical reaction is very slow. The very rapid oxidation of a substance is called combustion, or fire.

There are three basic theories that are used to describe the reaction known as fire. They are: the fire triangle, the tetrahedron of fire, and the life cycle of fire. Of the three, the first is the oldest and best known, the second is accepted as more fully explaining the chemistry of fire, while the third is a more detailed version of the fire triangle. Each is briefly described below.

The first of these theories, the fire triangle, is quite simplistic and provides a basic understanding of the three entities that are necessary for a fire. This theory states that there are three things necessary to have a fire: fuel, oxygen (or an oxidizer), and heat (or energy). It likens these three things to the three sides of a triangle, stating that as long as the triangle is not complete, that is, the legs are not touching each other to form the closed or completed triangle, combustion cannot take place.

The theory, as stated, is still correct. Without fuel to burn, there can be no fire. If there is no oxygen present, there can be no fire (technically, this is not correct, but we can make the fire triangle theory technically correct by changing the oxygen leg to an oxidizer leg). Finally, without heat, there can be no fire. This last statement must also be brought up to date. The fact is that heat is just one form of energy; it is really energy that is necessary to start a fire. This difference is mentioned, because there are some instances, where light or some other form of energy may be what is needed to start the combustion reaction. It is best to change the heat leg of the fire triangle to the energy leg. Therefore, our updated fire triangle now has three sides representing fuel, oxidizer, and energy.

Energy may also be generated by mechanical action; that is, the application of physical force by one body upon another. Examples of this are the energy created by the friction of one matter upon another or the compression of a gas. The force of friction in one case may produce energy that manifests itself as heat, while friction in the other case may result in a discharge of static electricity. Static electricity is created whenever molecules move over and past other molecules. This happens whether the moving molecules are in the form of a gas, a liquid, or a solid. (This is the reason why leaking natural gas under high pressure will ignite. This is also the reason why two containers must be bonded - connected by an electrical conductor - when you are pouring flammable liquids from one container to another. In any case, the amount of energy present and/or released could be more than enough to start the combustion reaction.) A third method of generation of energy is electrical - much like the discharge of static electricity. This method may manifest itself as heat, as produced in an electrical heater, as arcing in an electrical motor or in a short circuit, or as the tremendous amount of energy released as lightning. The fourth method of generation of energy is nuclear. Nuclear energy may be generated by the fission (splitting) of the atoms of certain elements and by the fusion (or joining together) of the nuclei of certain elements.

Once the energy - in many cases, heat - is generated, it must be transmitted to the fuel (the touching of the fuel and energy legs). This process is accomplished in three ways: conduction (the transfer of heat through a medium, such as a pan on a stove's heating element), convection (the transfer of heat with a medium, such as the heated air in a hot-air furnace), and radiation (the transfer of heat which is not dependent on any medium).

These three entities (fuel, oxidizer, and energy) make up the three legs of the fire triangle. It is a physical fact, a law of nature that cannot be repealed, that when fuel, oxidizers, and energy are brought together in the proper amounts, a fire will occur. If the three are brought together slowly, and over a long period of time, the oxidation will occur slowly, as in the rusting of iron. If the three are of a particular combination, the resulting oxidation reaction might even be an explosion. Whatever form the final release of energy takes, the thing that cannot be changed is that the chemical reaction will occur.

The second popular explanation of fire is the tetrahedron theory which is illustrated in Figure 1. This theory encompasses the three concepts in the fire triangle theory but adds a fourth side to the triangle, making it a pyramid, or tetrahedron; this fourth side is called the chain reaction of burning. This theory states that when energy is applied to a fuel like a hydrocarbon, some of the carbon-to-carbon bonds break, leaving an unpaired electron attached to one of the molecular fragments caused by the cleavage of the bond, thus creating a free radical. This molecular fragment with the unpaired electron, or dangling bond, is highly reactive and will therefore seek out some other material to react with in order to satisfy the octet rule. The same energy source that provided the necessary energy to break the carbon-to-carbon bond may have also broken some carbon-to-hydrogen bonds, creating more free radicals, and also broken some oxygen-to-oxygen bonds, creating oxide radicals. This mass breaking of bonds creates the free radicals in a particular space, and in a number large enough to be near each other, so as to facilitate the recombining of these free radicals with whatever other radicals or functional groups may be nearby. The breaking of these bonds releases the energy stored in them, so that this subsequent release of energy becomes the energy source for still more bond breakage, which in turn releases more energy. Thus the fire feeds upon itself by continuously creating and releasing more and more energy (the chain reaction), until one of several things happens: either the fuel is consumed, the oxygen is depleted, the energy is absorbed by something other than the fuel, or this chain reaction is broken. Thus, a fire usually begins as a very small amount of bond breakage by a relatively small energy (ignition) source and builds itself up higher and higher, until it becomes a raging inferno, limited only by the fuel present (a fuel-regulated fire) or the influx of oxygen (an oxygen-regulated fire). The earlier in the process that the reaction can be interrupted, the easier the extinguishment of the fire will be. This theory claims that the propagation of all hydrocarbon fires (or fires involving hydrocarbon derivatives) depends upon the formation of the hydroxyl (-OH) radical, which is found in great quantities in all such fires.

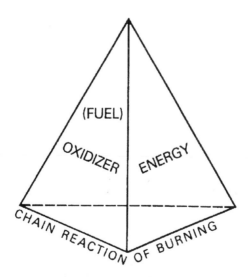

Figure 1. Illustrates the theory of the fire tetrahedron.

The third theory of fire is the life cycle theory, which is illustrated in Figure 2. According to this theory, the combustion process can be categorized by six steps, rather than the three of the fire triangle or the four of the tetrahedron of fire theory. Three of the steps in this theory are the same as the only three steps in the fire triangle theory. In the life cycle of fire theory, the first step is the input heat, which is defined as the amount of heat required to produce the evolution of vapors from the solid or liquid. The input heat will also be the ignition source and must be high enough to reach the ignition temperature of the fuel; it must be continuing and the entire piece must be raised to the proper temperature before it will begin to burn. The third part is oxygen in which the classical explanation of this theory only concerns itself with atmospheric oxygen, because the theory centers around the diffusion flame, which is the flame produced by a spontaneous mixture (as opposed to a pre-mixed mixture) of fuel gases or vapors and air self-generating and must heat enough of the fuel to produce the vapors necessary to form an ignitable mixture with the air near the source of the fuel. The second part of the life cycle of fire theory is the fuel, essentially the same as the fuel in the tetrahedron of fire and the fire triangle. It was assumed without so stating in the fire triangle theory, and is true in all three theories, that the fuel must be in the proper form to burn; that is, it must have vaporized.

This theory concerns itself with air-regulated fires, so airflow is crucial to the theory; this is why only atmospheric oxygen is discussed. Ignoring oxygen and the halogens that are generated from oxidizing agents should be viewed as a flaw in this theory. The fourth part of the theory is proportioning, or the occurrence of intermolecular collisions between oxygen and the hydrocarbon molecule (the "touching" together of the oxidizer leg and the fuel leg of the fire triangle). The speed of the molecules and the number of collisions depend on the heat of the mixture of oxygen and fuel; the hotter the mixture, the higher the speed.

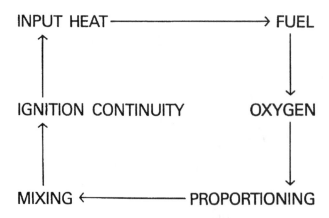

Figure 2. Illustrates the life cycle theory of fire.

A rule of thumb is used in chemistry that states the speed of any chemical reaction doubles for roughly every 10° C rise in temperature. The fifth step is mixing; that is, the ratio of fuel to oxygen must be right before ignition can occur (flammable range). Proper mixing after heat has been applied to the fuel to produce the vapors needed to burn is the reason for the "backdraft" explosion that occurs when a fresh supply of air is admitted to a room where a fire has been smoldering.

The sixth step is ignition continuity, which is provided by the heat being radiated from the flame back to the surface of the fuel; this heat must be high enough to act as the input heat for the continuing cycle of fire. In a fire, chemical energy is converted to heat; if this heat is converted at a rate faster than the rate of heat loss from the fire, the heat of the fire increases; therefore, the reaction will proceed faster, producing more heat faster than it can be carried away from the fire, thus increasing the rate of reaction even more. When the rate of conversion of chemical energy falls below the rate of dissipation, the fire goes out. That is to say, the sixth step, ignition continuity, is also the first step of the next cycle, the input heat. If the rate of generation of heat is such that there is not enough energy to raise or maintain the heat of the reaction, the cycle will be broken, and the fire will go out. The life cycle of fire theory adds the concepts of flash point and ignition point (heat input) and flammable range (mixing).

Given sufficient oxygen and heat input, all organic polymers will burn. Absolute fire safety of polymeric materials does not exist. There are always trade-offs in safety, utility, and costs.

Billions of pounds of synthetic polymers are used annually in the United States without presenting unmanageable fire safety problems. However, some uses of polymeric materials have seriously augmented the fire hazard. With increased volume and diversity of uses, such hazards could increase further. Many synthetic organic polymers burn in a manner differing from that of the more familiar natural polymers such as wood, paper, cotton, or wool. Some synthetics burn much faster, some give off much more smoke, some evolve potentially noxious and toxic gases, and some melt and drip. Others burn less readily than the natural polymers.

As the diversity and amount of polymeric materials used in confined areas (dwellings, vehicles, etc.) increases, the problems presented by the generation of smoke and toxic gases in a fire also increase. The fire safety of many polymers has been improved by the incorporation of fillers and/or compounds containing halogens, phosphorus, stabilizers, various fire retarders and/or antimony. In general, this is effective in resisting small ignition sources or low thermal fluxes in large-scale fires. Most of these systems burn readily and may lead to increased smoke generation or increased production of toxic and/or corrosive combustion products. Refer to *Flash Point*. (Source: N. P. Cheremisinoff, *Handbook of Industrial Toxicology and Hazardous Materials*, Marcel Dekker, Inc., New York, 1999).

B

BATCH INTENSIVE MIXING

Batch intensive mixing is generally carried out using two specially designed blades inside a temperature controlled chamber. It is most commonly used for compounding and mixing of rubber formulations. Its most important commercial use is the incorporation of carbon black and other additives into rubber for the manufacture of automobile tires. Laboratory size batch intensive mixers are extensively used for characterization of materials and processes. The following book provides several excellent chapters relevant to batch intensive mixing: *Mixing and Compounding of Polymers*, edited by I. Manas-Zloczower and Z. Tadmor; Hanser Publishers, New York (1994) ISBN 3-446-17368-4.

BASE

Term refers to any of a broad class of compounds, including *alkalis,* that react with *acids* to form salts, plus water. Also known as hydroxides. Hydroxides ionize in solution to form hydroxyl ions (OH^-); the higher the concentration of these ions, the stronger the base. Bases are used extensively in petroleum refining in *caustic washing* of *process streams* to remove acidic impurities, and are components in certain *additives* that neutralize weak acids formed during oxidation.

BENZENE

An aromatic hydrocarbon consisting of six carbon atoms and six hydrogen atoms arranged in a hexagonal ring structure. It is used extensively in the *petrochemical* industry as a chemical intermediate and reaction *diluent* and in some applications as a *solvent*. Benzene is a toxic substance, and proper safety precautions should be observed in handling it.

BIREFRINGENCE

Birefringence is a measure of optical anisotropy. It is defined as the maximum algebraic difference between two refractive indices measured in two perpendicular directions. Some materials are isotropic until stressed elastically, others have permanent birefringence induced by processing, e.g. drawing or rolling. The most convenient way of measuring birefringence is by means of a polarizing microscope. When a beam of light falls on an isotropic material it is resolved into two components. A polarizer is a device which is designed so that one of these components is totally absorbed (or internally reflected in the case of a Nicol prism). Thus light emerging from the polarizer is plane polarized. A similar sheet of polarising material mounted with its plane of polarization at right angles to the first will absorb all of the light. Only if the two directions of polarization are parallel will light pass through both filters. This second piece of polarizing material is known as the analyzer. The two arrangements described above are referred to as crossed polars and open polars, respectively.

BLOCK COPOLYMERS

Block copolymers display the unique behavior of being able to function as conventionally crosslinked elastomers over a certain temperature range, but they soften reversibly at high temperatures. As such, they can be processed as a thermoplastic. These polymers are macromolecules comprised of chemically dissimilar, terminally connected segments. According to the components that make up the segments, the thermoplastic rubbers can be classified according to Table 1.

Table 1.
Classification of Block Copolymers

Type	Specific Gravity Range	Shore Hardness Range
Styrenic	0.90-1.14	45A-53D
Olefinic	0.89-1.25	60A-60D
Polyurethane	1.10-1.34	70A-75D
Copolyester	1.13-1.39	35A-72D
Polyamide	1.01-1.14	75A-63D

The most commercially common among these is the styrenic group in which the thermoplastic blocks are comprised of polystyrene chemically linked to a rubbery block (polybutadiene or polyisoprene).

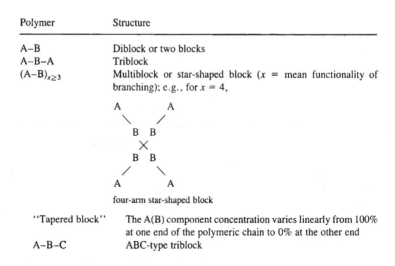

Figure 1. Structures of block copolymers.

Copolymers can be classified according to structural arrangement as follows:
1.   Random, in which the disposition of the components in the polymeric chain is irregular;
2.   Graft, in which branches of one homopolymer are grafted (linked) to the main chain of another homopolymer;
3.   Block, in which the polymeric chain is built up of blocks of different structures.

Refer to the structures in Figure 1, where the letters A, B, and C represent the various homopolymer blocks (or monomer in the tapered block case.

BOILING POINT

The value is the temperature of a liquid when its vapor pressure is 1 atm. For example, when water is heated to 100° C (212° F) its vapor pressure rises to 1 atm and the liquid boils. The boiling point at 1 atm indicates whether a liquid will boil and become a gas at any particular temperature and sea-level atmospheric pressure.

BULK POLYMERIZATION

Bulk polymerization is carried out in the absence of any solvent or dispersant and is thus the simplest in terms of formulation. It is used for most step-growth polymers and many types of chain-growth polymers. In the case of chain-growth reactions, which are generally exothermic, the heat evolved may cause the reaction to become too vigorous and difficult to control unless efficient cooling coils are installed in the reaction vessel. Bulk polymerizations are also difficult to stir because of the high viscosity, associated with high-molecular-weight polymers.

BLOWN-FILM EXTRUSION

The blown-film process is used in the manufacture of polyethylene and other plastic films. The molten polymer from the extruder head enters the die, where it flows round a mandrel and emerges through a ring-shaped opening in the form of a tube. The tube is expanded into a bubble of the required diameter by the pressure of internal air admitted through the center of the mandrel. The air contained in the bubble cannot escape because it is sealed by the die at one end and by the nip (or pinch) rolls at the other, so it acts like a permanent shaping mandrel once it has been injected. An even pressure of air or nitrogen is maintained to ensure uniform thickness of the film bubble. The film bubble is cooled below the softening point of the polymer by blowing air on it from a cooling ring placed round the die. When the polymer cools below the softening point, the crystalline material is cloudy compared with the clear amorphous melt. The transition line, which coincides with this transformation is therefore called the frost line. The ratio of bubble diameter to die diameter is called the blowup ratio. It may range as high as 4 or 5, but 2.5 is more typical. Molecular orientation occurs in the film in the hoop direction during blowup, and orientation in the machine direction, that is, in the direction of the extrudate flow from the die, can be induced by tension from the pinch rolls. The film bubble after solidification (at frost line) moves upward through guiding devices into a set of pinch rolls which flatten it. It can then be slit, gusseted, and surface-treated in line. A vertical extrusion scheme shown in Figure 1 is most common. (Source: Chanda, M. and S. K. Roy, Plastics Technology Handbook, Marcel Dekker, Inc., New York, NY, 1987). *See also Plunger-Type Transfer Molding, Compression Molding and Screw Transfer Molding, Extruders and Extrusion.*

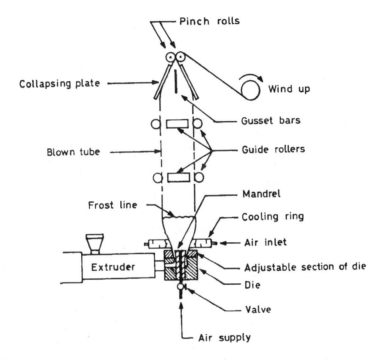

Figure 1. Vertical blown-film extrusion.

BLOW MOLDING

Blow molding is a fabrication process used for forming hollow plastic articles. There are five basic types of blow molding operations:

- Continuous extrusion blow molding
- Intermittent extrusion blow molding
- Injection blow molding
- Injection stretch blow molding
- Extrusion stretch blow molding

The continuous extrusion blow molding process is illustrated in Figure 1. The basic process steps involve:

- The extruder pushes the polymer melt through an annular die to form a cylinder of molten polymer, called a parison;

- The mold halves close on the parison, pinching at the top around a "blow pin" and the bottom;

- Air is passed through the blow pin and into the sealed parison to force the melt against the cold metal surfaces;

- The mold forms a part. The part ejects and the trimmer removes excess material.

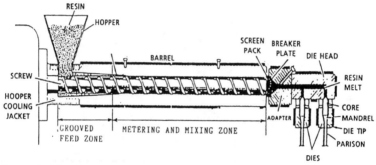

Figure 1. Continuous extrusion blow molding.

In order to obtain a uniform wall thickness for complex shaped containers, it is necessary to be able to control the thickness of the parison as it extrudes. Wall thickness control is achieved by varying the die gap as the parison is extruded. There are a variety of different arrangements for the blowing step of the process. These include the use of single dies with multiple molds (e.g., vertical rotary wheel), and multiple dies with multiple wheels. For some examples, refer to the illustrations in Figures 2 through 4. For enhanced barrier properties of multilayer structures, combinations of polymer materials are used, such as:

- PP/Regrind/TIE/EVOH/TIE/PP

- HDPE/Regrind/TIE/Nylon/TIE/HDPE

- PC/TIE/EVOH/PP

- Nylon/TIE/Regrind/HDPE

The intermittent extrusion blow molding process is generally used in the fabrication of large volume articles (e.g., > 20 liter bottles). An accumulator is employed to increase the parison extrusion speed (see diagram below). After the mold closes there is no additional mold movement. Refer to *Injection Stretch Blow Molding*.

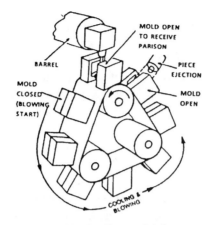

Figure 2. Single die, multiple
molds - vertical rotary wheel.

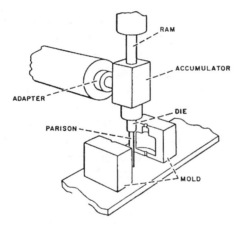

Figure 3. Extrusion blow molding with
accumulator.

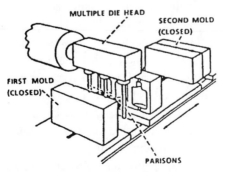

Figure 4. Shuttle mechanism for
multiple dies and multiple molds.

BROOKFIELD VISCOSITY

The apparent viscosity of an oil, as determined under test method ASTM D 2983. Since the apparent viscosity of a *non-Newtonian fluid* holds only for the *shear* rate (as well as temperature) at which it is determined, the Brookfield viscometer provides a known rate of shear by means of a spindle of specified configuration that rotates at a known constant speed in the fluid. The torque imposed by fluid friction can be converted to absolute viscosity units (centipoises) by a multiplication factor.

BUTANE

Gaseous paraffinic hydrocarbon (C_4H_{10}) is usually a mixture of iso- and normal butane, also called, along with *propane,* liquefied petroleum gas (LPG).

BUTYL (IIR)

A copolymer of isobutylene with 1-3% of isoprene to facilitate vulcanization. Its first application was as tire inner tubes because of its low permeability to air. It is used as a general purpose rubber. A readily distinguished characteristic is its very low resilience at 20° C. Available hardness range 35-85 IRHD. General properties include fairly good mechanical properties; low resilience at room temperature can lead to considerable heat buildup under the action of mechanical vibrations; good electrical properties; very low permeability to gases; high resistance to ozone and chemical attack and weathering; severe swelling in some hydrocarbons but good resistance to animal and vegetable oils and alcohols. Approximate working temperature range -50° C to +125° C. Typical uses include chemical tank linings, inner tubes, high voltage insulators, proofed goods.

Butyl rubber is produced by copolymerizing isobutylene with small amounts (1-4 per cent) of isoprene. This is in contrast to natural rubber which contains 100 percent isoprene, or SBR which is about 40 - 60 percent unsaturated. The isobutylene molecule contributes unusual properties such as outstanding damping and impermeability, while the low unsaturation leads to excellent heat and ozone resistance. Butyl's outstanding impermeability to gases quickly established it as the ideal elastomer for innertubes for many years. Low gas and moisture permeability; outstanding resistance to ozone, weather, abrasion, tear, flexing, heat aging, and chemical attack; excellent electrical insulation performance; and high shock absorption are among the properties that have given butyl rubber its wide appeal in a multitude of engineering applications. Due to its limited degree of chemical unsaturation it possesses greater chemical and heat resistance than natural rubber, especially in the presence of oxidizers. Butyl rubber should not be used in the presence of free halogens, petroleum oils, or halogenated or aromatic hydrocarbons.

C

CALENDERING

Calendering is a basic unit operation used in both plastics and rubber compounding. Rubber calenders consist of two or more hardened and accurately machined metal rolls rotating in bearing journal boxes which are set in rugged iron frames. At least one roll is equipped with screwdowns to control the thickness of the processed material. Adjacent pairs of rolls rotating in opposite directions form a "nip", in which the material being processed is squeezed into sheets or is laminated to form the desired product. The drives for the rolls include constant or variable-speed motors and reduction gearing to achieve roll surface speeds required by the processing requirements of the materials.

A calender, depending on the number and the design of its rolls, is capable of sheeting, frictioning, coating, profiling, and embossing. A variety of roll configurations, both horizontally and vertically, are available in sizes ranging from laboratory units to commercial designs weighing several tons. Three-roll vertical calender with 24-in. (diameter) x 68-in. (face length) rolls and the 4-roll "Z" and "L" with 28-in. x 78-in. rolls are typical of the machines used for mass production of tires, belting, sheeting, and similar articles. Three-roll calenders have been widely used for processing of mechanical rubber goods. Four-roll calenders are popular in tire plants. The four rolls permit simultaneous application of rubber compound on both sides of tire cord fabrics. Two-roll calenders are used to produce strips and profiles, often in combination with extruder feeding, in which case they are commonly referred to as *roller dies*.

Calenders are used in five separate operations in the manufacture of rubber products: sheeting, frictioning, coating, profiling, and embossing. Sheeting employs a two-roll calender in a horizontal or vertical configuration. The feed material, either in strip or *pig* form, is fed into one side of the nip and is flattened. The material emerges as a sheet that is pulled from the roll by some manual or mechanical means. Figure 1 illustrates basic calendering operations normally employed in processing. Thickness control is accomplished by use of the screwdowns and may be further refined by automatic control systems using sensors. The force required in the nip to flatten the feed material causes a slight deflection of the rolls. If some corrective steps are not taken, the product thickness will vary across the sheet, resulting in excessive variations of the product and possibly the production of scrap. In order to minimize air entrapment and blistering, the thickness of each sheet is generally limited. To build up the required thickness of the final sheet, two or more piles of calendered sheet are usually laminated on the bottom roll of a three-roll calender. The operation of frictioning involves rubbing or wiping an elastomeric compound into a substrate of textile or metallic cords, which may or may not be held together by *pick* threads or fill yarns, or the substrate may consist of a *square woven* fabric like *hose ducks* or *belt ducks*. Usually a three-roll calender is used. Rubber sheet is formed between the upper and middle rolls while the resulting sheet is simultaneously being frictioned into the substrate between the middle and bottom rolls. In-this operation the upper and middle rolls may be moving at "odd" or "even" surface speed, but the middle and bottom rolls will be run at "odd" or unequal surface speeds so that the rubber is effectively wiped into the substrate being carried on the bottom roll. A coating or skim-coating operation is similar to that described for frictioning except that the middle and bottom rolls are operated at "even" surface speed so that the rubber sheet is laid and pressed against the substrate. In a multipass operation the substrate will have been previously frictioned. The coating operation may produce a heavy deposit or a thin "skim" coat depending upon the product requirement. Generally, multipurpose calenders such as a three-roll unit are equipped with "even" and "odd" gearing arrangements so that a number of combinations on roll speed ratios are possible. A more complex form of coating calender is the four-roll "Z" or "V" arrangement. A four-roll calender can simultaneously apply a rubber coating onto both sides of a fabric. In effect, the no. 1 and 2 rolls and the no. 3 and 4 rolls form pairs from which two rubber sheets are produced. The sheets are then laminated to a substrate between the no. 2 and 3 rolls. Refer to *Profiling* and *Embossing*. (Source: Cheremisinoff, N.P., *An Introduction to Polymer Rheology and Processing*, CRC Press, Boca Raton, Florida, 1993).

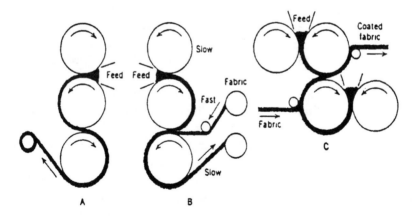

Figure 1. Shows the basic calendering operation.

CALORIE

Term applicable either to the gram calorie or the kilocalorie. The gram calorie is defined as the amount of heat required at a pressure of one atmosphere to raise the temperature of one gram of water one degree Celsius at 15° C. The kilocalorie is the unit used to express the energy value of food; it is defined as the amount of heat required at a pressure of one atmosphere to raise the temperature of one kilogram of water one degree Celsius; it is equal to 1000 gram calories.

CAPILLARY VISCOMETER

The capillary rheometer is one of the most common devices for measuring viscosity. Its main component is a straight tube or capillary. It has a pressure driven flow for which the velocity gradient or strain rate and also the shear component are maximum at the wall and zero at the center of the flow, making it a nonhomogeneous flow. Since a pressure driven viscometer employs non-homogeneous flows, it can only measure steady shear functions such as viscosity. However, it is widely used because it is inexpensive and simple to operate. Long capillary viscometers give the most accurate viscosity data available. Another major advantage is that the capillary rheometer has no free surfaces in the test region, unlike other types of rheometers such as the cone and plate rheometer. When the strain rate dependent viscosity of polymer melt is measured, capillary rheometers may be the only satisfactory method of obtaining such data at shear rates greater than 10 s^{-1}. This is important for processes with higher rates of deformation like mixing, extrusion and injection molding. Because its design is basic and it only needs a pressure head at its entrance, the device can easily be attached to the end of a screw- or ram-type extruder for on-line measurements. This makes the capillary viscometer an efficient tool for industry. The basic features of this instrument are shown in Figure 1. See *Cone and Plate Rheometer*. (Source: Osswald, T.A. and G. Menges, *Material Science of Polymers for Engineers*, Hanser Publishers, New York, 1996).

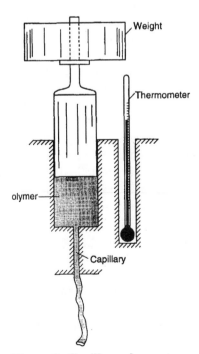

Figure 1. Capillary rheometer.

CARBON BLACK

Commercial carbon blacks used in rubber compounding applications are usually classified in terms of their morphology (i.e., particle size/ surface area, vehicle absorptive capacity). In the ASTM nomenclature system for carbon black the first digit is based on the mean particle diameter as measured with an electron microscope. Other commonly used methods for classifying blacks by size are iodine number, nitrogen adsorption, and tinting strength. Important also are the shapes of individual particles and of aggregates. DBP absorption is the principal technique employed for measuring the irregularity of the primary aggregates.

The term *structure* characterizes the bulkiness of individual aggregates. Often when carbon black is mixed into rubber, some of the aggregates fracture. In addition to fracturing, the separation of physically attracted or compacted aggregates (microagglomerates) can occur. Aggregates can be held together fairly rigidly because of the irregular nature of their structures. Microagglomerates can also be formed during mixing. That is, high shear forces can cause micro-compaction effects wherein several aggregates can be pressed together. Excessive micro-agglomeration can lead to persistent black network structures that can affect extrusion and ultimate vulcanizate properties. The relationship of carbon black morphology to the failure properties of rubber vulcanizates has been studied extensively. Strength properties are usually enhanced with increasing black surface area and loading. The upper limit of black loading for maximum tensile strength and tear resistance depends on carbon black fineness and structure, with the former usually being more important. Coarser, lower-structure blacks generally show peak strength properties at higher loadings. In terms of the ultimate level of strength reinforcement for different blacks, structure is significant as a dispersing aid, which may be attributable to better bonding between black and polymer. High structure is also important at lower black loadings, particularly for tear resistance.

There is some evidence that the strength reinforcing properties of fillers are directly related to modulus development. These properties are derived from the large stresses held by the highly extended polymer chains attached to the immobile particles. High tensile strength to energy dissipation can be related through a slippage mechanism at the filler surface. This is supported by the fact that high-surface-area, inactive (partially graphitized) carbon blacks give very high tensile strength under standard testing conditions. Under more severe conditions, however, adhesion between black and polymer becomes important.

Some investigators have shown that the path of rupture through a filled vulcanizate passes from one filler aggregate to another, which are sites of high stress concentration. One approach to increasing strength is to lengthen the overall rupture path. The finer the filler and the higher the loading, the greater the effective increase in total cross section. There is a limiting point, however, when the packing of the filler aggregates becomes critical and they are no longer completely separated by the polymer. Preferential failure paths between aggregates may then be formed. These, combined with the fact that a smaller amount of polymer is being strained, can result in failure. The level of filler-polymer adhesion is important, especially as the upper limits of loading (aggregate packing) are reached. Filler-polymer adhesion is important to the amount of elastic energy released by internal failure. Strongly adhering fillers enlarge the volume of rubber that must be highly strained during the process of rupture. It is worth noting that the relative importance of different carbon black parameters that influence vulcanizate properties greatly depends on the conditions of testing.

Carbon black fineness is the dominant factor governing vulcanizate strength properties. Tensile strength, tear resistance, and abrasion resistance all increase with decreasing black particle size. Cyclic processes such as cracking and fatigue are more complex because they may involve internal polymer degradation caused by heat buildup. It is important to note whether or not the end-use application represents constant energy input or constant amplitude vibrations. The latter mode of service greatly reduces the fatigue life of rubber compounds containing fine or high-structure blacks.

The relationship of carbon black fineness to failure properties is not straightforward since it is difficult to separate the effects of aggregate size and surface area. It is likely that both are important, since aggregate size relates to the manner in which the surface area is distributed. In general, decreasing aggregate size and increasing black loading reduce the average interaggregate spacing, thereby lowering the mean free rupture path. Increasing black fineness and loading can be viewed in terms of either increased interface between the carbon black and polymer, or increased total black cross section (lower aggregate spacing). Both parameters have a strong effect on tensile strength up to the maximum tensile value that can be achieved.

The ultimate value for tensile strength across all blacks appears to be determined by either aggregate size or specific surface area. Tensile strength can be improved to a limiting value by increasing the loading. It is not possible, however, to match the ultimate tensile of a fine black by increasing the loading of a coarse one, at least not without other modifications (e.g., improving the ultimate dispersion or the bonding between black and polymer).

Blends of two or more elastomers are used in a variety of rubber products to achieve different effects and end-use performance characteristics. The compatibility of elastomers in terms of their relative miscibility and response to different fillers and curing systems is of great importance to the rubber compounder. From the standpoint of carbon black reinforcement, certain combinations of polymers can give less than optimum performance if the black is not proportioned properly between them. Note that equivalent volume proportionality of the black is not always desirable. This depends on the nature of the polymers and their relative filler requirements in terms of strength reinforcement.

Compatibility has been studied employing a variety of different analytical techniques, including optical and electron microscopy, differential thermal analysis (DTA), gel permeation chromatography (GPC), solubility, thermal and thermochemical analysis, and x-ray analysis. Microscopic methods are especially useful in studying the phase separation (zone size) of different polymer combinations and in determining the relative amounts of filler in each blend component. Phase-contrast optical microscopy is particularly useful, where the contrast mechanism is based on differences in the refractive indexes of the polymers. Blends such as NR/SBR, NR/BR, and SBR/BR have been extensively studied. Such studies have indicated that few, if any, elastomers can be blended on a molecular scale. Such studies have also reported that fillers and curing agents do not necessarily distribute proportionately. (Sources: N. P. Cheremisinoff, *Product Design and Testing of Polymeric Materials*, Marcel Dekker, Inc., New York, 1990 and N. P. Cheremisinoff, *Polymer Mixing and Extrusion Technology*, Marcel Dekker, Inc., New York, 1987.)

CARBON TYPE ANALYSIS

Refers to an empirical analysis of rubber process oil composition that expresses the percentages of carbon atoms in aromatic, naphthenic, and paraffinic components, respectively. See *Rubber Oil, Aromatic, Naphthene, Paraffin*.

CAS REGISTRY NUMBERS

CAS stands for the Chemical Abstract Service registry number identifying numbers assigned to chemical substances by the Chemical Abstract Service of the American Chemical Society and used by the Environmental Protection Agency (EPA) to aid in registering chemicals under the federal Toxic Substances Control Act (TSCA) of 1976. Petroleum products containing additives are termed "mixtures" by the TSCA and, as such, do not have CAS numbers. All chemical substances used in such mixtures are assigned CAS numbers and must be listed with the EPA by the supplier.

CASTING

If the material to be molded is already a stable liquid, simply pouring (casting) the liquid into a mold may suffice. Since the mold need not be massive, even the cyclical heating and cooling for a thermoplastic is efficiently done. One example of a cast thermoplastic is a suspension of finely divided, low-porosity PVC particles in a plasticizer such as DOP. This suspension forms a free-flowing liquid (a plastisol) that is stable for months. However, if the suspension (for instance, 60 parts PVC and 40 parts plasticizer) is heated to 180° C for five minutes, the PVC and plasticizer will form a homogeneous gel that will not separate into its components when cooled back to room temperature. A realistic insect can be cast from a plastisol using inexpensive molds and a cycle requiring only minutes. In addition, when a mold in the shape of a hand is dipped into a plastisol and then removed, subsequent heating will produce a glove that can be stripped from the mold after cooling (a process called dipping). In casting a mixture of polymer and multifunctional monomers with initiators can be poured into a heated mold. When polymerization is complete, the article can be removed from the mold. A transparent lens can be formed in this way using a diallyl diglycol carbonate monomer and a free-radical initiator.

CAVITATION

Refers to the formation of an air or vapor pocket (or bubble) due to lowering of pressure in a liquid, often as a result of a solid body, such as a propeller or piston, moving through the liquid; also, the pitting or wearing away of a solid surface as a result of the collapse of a vapor bubble. Cavitation can occur in a hydraulic system as a result of low fluid levels that draw air into the system, producing tiny bubbles that expand explosively at the pump outlet, causing metal erosion and eventual pump destruction. Cavitation can also result when reduced pressure in lubricating grease dispensing systems forms a void, or cavity, which impedes suction and prevents the flow of greases.

CELLULAR PLASTICS

The term *Cellular Plastics* also refers to plastic foams and expanded or sponge plastics. These materials generally consist of a minimum of two phases, a solid polymer matrix and a gaseous phase derived from a blowing agent. There may be more than one solid phase present, as in the case of a blend or alloy of polymers (generally heterogeneous). Other solid phases may be present in the foams in the form of fillers, either fibrous or other shapes. The fillers may be of inorganic origin, of a material such as glass, ceramic, or metal, or polymeric in nature.

Foams may be flexible or rigid, depending upon whether their glass transition temperature is below or above room temperature, which in turn depends upon their chemical composition, the degree of crystallinity, and the degree of crosslinking. Intermediate between flexible and rigid foams are semirigid or semiflexible foams. The cell geometry may be open-(tunnels between the cells), or closed-cell (refer to Figures 1 and 2).

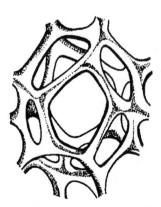

Figure 1. Open gas structural element.

Figure 2. Closed gas structural element.

Closed-cell foams are most suitable for thermal insulation and are generally rigid, while open-celled foams are best for car seating, furniture, bedding and acoustical insulation, among applications. These are generally flexible.

Plastic foams are manufactured in a variety of densities ranging from about 1.6 kg/m^3 to over 960 kg/m^3. The mechanical strength properties are generally proportional to the foam densities. The applications of these foams usually determine which range of foam densities should be produced. Rigid foams for load-bearing applications require high density, fiber reinforcement, or both, while low densities are usually used for thermal insulation. Low-density flexible foams (around 30 kg/m^3) are usually used in furniture and automotive seating, while somewhat higher densities are used for carpet backing and energy-absorbing applications.

The foaming of polymeric materials can be carried out by either mechanical, chemical, or physical means. Some of the most commonly used methods include: thermal decomposition of chemical blowing agents, generating either nitrogen or carbon dioxide or both, by application of heat or as a result of the exothermic heat of reaction during polymerization; mechanical whipping of gases (frothing) into a polymer system (melt, solution, or suspension), which hardens either by catalytic action or heat or both, thus entrapping the gas bubbles in the matrix; volatilization of low-boiling liquids (such as fluorocarbons or methylene chloride) within the polymer mass as a result of the exothermic heat of reaction or by application of heat; volatilization by the exothermic heat of reaction of gases produced during polymerization, such as occurs in the reaction of isocyanate with water to form carbon dioxide; expansion of gas dissolved in a polymer mass upon reduction of pressure in the system; incorporation of microspheres into a polymer mass (the hollow microspheres may consist of either glass or plastic beads); and expansion of gas-filled beads by thermal application, or the expansion of these beads in a polymer mass via heat of reaction (an example of this last case is the expansion of polystyrene beads in polyurethane or epoxy resin systems. The most common methods of foaming are:

1. production of continuous slab stock by pouring; multi-component foam machines using impingement mixing are employed;
2. compression molding of foams;
3. reaction injection molding (RIM);
4. foaming in place by pouring from a dual or multi-component head;
5. spraying of foams;
6. extrusion using expandable beads or pellets;
7. injection molding of expandable beads or pellets;
8. rotational casting;
9. frothing of foams by introducing either air or a low-boiling volatile solvent;
10. lamination of foams, known as foam board production;
11. manufacture of foam composites;
12. precipitation foam processes, in which a polymer phase forms by

polymerization or precipitation from a liquid that is later allowed to escape.

A good reference to consult is that of D. Klempner and K. C. Frisch, *Handbook of Polymeric Foams and Foam Technology*, Hanser Publihers, New York, 1991.

CELLULOSE ACETATE PROPIONATE

Table 1 provides a summary of the average properties of this material. In reviewing the general properties of this polymer, note the use of the following legend: A = amorphous - Cr = crystalline - C = clear - E = excellent - G = good - P = poor - O = opaque - T = translucent- R = Rockwell - S = Shore.

Table 1
Average Properties of Cellulose Acetate Proprionate.

STRUCTURE: A	FABRICATION		
	Bonding	**Ultrasonic**	**Machining**
	P	P	P
Specific Density: 1.22	**Deflection Temperature (° F)**		
	@ 66 psi	**@ 264 psi**	
	194	167	
Water Absorbtion Rate (%): 2	**Utilization Temperature (° F)**		
	min:	max:	
	-40	212	
Elongation (%): 50	**Melting Point (° F)**: 248		
Tensile Strength (psi): 6000	**Coefficient of Expansion**: 0.00007		
Compression Strength (psi): 8500	**Arc Resistance**: 180		
Flexural Strength (psi): 8500	**Dielectric Strength (kV/mm)**: 13		
Flexural Modulus (psi): 180000	**Transparency**: C		
Impact (Izod ft. lbs/in): 1.5	**UV Resistance**: G		
Hardness: R80	**CHEMICAL RESISTANCE**		
	Acids	**Alkalis**	**Solvents**
	P	P	P

CHEMICAL ANTIOZONANTS

Many compounds have been reported in the literature to be chemical antiozonants, and nearly all contain nitrogen. Compound classes include derivatives of 2,2,4-trimethyl-1, 2-dihydroquinoline, N-substituted ureas or thioureas, substituted pyrroles, and nickel or zinc dithiocarbamate salts. The most effective antiozonants, however, are derivatives of p-phenylenediamine (p-PDA). The commercial materials are grouped into three classes: N,N'-dialkyl-p-PDAs, N-alkyl-N'-aryl-p-PDAs, and NX-diaryl-p-PI)As. The NX-dialkyl-p-PDAs (where the alkyl group may be 1-methylheptyl, 1-ethyl-3-methylpentyl, 1,4-dimethylpentyl, or cyclohexyl) are the most effective in terms of their reactivity to ozone. These derivatives increase the critical stress required for the initiation of crack growth, and they also reduce the rate of crack growth significantly. The sec-alkyl group is most active, for reasons that are not yet completely clear. The drawbacks of these derivatives are:

1. their rapid destruction by oxygen and consequent shorter useful lifetimes,
2. their activity as vulcanization accelerators and hence increased scorchiness,
3. their tendencies to cause dark red or purple discoloration, and
4. their liquid nature and hence difficulty in handling.

The dialkyl-p-PDAs are seldom used alone in rubber compounds, although they can be used effectively when blended with alkyl/aryl-p-PDAs. The N-alkyl-N'-aryl-p-PDAs (where the aryl is phenyl and the alkyl may be isopropyl, sec-butyl, cyclohexyl, 1-methylheptyl, or 1,3-dimethylbutyl) are the most widely used p-PDAs. These derivatives reduce the rate of crack growth and also the number of cracks, but in general-purpose rubbers the critical stress is not appreciably raised unless the compound is combined with a wax or a dialkyl type. It is interesting that their behavior is synergistic with a wax under static conditions to a much greater extent than with dialkyl derivatives.

The alkyl/aryl-p-PDAs are in general excellent antiozonants, particularly in dynamic environments. These derivatives are destroyed only slowly by oxygen and increase the scorchiness of the stock only slightly. In addition to their antiozonant ability, they are also very effective as antifatigue (mechanostabilization) agents and antioxidants. Since they are low-melting solids, they are easily handled. A principal problem in selected applications is that these derivatives (like the dialkyls) are staining. The NX-diaryl-p-PI)As (where the aryl group may be phenyl, tolyl, or xylyl) are only moderately active antiozonants, which can only be used at low concentrations (generally less than 2 phr) because of their poor solubility. However, they have minimal scorching effects, are the most stable toward oxygen, and stain the least. Their main advantage is high resistance to loss by consumption and vaporization, and in combination with more reactive antiozonants they can offer a degree of increased protection in such longterm applications as radial passenger tires. Very few chemical antiozonants outside the class of p-PDAs have much commercial importance. One of the few exceptions to this rule is 6-ethoxy-2,2,4-trimethyl-1,2-dihydroquinoline, one of the first commercial antiozonants.

It should be clear from the foregoing discussion that the principal objection to p-PDA antiozonants is their staining characteristics. The lack of suitable alternative antiozonants for light-colored diene rubber articles is one of the major outstanding problems in rubber technology. A notable exception is chloroprene rubber, which has more natural ozone resistance than other diene rubbers. For example, pentaerythritol acetal derivatives have been shown to be effective nonstaining antiozonants for CR. These derivatives are reactive enough with ozone to protect CR vulcanizates, but they are not effective in other general-purpose diene rubbers. For natural rubber, two compound classes which have some promise as nonstaining antiozonants are substituted thioureas and metal dithiocarbamates. Tributylthiourea has been used as an NR antiozonant, but its activity is considerably less than that of p-PDAs and in addition it is very scorchy.

Zinc dialkyl dithiocarbamates reduce the rate of crack growth, but their scorch resistances are prohibitively low. A few other classes of compounds have been reported to react rapidly with ozone, such as phosphines and stibines. These materials also exhibit antiozonant activity when swollen into rubber vulcanizates after the cure. However, they cannot be used practically for two reasons:

- they react readily with oxygen, and
- they are destroyed during vulcanization.

Toxicity rules out selenium and tellurium compounds, which have also been reported to have antiozonant activity. Several theories have appeared in the literature regarding the mechanism of protection by p-PDA antiozonants. The *scavenger* theory states that the antiozonant diffuses to the surface and preferentially reacts with ozone, with the result that the rubber is not attacked until the antiozonant is exhausted.

The *protective* film theory is similar, except that the ozoneantiozonant reaction products form a film on the surface that prevents attack. The *relinking* theory states that the antiozonant prevents scission of the ozonized rubber or recombines severed double bonds. A fourth theory states that the antiozonant reacts with the ozonized rubber or carbonyl oxide to give a low molecular weight, inert *self-healing* film on the surface. The literature suggests that more than one mechanism may be operative for a given antiozonant and that different mechanisms may be applicable to different types of antiozonants. All of the evidence, however, indicates that the scavenger mechanism is the most important one.

All antiozonants react with ozone at a much higher rate than does the rubber which they protect. For example, selected antiozonants have been shown to react preferentially with ozone in rubber films as well as in solution. Antiozonant-ozone reaction rates are typically one to two orders of magnitude higher than the rates for diene rubbers.

Surface spectroscopy also supports preferential attack of ozone on the antiozonant. Although all antiozonants must react rapidly with ozone, not all highly reactive materials are antiozonants. Something else in addition to the scavenging effect is required. The protective film theory contends that ozonized products, to a considerable extent, prevent ozone from reaching the rubber. There is visual and microscopic evidence for formation of a protective film on the rubber surface. Spectroscopic characterization has shown that this film consists of unreacted antiozonant and many of the same components observed in ozonized liquid antiozonant.

The components of the film are polar and tend not to diffuse back into the rubber bulk. The surface film evidently acts as a secondary scavenger, as well as a partial physical barrier, for ozone. It has been assumed that the physical properties of these films play an important role in the functioning of an antiozonant. Differences in these properties have been used to explain why *N,N'-tri*substituted and NX-tetrasubstituted p-phenylenediamines are not as effective as their NX-disubstituted counterparts, even though all of these derivatives scavenge ozone at about the same rate.

The relinking and self-healing film theories require chemical interaction between the antiozonant, or ozonized antiozonant, and the rubber or ozonized rubber. The evidence for these interactions is sparse in the literature. The products of the ozone-antiozonant reaction are soluble in acetone. Thus if only the scavenger and protective film mechanisms are operative, no nitrogen from the antiozonant should be left in the rubber after ozonation and subsequent acetone extraction. Nitrogen analyses of extracted rubber showed, however, that some of the nitrogen was unextractable; this nitrogen was presumably attached chemically to the rubber network.

Experiments showing that antiozonants react readily with aldehydes led to the suggestion that the antiozonant may work by relinking aldehydic end groups of rubber chains that have been broken by ozonolysis. While it is confirmed that vulcanization and ozone aging can lead to attachment of p-PDA fragments to the rubber matrix, mechanisms other than antiozonant action can cause this unextractable nitrogen effect. In particular, it has been proposed that the principal reason for nitrogen becoming bound during aging is the trapping of macroalkyl radicals by aromatic amine-derived species (e.g., nitrones and nitroxides). See *Antiozonants* and *Ozone Cracking*. (Source: *Handbook of Polymer Science and Technology, Volume 2: Performance Properties of Plastics and Elastomers,* N. P. Cheremisinoff, Macrel Dekker Publishers, New York, 1989).

CHEMICAL RESISTANCE

Table 1 has been compiled from product literature and an overall qualitative rating developed for plastics in common use throughout industry. This chemical resistance chart can be used as a general selection guide for polymers. The reader should consult grade specific data, especially with regard to the effects of material exposures to temperature and immersion times to different chemicals. Under severe environmental or process conditions, it is advisable to perform accelerated chemical exposure testing. Although there are no strict guidelines for these types of tests, both suppliers and the general literature can be of assistance.

Table 1. Chemical Resistance Chart for Common Polymers

A = Very good B = Good C = Moderate D = Not Recommended							
Chemicals	**CPCV**	**Epoxy**	**PP**	**PE**	**PVC**	**Phenolic**	**PA**
Acetaldehyde	D	A	A	C	D	A	A
Acetamide	-	A	A	A	D	D	A
Acetic Acid, 80%	C	C	A	D	D	D	D
Acetone	D	D	A	B	D	D	A
Acetylene	C	A	A	A	A	A	A
Amyl Alcohol	A	D	B	B	A	A	A
Benzyl Alcohol	A	A	A	D	D	A	A
Butyl Alcohol	A	A	A	A	A	C	A
Ethyl Alcohol	A	A	A	B	C	A	A
Isopropyl Alcohol	A	A	A	A	A	A	B
Methyl Alcohol	A	B	A	A	A	A	A
Aluminum Sulfate	A	A	A	A	A	A	A
Ammonia	A	A	A	C	A	A	B
Ammonia Nitrate	B	A	A	-	B	A	D
Aniline	C	C	C	B	C	D	C
Anti-freeze	A	A	D	-	A	A	D
Aromatic HCs	D	A	D	C	D	A	-
Arsenic Acid	A	A	A	B	A	A	C
Barium Carbonate	A	A	A	B	A	-	A
Barium Sulfate	B	C	B	B	B	-	A
Benzaldehyde	D	A	A	A	D	D	C
Benzene	C	C	C	C	C	A	A
Benzoic Acid	A	A	C	B	A	C	C
Benzol	-	A	A	C	-	A	D
Borax	A	A	B	A	B	D	B

Table 1 continued

Chemicals	CPCV	Epoxy	PP	PE	PVC	Phenolic	PA
Boric Acid	A	A	A	A	A	A	B
Butadiene	A	A	D	D	C	-	A
Butane	C	A	C	C	C	-	A
Butylene	A	A	-	B	C	A	B
Calcium Stearate	A	A	A	B	A	D	D
Carbon Bisulfide	A	A	C	-	D	-	A
Carbon Dioxide	A	A	A	C	A	A	A
Carbon Disulfide	C	C	D	C	D	A	B
Carbonic Acid	A	B	B	B	A	A	A
Chloric Acid	A	-	-	-	A	-	D
Chlorine, Anhydrous	B	C	B	B	C	D	D
Chloroform	D	C	C	C	D	A	D
Chromic Acid 50%	C	D	B	A	C	C	D
Citric Acid	B	A	A	A	B	A	A
Clorox (Bleach)	A	A	D	-	B	A	A
Copper Sulfate	A	A	A	B	A	A	C
Cyanic Acid	-	A	-	-	-	D	-
Diesel Fuel	A	A	A	C	A	-	A
Ethane	-	A	C	-	D	-	D
Ethylene Glycol	A	C	A	A	A	A	B
Fatty Acids	B	A	B	A	B	D	A
Ferric Chloride	A	A	B	A	A	A	C
Ferric Sulfate	A	A	B	A	A	D	A

Table 1 continued

Chemicals	CPCV	Epoxy	PP	PE	PVC	Phenolic	PA
Flourine	A	D	D	C	D	-	D
Formaldehyde	A	A	C	B	A	B	D
Formic Acid	A	C	A	B	A	C	C
Gasoline	C	A	C	C	C	A	A
Grease	-	A	-	-	A	D	-
Heptane	A	A	C	B	C	D	A
Hydrazine	D	A	C	-	-	A	-
Hydrochloric Acid 20%	A	A	B	A	A	A	D
Hydrocloric Acid 100%	A	-	B	-	B	-	D
Hydrogen Peroxide 30%	A	B	B	C	A	D	D
Hydrogen Peroxide 100%	A	A	B	C	C	D	D
Iodine	D	C	C	A	D	-	D
Magnesium Hydroxide	A	A	A	A	A	D	B
Mercury	A	A	B	A	B	D	A
Oleum 100%	D	D	D	D	D	C	D
Petrolatum	-	A	D	B	B	-	D
Phenol	A	C	B	B	C	D	C
Phosphoric Acid	A	B	B	B	B	D	B
Picric Acid	D	A	B	-	D	D	C
Potassium Carbonate	A	A	A	A	A	A	A

Table 1 continued

Chemicals	CPCV	Epoxy	PP	PE	PVC	Phenolic	PA
Silver Nitrate	A	A	A	B	A	A	A
Sodium Bicarbonate	A	A	A	A	A	A	A
Stearic Acid	B	B	A	B	B	D	A
Sulfuric Acid 10%	A	A	A	A	A	D	D
Sulfuric Acid > 75%	C	C	C	B	D	D	D
Tannic Acid	A	A	A	B	A	-	C
Toluene	D	B	C	C	D	A	A
Zinc Sulfate	A	A	A	A	A	A	A

CPCV = Chlorinated Polyvinyl Chloride; PP = Polypropylene Homopolymer, PE = Polyethylene; PVC = Polyvinyl Chloride; PA = Polyamide (Nylon).

CHEMICAL TRANSPORTATION EMERGENCY CENTER

The *Chemical Transportation Emergency Center* is more commonly known as CHEMTREC. In the United States, the Manufacturing Chemists Association operates CHEMTREC 24 hours a day. By calling the appropriate toll-free number listed below, one can consult experts on chemicals and spill response.

- Continental United States (except Alaska & District of Columbia) 800-424-9300.
- Alaska, Hawaii, and District of Columbia 202-483-7616.

Some useful references that the reader can refer to on chemical safety are as follows:

1. *North American Emergency Response Guidebook*, J. J. Keller, Wisconsin, 1996.
2. Cheremisinoff, N.P., *Handbook of Industrial Toxicology and Hazardous Materials*, Marcel Dekker Publishers, New York and Basel (1999).
3. Cheremisinoff, N.P., *Handbook of Pollution and Hazardous Materials Compliance*, Marcel Dekker Publishers, New York and Basel (1996).
4. Cheremisinoff, N.P., *Handbook of Emergency Response to Toxic Chemical Releases: A Guide to Compliance*, Noyes Publishers, Westwood, New Jersey (1995).
5. Cheremisinoff, N.P., *Environmental Health & Safety Management: A Guide for the Professional Hazards Manager*, Noyes Publishers, Westwood, New Jersey (1995).
6. Cheremisinoff, N.P., *Transportation of Hazardous Materials: A Guide to*

Compliance, Noyes Publishers, Westwood, New Jersey (1994).

7. Cheremisinoff, N.P., *Hazardous Chemicals in the Polymers Industry*, Marcel Dekker Publishers, New York and Basel (1995).

8. Cheremisinoff, N. P., *Practical Guide to Industrial Safety: Methods for Process Safety Professionals*, Marcel Dekker, Inc., New York, (2000).

CHLORINATED POLYVINYL CHLORIDE (CPVC)

Table 1 provides a summary of the average properties of this material. In reviewing the general properties of this polymer, note the use of the following legend: A = amorphous - Cr = crystalline - C = clear - E = excellent - G = good - P = poor - O = opaque - T = translucent- R = Rockwell - S = Shore - ND = no data.

Table 1

Average Properties of CPCV

STRUCTURE: A	FABRICATION		
	Bonding	**Ultrasonic**	**Machining**
	G	P	P
Specific Density: 1.56	**Deflection Temperature (° F)**		
	@ 66 psi	**@ 264 psi**	
	162:	140	
Water Absorption Rate (%): 0.15	**Utilization Temperature (° F)**		
Elongation (%): 40	min	max	
	ND	178	
	Melting Point (° F): 175		
Tensile Strength (psi): 8700	**Coefficient of Expansion:** 0.00004		
Compression Strength (psi): 14500	**Transparency:** C		
Flexural Strength (psi): 15500	**UV Resistance:** P		
Flexural Modulus (psi): 435000	**CHEMICAL RESISTANCE**		
Impact (Izod ft. lbs/in): 12	**Acids**	**Alkalis**	**Solvents**
Hardness: R110	E	E	G

CHLOROPRENE OR NEOPRENE, AND DESIGNATED CR

Also called Polychloroprene. A versatile special purpose oil resistant rubber. Compared with NR most of the physical properties are stable when in a hostile environment. General properties include good mechanical and electrical properties; very good resistance to ozone, oxidation, heat and flame; good resistance to swelling in oils; less permeable to most gases than NR or SBR. Approximate working temperature range -20° C to +130° C.

CHLOROSULPHONATED POLYETHYLENE

More commonly known as Hypalon. This is a special purpose rubber formed by substituting chlorine and sulphonylchloride groups into polyethylene. General properties include moderate mechanical properties; excellent resistance to ozone, oxidation, weathering, and oxidizing chemicals. Its upper temperature limit is ca. +150° C. Hypalon® or Chlorosulfonated Polyethylene Rubber is made by DuPont-Dow. It is an elastomer and is a thermoset material. There are many grade variations of this specialty elastomer. A typical grade is 29 % chlorine content and 1.4% sulfur content. Its neat product form is white, odorless chips. A distinguishing feature is that it is readily soluble in common solvents. The polymer has the following characteristics:

1.   good low temperature flexibility,

2.   very good abrasion resistance,

3.   good chemical resistance,

4.   fair compression set resistance,

5.   fair flame resistance,

6.   very good heat resistance,

7.   excellent ozone resistance,

8.   fair petroleum oil resistance,

9.   excellent weather resistance,

10.   good low temperature properties, and

11.   fair tear strength.

In terms of processability, it is described as demonstrating fair extrusion, good molding, and fair calendering. Some general physical properties are reported in Table 1, however the reader should examine individual grades. The physical property data below (except viscosity and density) are for vulcanizate.

Vulcanizates of this chlorosulfonated polyethylene synthetic rubber are highly resistant to ozone, oxygen, weather, heat, oil, and chemicals. Hypalon resists discoloration on exposure to light and is widely used in light-colored vulcanizates.

It can be compounded to give excellent mechanical properties. Several grades are available, all of which may be processed and used in the usual manner for solid elastomeric vulcanizates. Various grades of Hypalon have been used in single-ply roofing systems; auto power steering and oil cooler hoses; chemical-resistant liners; cable sheathing; and other coatings.

Table 1

Average Physical Properties of a Hypalon® Grade

Properties	Values	Comments	US/other units
Density, g/cc	1.12		1.12 g/cc
Hardness, Shore A	70	45-95. ASTM D2240-81	70
Viscosity	28	Mooney Viscosity ML 1+4 at 100° C. ASTM D1646-81	28
Tensile Strength, Ultimate, MPa	20.6	Upper limit with carbon black stocks. ASTM D412-80	2,987 psi

CHROMATOGRAPHY

Chromatography is the separation of molecular mixtures by distribution between two or more phases, one phase being essentially two-dimensional (a surface) and the remaining phase, or being a bulk phase brought into contact in a counter-current fashion with the two dimensional phase. Various types of physical states of chromatography are possible, depending on the phases involved.

Chromatography is divided into two main branches. One branch is gas chromatography, the other is liquid chromatography. The sequence of chromatographic separation is as follows: A sample is placed at the top of a column where its components are sorbed and desorbed by a carrier. This partitioning process occurs repeatedly as the sample moves towards the outlet of the column.

Each solute travels at its own rate through the column, consequently, a band representing each solute will form on the column. A detector attached to the column's outlet responds to each band. The output of detector response versus time is called a chromatogram.

The time of emergence identifies the component, and the peak area defines its concentration, based on calibration with known compounds. Refer to *Gas Chromatography* and *Liquid Chromatography*.

CLEVELAND OPEN CUP (COC)

Test (ASTM D 92) for determining the *flash point* and *fire point* of all petroleum and rubber products except fuel oil and products with flash points below 79° C (175° F). The oil sample is heated in a precisely specified brass cup containing a thermometer. At specified intervals a small flame is passed across the cup. The lowest temperature at which the vapors above the cup briefly ignite is the flash point; the temperature at which the vapors sustain combustion for at least five seconds is the fire point. Refer to *Closed Cup Method*.

CLOSED CUP METHOD

Method for determining the *flash point* of fuels, solvents, resins and cutback asphalts, utilizing a covered container in which the test sample is heated and periodically exposed to a small flame introduced through a shuttered opening. The lowest temperature at which the vapors above the sample briefly ignite is the flash point.

CLOUD POINT TEMPERATURE

The temperature at which a cloud or haze of wax crystals appears at the bottom of a sample of lubricating oil in a test jar, when cooled under conditions prescribed by test method ASTM D 2500. Cloud point is an indicator of the tendency of the oil to plug filters or small orifices at cold operating temperatures. It is very similar to wax appearance point.

COATING

Although color may be added in the form of a pigment or dye throughout a plastic article, there are many applications where a surface coating is valuable for protective or decorative purposes. The automobile bumpers produced by reaction injection molding can be painted to match the rest of the body. It is important in applying coatings to plastics that the solvent used does not cause swelling of the underlying substrate. For this reason, latex dispersion paints have found favor, although surface treatment is necessary to provide good bonding with these materials.

CODE OF FEDERAL REGULATIONS

The hazard class specified in the Code of Federal Regulations, Title 49, Part 172. Chemicals not specifically listed therein have been classified as "Flammable", if their flash point (closed cup) is below 100° F. Users of compounding ingredients or monomers should always refer to CFR when transporting, packaging or storing the chemicals. The CFR provides valuable safety and regulatory information. In addition to Title 49, readers should refer to 29CFR 1910.120 - The OSHA Safety Regulations.

COEXTRUSION

Coextrusion is a technique by which different polymers are joined together in the melt phase. This technique is applicable to various processing methods such as cast and blown film, extrusion coating and even blow molding or cable coating. Die designs are usually very complex. Rheological compatibility and interlayer adhesion are critical. To join incompatible polymers, tie layers have to be used. Coextruslon allows one to combine interesting properties of different materials into one product, e.g., Barrier and sealing properties (nylon/tie layer/PE, PF,/tie I./EVOH/tie I./PE); Single-side tack (PF/PE plus tackifier, PE/HEVA); different in/outside color (black/white hydroculture film) for opacity plus light reflection. A blown film coextrusion line (3 layers) is illustrated in Figure 1.

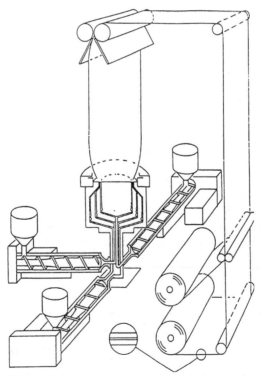

Figure 1. Blown film coextrusion (3 layers).

COKNEADER

A cokneader is a single screw extruder with pins on the barrel and a screw that oscillates in the axial direction. Figure 1 shows a schematic diagram of a cokneader. The pins on the barrel practically wipe the entire surface of the screw, making it a self-cleaning single-screw extruder. This results in a reduced residence time, which makes it appropriate for processing thermally sensitive materials. The pins on the barrel also disrupt the solid bed creating a dispersed melting which improves the over melting rate while reducing the overall temperature in the material.

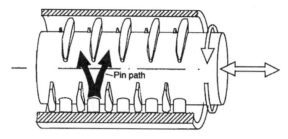

Figure 1. Shows features of a cokneader.

COLD-FLOW IMPROVER

An additive to improve flow of diesel fuel in cold weather. In some instances, a cold-flow improver may improve operability by modifying the size and structure of the wax crystals that precipitate out of the fuel at low temperatures, permitting their passage through the fuel filter.

In most cases, the additive depresses the *pour point,* which delays agglomeration of the wax crystals, but usually has no significant effect on diesel engine performance. A preferred means of improving cold flow is to blend kerosene with the diesel fuel, which lowers the wax *appearance point* by about 10° C for each 10% increment of kerosene added.

COLD MOLDING

Even without heating, some thermoplastics can be formed into new shapes by the application of sufficient pressure. This technique, called cold molding, has been used to make margarine cups and other refrigerated food containers from sheets of acrylonitrile-butadiene-styrene copolymer.

COLORANTS

For most consumer applications, plastics and elastomers are colored. The ease with which color is incorporated throughout a molded article is an advantage of plastics over metals and ceramics, which depend on coatings for color. Popular pigments for coloring plastics and elastomers include titanium dioxide and zinc oxide (white), carbon (black), and various other inorganic oxides such as iron and chromium. Organic compounds can be used to add color either as pigments (insoluble) or as dyes (soluble).

COMPOSITES

A simple definition for a composite is two or more materials, that when combined, will retain their original identities but do interface chemically and physically when subjected to specific conditions of either time, temperature, pressure, or chemical catalyzation. Composites are widely available with different resin systems and reinforcement combinations. They are produced in several physical forms and can be shaped with conventional molding methods as well as a variety of other less conventional methods such as resin transfer molding (RTM), resin injection molding (RIM), structural resin injection molding (SRIM), hand-lay-up, and autoclave molding(i.e., vacuum bag technique).

COMPOUNDED OIL

Refers to a mixture of a petroleum oil with animal or vegetable fat or oil. Compounded oils have a strong affinity for metal surfaces; they are particularly suitable for wet-steam conditions and for applications where lubricity and extra load-carrying ability are needed. They are not generally recommended where long-term oxidation stability is required.

COMPOUNDING

The first step in most plastic or polymer fabrication procedure is compounding, the mixing together of various raw materials in proportions according to a specific recipe. Most often the plastic resins are supplied to the fabricator as cylindrical pellets (several millimeters in diameter and length) or as flakes and powders. Other forms include viscous liquids, solutions, and suspensions. Elastomers are usually in the form of crumb, bales or pellets. Mixing liquids with other ingredients may be done in conventional stirred tanks, but certain operations demand special machinery. Dry blending refers to the mixing of dry ingredients prior to further use, as in mixtures of pigments, stabilizers, or reinforcements. However, PVC as a porous powder can be combined with a liquid plasticizer in an agitated trough called a ribbon blender or in a tumbling container. This process also is called dry blending, because the liquid penetrates the pores of the resin, and the final mixture, containing as much as 50 percent plasticizer, is still a free-flowing powder that appears to be dry. The workhorse mixer of the plastics and rubber industries is the internal mixer, in which heat and pressure are applied simultaneously. The Banbury mixer resembles a robust dough mixer in that two interrupted spiral rotors move in opposite directions at 30 to 40 rotations per minute. The shearing action is intense, and the power input can be as high as 1,200 kilowatts for a 250-kilogram (550-pound) batch of molten resin with finely divided pigment. In some cases, mixing may be integrated with the extrusion or molding step, as in twin-screw extruders.

COMPRESSION SET

Compression set is a measure of a compound's resistance to flow while in a compressed state. Unreacted curative will contribute to poor compression set resistance because it continues to react while the sample is in the compressed state. These new cross-links then prevent the stock from returning to its original shape. If peroxide formulations are cured to the cure plateau, 7 to 10 half-lives, no further curing is possible and compression set resistance is at a maximum for that particular compound. Compression set resistance in stocks cured in this manner varies directly with state of cure (or the quantity of peroxide in the formulation). Figure 1 shows the typical response of an elastomer fully cured with various levels of peroxide. The isolated point on the curve shows the result of undercuring a stock and utilizing only 50% of the peroxide added.

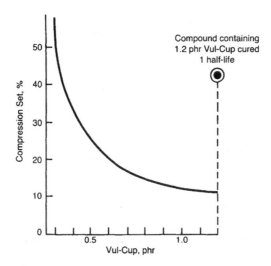

Figure 1. Shows compression set for a peroxide cure vulcanizate.

CONDENSERS

Condensation is the process of reduction of matter into a denser form, as in the liquefaction of vapor or steam. Condensation is the result of the reduction of temperature by the removal of the latent heat of evaporation. The removal of heat shrinks the volume of the vapor and decreases the velocity of, and the distance between, molecules. The process can also be thought of as a reaction involving the union of atoms in molecules. The process often leads to the elimination of a simple molecule to form a new and more complex compound.

Condensation heat transfer is a vital process in the process and power generation industries, as well as polymerization reactions and laboratory characterization studies of polymers. The existing modes of condensation are **filmwise** and **dropwise**. Filmwise is currently used by industry, while dropwise is an alternative which is under development because it offers attractive higher rates of heat transfer by preventing the buildup of the insulating liquid layer found in filmwise condensation. All but a few precious metals will in an untreated state tend to condense filmwise: this is why industrial condensers operate filmwise. The type of condensation behavior which a metal displays is related to it's surface energy. Materials with a high surface energy condense filmwise while those with a low surface energy condense dropwise. With suitable promoters or surface treatments, most metals, including those with high surface energies, can promote dropwise condensation.

Non-azeotropic mixtures have been utilized in refrigeration systems for several direct and indirect advantages like, enhanced coefficient of performance, lower power consumption, reduced thermal irreversibility, increased chemical stability, improved oil miscibility, varying condensation temperatures and variable capacity refrigeration systems. All these merits offer rich prospects for the use of mixed component working fluids in heat pumps, power cycles and refrigeration systems.

In the mid 1980s, a new thermodynamic power cycle using a multicomponent working fluid as ammonia-water with a different composition in the boiler and condenser was proposed (known as the Kalina cycle). The use of a non-azeotropic mixture decreases the loss of availability in a heat recovery boiler when the heat source is a sensible heat source, and in a condenser when the temperature decreases with heat exchange. Most heat input to a plant's working fluid is from variable temperature heat sources.

Due to its variable boiling temperature, the temperature rise in an ammonia-water mixture in a counterflow heat-exchanger closely follows the straight line temperature drop of the heat source. Although this is an advantage over the conventional single component Rankine cycle, given equal condenser cooling temperatures, the ammonia-water mixture will have a significantly higher pressure and temperature than the steam at the condensing turbine outlet. The higher pressure is a result of ammonia being more volatile than water. The higher temperature is a result of variable condensing temperature of the ammonia-water mixture.

In the Rankine cycle much of the heat (almost 65%) from a turbine exhaust cannot be recuperated because there is no temperature difference between the steam at the turbine exhaust and the water at the condenser outlet. However, in the Kalina cycle, much of this latent heat can be extracted due to the higher temperature of the turbine exhaust over the ambient coolant temperature. The need for effectively extracting latent heat from the turbine exhaust in the Kalina cycle is the motivation behind numerous studies of convective condensation of non-azeotropic vapor mixtures. In condensers operating with pure vapors, the vapor pressure generally remains constant during the process of phase change. Therefore, it implies that the temperature difference between the vapor and the coolant increases along the direction of vapor flow in a counterflow type of heat exchanger. Thus, a situation is created in which the available excess energy is maximum at the exit of the condensate and minimum at the entrance of the vapor. As a result, all the available energy is not utilized in pure vapor condensation.

The utilization of availability can be enhanced by maintaining a constant temperature difference between the vapor and coolant, all along the heat exchanger. This can be achieved by using a certain non-azeotropic vapor mixture which can maintain a constant temperature difference due to its variable boiling temperature characteristics. The introduction of another condensable vapor may alter the composition of the vapor and decrease the heat and mass transport in the condenser. Furthermore, the orientation of the condenser can affect the flow regime in the condenser, and hence alter the performance of the condenser.

Condensers are a standard piece of equipment found in any solution polymerization process. They are often used alongside heat exchanger units. In surface and contact condensers, the vapors can be condensed either by increasing pressure or extracting heat. In practice, condensers operate through removal of heat from the vapor. Condensers differ principally in the means of cooling. In surface condensers, the coolant does not contact the vapors or condensate. In contact condensers, coolant, vapors, and condensate are intimately mixed.

Most surface condensers are of the tube and shell type shown in Figure 1. Water flows inside the tubes, and vapors condense on the shell side. Cooling water is normally chilled, as in a cooling tower, and reused. Air-cooled surface condensers and some water-cooled units condense inside the tubes. Air-cooled condensers are usually constructed with extended surface fins. Most vapors condense inside tubes cooled by a falling curtain of water. The water is cooled by air circulated through the tube bundle. The bundles can be mounted directly in a cooling tower or submerged in water. Contact condensers employ liquid coolants, usually water, which come in direct contact with condensing vapors. These devices are relatively uncomplicated, with typical configurations illustrated below. Some contact condensers are simple spray chambers, usually with baffles to ensure adequate contact. Others incorporate high-velocity jets designed to produce a vacuum. In comparison to surface condensers, contact condensers are more flexible, are simpler, and are considerably less expensive to install. On the other hand, surface condensers require far less water and produce 10 to 20 times less condensate than contact type condensers.

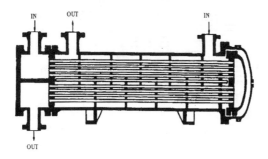

Figure 1. Shell and tube condenser.

Condensate from contact units cannot be reused and may constitute a waste disposal problem. Surface condensers can be used to recover valuable condensate. Surface condensers must be equipped with more auxiliary equipment and generally require a greater degree of maintenance. In general, subcooling requirements are more stringent for surface units than for contact condensers, where dilution is much greater. Nevertheless, many surface condenser designs do not permit adequate condensate cooling. In the typical water-cooled, horizontal, tube-and-shell condenser (Figure 2), the shell side temperature is the same throughout the vessel. Vapors condense, and condensate is removed at the condensation temperature, which is governed by pressure. In a horizontal-tube unit of this type, condensate temperature can be lowered by:

1. reducing the pressure on the shell side,

2. adding a separate subcooler, or

3. using the lower tubes for subcooling.

Vertical-tube condensers provide some degree of subcooling even with condensation on the shell side. With condensation inside the tubes, subcooling occurs in much the same manner, whether tubes are arranged vertically or horizontally. With inside-the-tube condensation, both condensate and uncondensed vapors pass through the full tube length. A separate hot well is usually provided to separate gases before the condensate is discharged. Water requirements for contact condensers can be calculated directly from the condensation rate, by assuming equilibrium conditions. Refer to Figure 3 for other common examples. (Source: N. P. Cheremisinoff, *Handbook of Chemical Processing Equipment*, Butterworth-Heinemann Publishers, U.K., 2000).

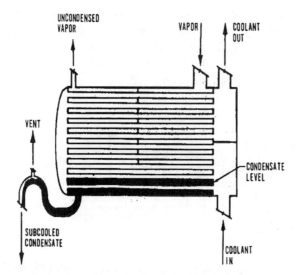

Figure 2. Subcooling arrangement in a horizontal tube and shell condenser.

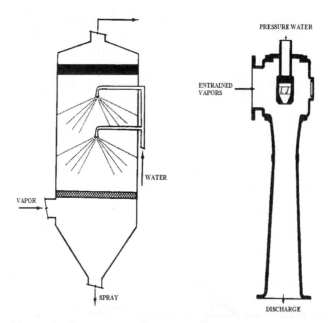

Figure 3. Common condenser configurations.

CONE-PLATE RHEOMETER

The cone-plate rheometer is used when measuring the viscosity and the primary and secondary normal stress coefficient functions as a function of shear rate and temperature. The geometry of a cone-plate rheometer is shown in Figure 1. A detailed explanation of this instrument can be found in the following reference: N. P. Cheremisinoff, *An Introduction to Polymer Rheology and Processing*, CRC Press, Boca Raton, FL, 1993.

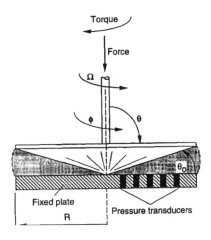

Figure 1. Cone and plate rheometer.

CONGEALING POINT

The temperature at which molten wax ceases to flow, as measured by test method ASTM D 938; of importance where storage or application temperature is a critical factor.

CONTROL CHARTS

Control charts are a fundamental tool in statistical process control (SPC) methods, which has become the foundation of quality control. Quality control is an essential issue throughout every aspect of a polymer's life - ranging from polymerization, to polymer finishing operations, to compounding and processing. Industrial processes are inherently unstable. The concept of an unstable process is illustrated conceptually in Figure 1. In such a process there are three types of variations: (1) *Drift* - the gradual shift of the process mean overtime; (2) *Fluctuations* - the short term variations about the current process mean; and (3) *Interruptions* - random shocks to the process. Since product quality is directly related to process consistency, in order to improve quality we must reduce all these three types of variations. Improvements can be made in both the short term and in the long term. By persistently steering the process mean towards its target, variation will be reduced. This is the basic function of process control. Yet quality improvement goes beyond effectively employing process control and developing better process control methods. Quality improvement also includes:

1.   discovering why the process tends to drift and then implementing programs that can eliminate the root causes;

2.   identifying the causes of the interruptions and taking measures to prevent reocurrences;

3.   conducting designed experiments to identify the variables that contribute to process fluctuations and then modifying the process to avoid future fluctuations.

While an unstable process in its conceptual form is complicated, the actual measurements of the outcome of an unstable process can be even more perplexing. These are some of the possible reasons: (1) the measured variables may be downstream output variables rather than true process variables; (2) the measurements may be periodic rather than continuous; (3) there may be considerable measurement errors.

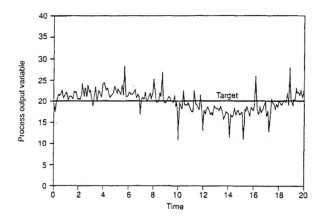

Figure 1. The inherent instability of a process.

The measured output of an unstable process is illustrated in Figure 2. The pattern of variation is no longer as obvious as in the case of the conceptual process, yet in order to monitor and improve quality one must extract from such data, clues concerning the three types of variations. This is accomplished with the help of SPC methods. Statistical Process Control methods are a group of procedures which allows us to use statistics to analyze quality data and to identify the types of variations. One simple method is to fit a polynomial to the data. The fitted curve is then an estimate of the process drift.

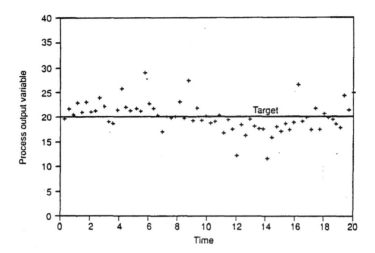

Figure 2. Shows measured output of an unstable operation.

The distances between the individual data points and the fitted curve can now be measured. The variance of these distances is an estimate of the average magnitude of the process fluctuations. Once we have quantified fluctuations in the form of a *variance*, we can calculate the standard deviation representing the degree of fluctuation. In fact, one can draw a pair of parallel lines at a distance of 3a from the original curve. Between these two new curves, we would expect to find 99.7% of the data points. These two lines are usually called the *control limits*. In reality, though, they are simply arbitrary boundaries used to spot the third type of variation, the interruptions. With the polynomial model one can monitor all three types of variations: drift, fluctuations, and interruptions. A third order polynomial is quite common in many applications, but models of other orders can be used to achieve different degrees of *goodness of fit*. Refer to *Shewhart Control Chart* and the *CUSUM Chart*.

COPOLYMERS

Copolymers are polymeric materials with two or more monomer types in the same chain. A copolymer that is composed of two monomer types is referred to as a *bipolymer,* and one that is formed by three different monomer groups is called a *terpolymer*. Depending on how the different monomers are arranged in the polymer chain one distinguishes between *random, alternating, block* or *graft copolymers*. The four types of copolymers are schematically represented in Figure 1. A common example of a copolymer is an ethylene-propylene copolymer. Although both monomers would result in semi-crystalline polymers when polymerized individually, the melting temperature disappears in the randomly distributed copolymer with ratios between 35/65 and 65/35, resulting in an elastomeric material. EPDM rubbers are widely used because of their resistance to weathering. On the other hand, the ethylene-propylene block copolymer maintains a melting temperature for all ethylene/propylene ratios. Another widely used copolymer is high impact polystyrene (PS-HI), which is formed by grafting polystyrene to polybutadiene. If styrene and butadiene are randomly copolymerized, the resulting material is an elastomer called styrene-butadiene-rubber (SBR). Another example of copolymerization is the terpolymer acrylonitrilebutadiene-styrene (ABS). (Source: Osswald, T.A. and G. Menges, *Material Science of Polymers for Engineers*, Hanser Publishers, New York, 1996).

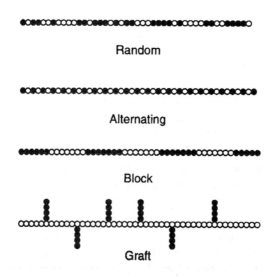

Figure 1. Representation of common copolymers.

COMPRESSION MOLDING

Compression molding is the most common method by which thermosetting plastics are molded. The plastic is introduced to the machine in the form of powder, pellet, or disc, is dried by heating and then further heated to near the curing temperature; this heated charge is loaded directly into the mold cavity. The temperature of the mold cavity is held at 150 to 200° C, depending on the material. The mold is then partially closed, and the plastic, which is liquefied by the heat and the exerted pressure, flows into the recess of the mold. At this stage the mold is fully closed, and the flow and cure of the plastic are complete. During the final stage, the mold is opened, and the completely cured molded part is ejected. Compression-molding equipment consists of a matched mold, a means of heating the plastic and the mold, and some method of exerting force on the mold halves. For severe molding conditions molds are usually made of hardened steel. Brass, mild steel, or plastics are used as mold materials for less severe molding conditions or short-run products. In compression molding a pressure of 2250 psi to 3000 psi is suitable for phenolic materials. The lower pressure is adequate only for an easy-flow material and a simple uncomplicated shallow molded shape. For a medium-flow material and where there are average-sized recesses, cores, shapes, and pins in the molding cavity, a pressure of 3000 psi or above is required.

For molding urea and melamine materials, pressures of approximately one and one-half times that needed for phenolic material are necessary. In compression molding of thermosets the mold remains hot throughout the entire cycle. After the molded part is ejected, a new charge of molding powder is introduced. Unlike thermosets, thermoplastics must be cooled to harden. So before a molded part is ejected, the entire mold must be cooled, and as a result, the process of compression molding is slow with thermoplastics.

Compression molding is commonly used for thermosetting plastics such as phenolics, urea, melamine, and alkyds. In special cases, such as when extreme accuracy is needed, thermoplastics are also compression molded. Compression molding is applied for such products as electrical switch gear and other electrical parts, plastic dinnerware, radio and television cabinets, furniture drawers, buttons, knobs, handles, etc. Compression molds can be divided into hand molds, semiautomatic molds, and automatic molds. The design of any of these molds must allow venting to provide for escape of steam, gas, or air produced during the operation. After the initial application of pressure the usual practice is to open the mold slightly to release the gases. This procedure is known as *breathing*. Refer to *Compression Molds*.

COMPRESSION MOLDS

There are four basic compression mold designs. These are flash type, positive type, semipositive, and the semipositive subcavity gang type. *Flash* type is the simplest of all the thermoset molds and is still widely used in the production of simple shapes such as molded dinnerware, ashtrays, closures, or knobs. The mold is constructed with a retainer set for the force and cavity blocks and is provided with guide pins and bushings for positive mold alignment when opened and closed. Holes for heaters are drilled into the clamping plates. The force and cavity should be built with a good tool steel, through hardened to resist the compression load on the cut-off lands of the cavity. Both units are highly polished and hard chrome plated for easy mold release and a superior surface finish.

In *positive* type mold makes, the force in the upper mold half is allowed to enter the cavity in the lower half with a tight fit between the two components, thus providing direct and positive pressure on the compound in the mold cavity as well as correct alignment of the force and cavity during mold closing. The mold charge will be in excess of the part weight, and the excess compound is allowed to escape in a vertical direction through "bleed-outs" or flats that are ground deep in strategic locations on the side of the force that enters the cavity. The excess compound is typically referred to as the "flash". It is easily removed by manually pressing the upper face of the mold part against the abrasive surface and rotating it until the flash is entirely vacated. A positive mold tends to promote excellent mold density while the hardened pressure pads at the parting line of the mold are closed with no compound in the cavity.

A positive type of mold is often multicavitied, and is generally run semi-automat, using preheated preforms that are fed into each cavity manually or with the use of a "loading board" that will feed all cavities at one time, thus reducing the mold open time and cycle time. The preforms are placed in the holes in a plate, and are in alignment with the mold cavities. The loading is done during the previous molding cycle. With the mold open and ready to receive the next load of preforms, the board is properly located over the cavities and the plate is moved to bring its holes directly under those of the plate allowing the preforms to fall into the cavities. There are loading boards designs that are capable of being placed within a dielectric preheater so the preforms will be of the proper temperature when fed to the mold.

The *semipositive* mold resembles the positive mold except that the force actually "lands" within the cavity well, producing a horizontal as well as a vertical flash. The effect of landing the force provides a better control of the flash thickness as well as the part dimensions that are determined from the parting line of the mold. Again, hardened pressure pads are mounted on the mold at the parting line to absorb the high molding pressure and prevent damage to the land areas from the constant opening and closing of the mold when in production. In both positive and semipositive molding, the telescoping of the force into the cavity serves the important function of maintaining precise mold alignment and will produce parts with superior molded density.

Semipositive Gang type molds utilize the benefits derived from the positive and semipositive. designs along with the ability to load the compound into a single, centrally located area instead of having to feed each individual cavity. The cavities are mounted into a hardened retainer, with the compound being fed into the cavity area and the mold closed under pressure. The compound will flow into each separate cavity and the excess compound will serve to connect all the cavities at the parting line. It should be noted that the compression molding process requires high molding pressures, and it is imperative that the mold components be of high grade tool steels and the mold built with ample support throughout to prevent any plate deflection over the life of the mold.

COUETTE RHEOMETER

The concentric cylinder or Couette flow rheometer is schematically depicted in Figure 1. The torque, T, and rotational speed, Q, can easily be measured. The torque is related to the shear stress that acts on the inner cylinder wall. The major sources of error in a concentric cylinder rheometer are the end-effects. One way of minimizing these effects is by providing a large gap between the inner cylinder end and the bottom of the closed end of the outer cylinder.

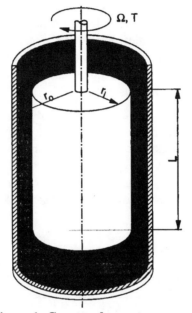

Figure 1. Couette rheometer.

CRACKING

Petroleum refining process in which large-molecule liquid hydrocarbons are converted to small-molecule, lower-boiling liquids or gases; the liquids leave the reaction vessel as unfinished gasoline, kerosene, and gas oils. At the same time, certain unstable, more reactive molecules combine into larger molecules to form tar or coke bottoms. The cracking reaction may be carried out under heat and pressure alone (thermal cracking), or in the presence of a catalyst (catalytic cracking).

CROSS-LINKED POLYMERS

Cross-linked polymers, such as thermosets and elastomers, behave completely different than their counterparts, thermoplastic polymers. In cross-linked systems, the mechanical behavior is also best reflected by the plot of the shear modulus versus temperature. Figure 1 compares the shear modulus between highly cross-linked, coarsely cross-linked and uncross-linked polymers. The coarse cross-linked system, typical of elastomers, has a low modulus above the glass transition temperature. The glass transition temperature of these materials is usually below -50°C, so they are soft and flexible at room temperature. In contrast, highly cross-linked systems, typical in thermosets, show a smaller decrease in stiffness as the material is raised above the glass transition temperature; the decrease in properties becomes smaller as the degree of cross-linking increases.

Figure 2 shows ultimate tensile strength and strain curves plotted versus temperature. The strength remains fairly constant up to the thermal degradation temperature of the material. The failure of a fiber filled material begins at the interface between filler or reinforcement and the matrix. A micrograph of a glass fiber-filled polyester specimen placed under loading will show the breakage of the adhesion between imbedded glass fibers and their matrix. This breakage is generally referred to as *debonding*.

This initial microcrack formation is reflected in a stress-strain curve by the deviation from the linear range of the elastic constants. In fact, the failure is analogous to the microcracks that form between spherulites when a semi-crystalline polymer is deformed. (Source: Osswald, T.A. and G. Menges, *Material Science of Polymers for Engineers*, Hanser Publishers, New York, 1996). Refer also to *Vulcanization*, *Peroxides*, *Peroxide Cross-Linking*, *Sulfur Vulcanization*, and *Vulcanizing Agents*.

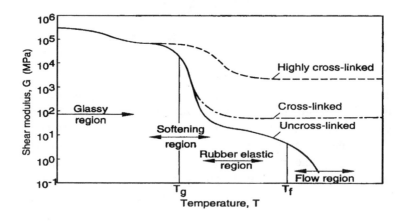

Figure 1. Shows shear modulus and behavior of cross-linked and uncross-linked polymers.

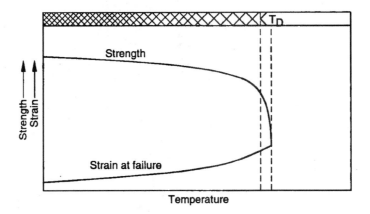

Figure 2. Tensile strength and strain at failure as function of temperature for thermosets.

CREEP RUPTURE

During creep, a loaded polymer component will gradually increase in length until fracture or failure occurs. This phenomenon is usually referred to as *creep rupture* or, sometimes, *as static fatigue*.

During creep, a component is loaded under a constant stress, constantly straining until the material cannot withstand further deformation, causing it to rupture.

At high stresses, the rupture occurs sooner than at lower stresses. However, at low enough stresses failure may never occur. The time it takes for a component or test specimen to fail depends on temperature, load, manufacturing process, environment, etc. It is important to point out that damage is often present and visible before creep rupture occurs.

Results from creep rupture tests are usually presented in graphs of applied stress versus the logarithm of time to rupture. Although the creep test is considered a long-term test, in principle it is difficult to actually distinguish it from monotonic stress strain tests or even impact tests. In fact, one can plot the full behavior of the material, from impact to creep, on the same graph under tensile loads at room temperature. Creep tests can be used effectively in product development efforts, especially when developing tailored molecular weight polymer grades. Figure 1 shows the typical apparatus used for a creep rupture test.

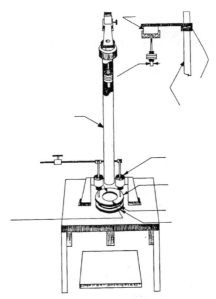

Figure 1. Creep test apparatus.

CRYSTALLINE POLYMER

The term refers to a material which shows regular order in its solid state. Examples are PE, PP, PA. Some polymers, such as PET, which are normally amorphous when quenched in a cold mold, can be nucleated and thermally conditioned to CPET which demonstrates a degree of crystallinity useful in ovenable food containers. See *Semi-crystalline Thermoplastics* and *Crystallization*.

CRYSTALLIZATION

Crystallization is characterized by several physical phenomena. With regard to rubbers, a crystal unit cell is of higher density (1.0 g/cm^3 for NR) than is amorphous unfilled rubber (0.91 g/cm^3 for NR). Crystallization is thus accompanied by an increase in density (i.e., decrease in volume) of up to 10%. Small volume changes (less than 3% for NR) have been measured by both hydrostatic balance and dilatometric techniques. Crystallite formation also involves the evolution of heat of fusion, and calorimetry has been used to derive values for the rate and degree of crystallization in NR in broad agreement with volume change data.

Differential thermal analysis (DTA) has also been exploited, mainly to determine polymer crystal melting temperatures but also (less frequently) to determine crystallization kinetics. Crystallite formation also changes the optical properties of rubber. Birefringence and low-angle light-scattering techniques to provide evidence for crystal morphology have been used in the past. When rubber is held at a fixed strain, crystallization is accompanied by a progressive relaxation of stress - in many cases to and beyond zero. Use of both stress relaxation and volume change measurements in investigating the kinetics of crystallization of various vulcanized natural rubbers at different temperatures and strains are reported in the literature. This contrasts with other stress changes reported for other polymers. As stress relaxes toward zero the elastic modulus increases dramatically (by up to 2 orders of magnitude).

The process of crystallization may be considered in three stages: crystal nucleation, crystal growth, and the equilibrium partially crystalline state. Embryo nuclei may continuously form and grow even in amorphous rubber. However, below a certain critical size they are unstable enough to disappear due to random thermal motions of the molecular chain and may not have measurable macroscopic consequences.

Nucleation may occur spontaneously and homogeneously throughout the amorphous phase (most likely at low temperatures) or be "seeded" by foreign surfaces or structural discontinuities. The type and density of nucleation are likely to strongly influence the character of subsequent crystal growth. Above the critical size, nuclei become stable, permitting crystal growth to occur.

Crystal growth involves physical consequences and its kinetics may be characterized in terms of progressive changes in almost any one of them. The ultimate degree of crystallinity is below unity and characterizes an equilibrium state, which is a function of the rubber cross-link density, temperature, and strain. Thermodynamic arguments have been used to derive theoretical expressions for the relaxation of stress due to crystallization (occurring under fixed strain) for the equilibrium degree of crystallinity in terms of strain and temperature, and for the melting temperature as a function of the crystallization temperature. Crystallization becomes theoretically possible whenever the melt free energy exceeds the crystal free energy, enabling a reduction in free energy to result from the crystallization process. The equilibrium melting temperature is the characteristic temperature at which all crystallites melt when heated very slowly (i.e., under equilibrium conditions).

The term *crystalline,* as used here, refers to semicrystalline bulk polymer specimens that are also called *polycrystalline.* Bulk polymers are, as a rule, only partly crystalline and are partly amorphous. The degree of crystallinity of a polymer (weight percent of the total polymer that is in the crystalline state) is typically 40%.

While polymer crystals have different physical properties in different crystal directions, in a bulk specimen the individual microscopic crystals are oriented at random so that the macroscopic properties are the same in every direction and the physical properties are said to be isotropic. It will be shown that yield properties depend on the degree of crystallinity and the polymer morphology so that the designation of a polymer by a single term, such as polyethylene, does not fully specify the material. There are significant variations among different polyethylenes. (Source: Handbook of Polymer Science and Technology, Volume 2: Performance Properties of Plastics and Elastomers, N. P. Cheremisinoff, Macrel Dekker Publishers, New York, 1989).

CURE RHEOMETER

Cure of rubber vulcanizates is commonly measured using a Monsanto cure rheometer, which plots the change in torque as a function of time as is shown in Figure 1. From the percentage rise of the graph, the percent cure can be calculated.

This figure also illustrates the three main regions of rubber vulcanization. The first regime is the induction period, or scorch delay, during which accelerator complex formation occurs. The second time period is the cure period, in which the network or sulfurization structures are formed. The network structures can include crosslinks, cyclics, main chain modification, isomerization, etc. The third regime is the overcure, or reversion regime.

During this time period network maturation occurs. The three different curves in this region represent continued crosslinking (top curve), which is also referred to as "creeping modulus"; no change, which is the middle curve; and reversion (bottom curve), which is a degradation of the network.

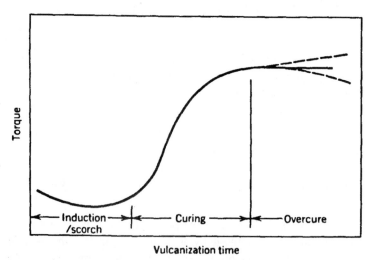

Figure 1. Cure rheometer trace for rubber vulcanization.

CURE TIME

Cure time is the time required to harden thermosetting materials. Depending on the type of molding material, preheating temperature, and the thickness of the molded article, the cure time may range from seconds to several minutes. It is important to note that cure time can be a function of the medium. Consider peroxide vulcanization by way of an example. Theoretically, the only factor affecting the rate of peroxide decomposition, and therefore the rate of cure, is temperature. In practice, it has been found that the medium (i.e., polymer) within which the peroxide decomposes has a pronounced effect on the rate of peroxide decomposition. Half-life curves describing cross-linking rates for one peroxide, Vul-Cup, in polyethylene, EP rubbers, BR, NBR, SBR, NR, and IR rubbers, and in solution are shown in Figure 1. Half-life curves for Di-Cup are shown in Figure 2. Other commercially available peroxides have been found to behave similarly. These plots show that the polymer strongly affects the rate of peroxide cross-linking. The reasons for this variation in rate are not known; however, they are undoubtedly related to the ease of hydrogen abstraction from the polymer, viscosity, and solvent characteristics of the polymer.

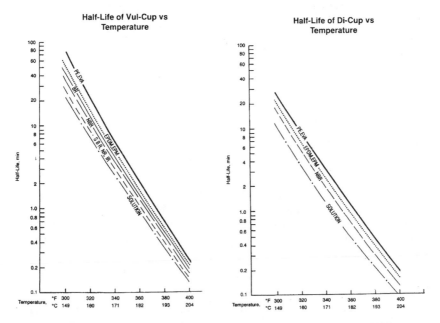

Figure 1. Peroxide cure - Vul-Cup.   **Figure 2. Peroxide cure - Di-Cup.**

Figure 3 shows typical 300% modulus, tensile strength, and compression set resistance response to cure time. Tensile strength rapidly attains a satisfactory value. Modulus rises rapidly at first and slowly comes to a plateau when all of the curative has decomposed. Compression set follows the same pattern as modulus. There appears to be very little gain from t_{95} (time to 95% cure) to the cure plateau. The major benefit of curing to the cure plateau vs curing to t_{95} appears to be a minor gain in compression set resistance. If optimum compression set resistance is not required, the cure for many polymers can be terminated between t_{90} and t_{95} and the stock can be subjected to a postcure. Refer to *Postcuring*.

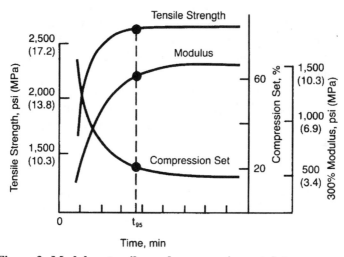

Figure 3. Modulus, tensile, and compression set data.

CUSUM CHART

The CUSUM (or Cumulative Sum) Control Chart is used to avoid the complicated calculations required by sequential analysis in SPC control charting techniques. A V-mask is superimposed onto a chart representing the accumulated variations from the process target, to determine whether the process has drifted away from the aiming point. There are two basic steps in using the CUSUM Chart: *Step (1)* - constructing the V-mask; and *Step (2)* updating the cumulative function and testing for significance with the use of the V-mask. The V-mask has two calculated parameters: θ the half angle of the V-mask; P-O the leading distance between the current data point and the apex of the V-mask. These calculated parameters are the geometric representation of the balancing of the original set of design parameters: σ - the standard deviation of the process; d - the size of a significant change; α - type 1 error, the risk of failing to act; β - type II error, the risk of over-reacting. The CUSUM chart is illustrated in Figure 1. The updating of the cumulative function involves the following steps:

1.   finding the difference between the current data and the target;

2.   adding the difference to the total of the previous differences;

3.   plotting the new total according to its sample number.

The V-mask can now be placed over the graph with P-O parallel to the x-axis and the point P over the current data point. If any part of the cumulative function protrudes the boundaries prescribed by the V-mask, we would conclude that the current mean of the process has deviated from the target. The CUSUM chart, in general, is more effective than the Shewhart Control Chart when used to monitor continuous processes that tend to drift over time. The CUSUM Chart, however, is quite vulnerable to the impact of process interruptions. Another drawback of the CUSUM Chart is that its direct relationship to the actual time variation of the process is not always clear, making it rather difficult for us to analyze and to improve the process. Refer to *Stepwise SPC Chart*.

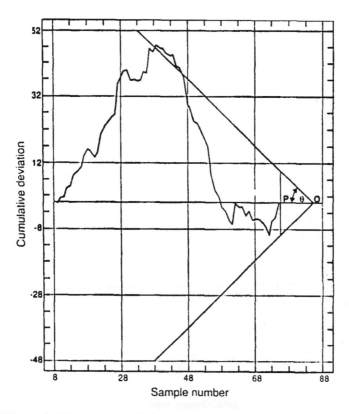

Figure 1. Shows a cumulative sum chart.

D

DEGRADABLE PLASTICS

None of the commodity plastics degrades rapidly in the environment. Nevertheless, environmentalists have seen biodegradable and photodegradable plastics as a solution to the problem of litter. Such materials have been developed, but they have not been successful on a large scale primarily because of high production costs and problems of stability during their processing and use. On the other hand, the plastic rings that hold six-packs of soft-drink and beer cans together represent an application where photodegradation has been used effectively. A copolymer of ethylene with some carbon monoxide contains ketone groups that absorb sufficient energy from sunlight to cause extensive scissioning of the polymer chain. The photodegradable plastic, very similar in appearance and properties to low-density polyethylene (LDPE), decomposes to a powder within a few months of exposure in sunny climates. Refer to *Recycling*.

DEMULSIBILITY

Refers to the ability of an oil to separate from water, as determined by test method ASTM D 1401 or D 2711. Demulsibility is an important consideration in lubricant maintenance in many circulating lubrication systems. *Refer to Demulsifier* and *Dynamic Demulsibility*.

DEMULSIFIER

The term refers to an additive that promotes oil-water separation in lubricants that are exposed to water or steam. Refer to *Demulsibility and Dynamic Demulsibility*

DENATURING OIL

An unpalatable oil, commonly kerosene or No. 2 heating oil, required to be added to food substances condemned by the U.S. Department of Agriculture, to ensure that these substances will not be sold as food or consumed as such.

DE-OILING

Refers to the removal of oil from petroleum wax; a refinery process usually involving filtering or pressing a chilled mixture of slack wax and a solvent that is miscible in the oil, to lower the oil content of the wax.

DEPARTMENT OF TRANSPORTATION (DOT)

The DOT provides guidelines and mandatory requirements for the safe transportation of hazardous materials. This information can be found in Title 49 of the Code of Federal Regulations (CFR).

DETERGENT

An important component of engine oils that helps control varnish, ring zone deposits, and rust by keeping insoluble particles in colloidal suspension and in some cases, by neutralizing acids. A detergent is usually a metallic (commonly barium, calcium, or magnesium) compound, such as a sulfonate, phosphonate, thiophosphonate, phenate, or salicylate. Because of its metallic composition, a detergent leaves a slight ash when the oil is burned. A detergent is normally used in conjunction with a dispersant.

DEWAXING

Refers to the removal of paraffin wax from lubricating oils to improve low temperature properties, especially to lower the cloud point and pour point.

DIE SWELL

This effect is well known to anybody who has operated an extruder, where it is observed that the extrudate cross section is greater than the die cross section (providing no drawing down has occurred). Sometimes known as the Barus effect, this contrasts with the behavior of Newtonian liquids and similar real materials where there is in fact a contraction in cross sections. As a general guide we can assume that die swell will normally only occur with elastic materials. A number of theories have been proposed to explain die swell but perhaps the most useful working hypothesis is to assume that the molecules are oriented in the die and on emergence into the atmosphere recoiling takes place with a contraction in the direction of flow being offset by lateral expansion. If the leading ends of samples cut at the die face are examined it is seen that they have a convex profile. This is consistent with the fact that it is at the die wall that the greatest shear has occurred and hence on the emergence from the die the greatest longitudinal shrinkage will take place on the (outside) surfaces of extrudates. The following useful conclusions may be made by studying die swell behavior:

1. Die swell increases with shear up to a limit which is near to the *critical shear rate;* beyond this die swell decreases.
2. At a fixed shear rate (or output rate for a given die) die swell decreases with temperature but the maximum swelling ratio increases with temperature.
3. At a fixed shear stress, die swell is very little affected by temperature (at shear stresses well below the critical shear stress).
4. At a fixed shear rate die swell decreases with the length of the die.
5. The greater the residence time in the capillary the less is the die swell.

The greater the shear strain that occurs in the capillary the less the die swell. This is one of the most important factors determining the die swell of a given polymer. It indicates that continuing strain causes disentanglement, an effect which has been noted in rotational viscometers. It is also possible to make some comments concerning the molecular factors influencing swell. In general, it is found that molecular features that tend to increase entanglements also tend to increase swelling. Hence, an increase in molecular weight increases the amount of swell; long-chain branched molecules are more compact than linear polymers of similar weight and hence show less swelling. Refer to *Melt Fracture*.

DIFFERENTIAL SCANNING CALORIMETRY (DSC)

The DSC measures the power (heat energy per unit time) differential between a small weighed sample of polymer (ca. 10 mg) in a sealed aluminum pan referenced to an empty pan in order to maintain a zero temperature differential between them during programmed heating and cooling temperature scans. The technique is most often used for characterizing the Tg, Tm, Tc, and heat of fusion of polymers.

The technique can also be used for studying the kinetics of chemical reactions, e.g., oxidation and decomposition. The conversion of a measured heat of fusion can be converted to a % crystallinity provided, of course, the heat of fusion for the 100% crystalline polymer is known. An example of a thermogram is shown in Figure 1. Refer to Thermal Analysis and *Thermogravirnetric Analysis*. (Source: Cheremisinoff, N.P. *Polymer Characterization: Laboratory Techniques and Analysis*, Noyes Publishers, New Jersey, 1996).

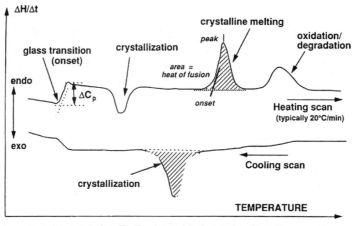

Important characteristics: T_g, T_m, heat of fusion on heating; T_c on cooling

Figure 1. Example of DSC thermogram.

DILUENT

An usually inert (unreactive) liquid or solvent, used to dilute, carry, or increase the bulk of some other substance. Petroleum oils and solvents are commonly used as diluents in such products as paints, pesticides, and additives. In polymerization processes, the diluent phase is normally a hydrocarbon, but may also be water. Refer to *Solution Polymerization*.

DISTILLATE

Refers to any of a wide range of petroleum products produced by distillation, as distinct from bottoms, cracked stock, and natural gas liquids. In fuels, a term referring specifically to those products in the mid-boiling range, which include kerosene, turbo fuel, and heating oil--also called middle distillates and distillate fuels. In lubricating oils, a term applied to the various fractions separated under vacuum in a distillation tower for further processing (lube distillate).

DISTILLATION

Another term for distillation is fractionation - the primary refining step, in which crude is separated into fractions, or components, in a distillation tower, or pipe still. Distillation methods are quite common in polymer characterization laboratories for separation of oligomers and oily materials. Heat, usually applied at the bottom of the tower, causes the low boiling solvent vapors to rise through progressively cooler levels of the tower, where they condense onto plates and are drawn off in order of their respective condensation temperatures, or boiling points --the lighter-weight, lower-boiling point fractions, exiting higher in the tower.

Heavy materials remaining at the bottom are called the bottoms or residuum. Those fractions taken in liquid form from any level other than the very top or bottom are called sidestream products; a product. Product removed in vapor form from the top of the distillation tower is called overhead product.

In petroleum refining, distillation may take place in two stages:

- first, the lighter fractions--gases, naphtha, and kerosene--are recovered at essentially atmospheric pressure;
- next, the remaining crude is distilled at reduced pressure in a vacuum tower, causing the heavy lube fractions to distill at much lower temperatures than possible at atmospheric pressure, thus permitting more lube oil to be distilled without the molecular cracking that can occur at excessively high temperatures.

DISTILLATION TEST

A method for determining the full range of volatility characteristics of a liquid by progressively boiling off (evaporating) a sample under controlled heating. Initial boiling point (IBP) is the fluid temperature at which the first drop falls into a graduated cylinder after being condensed in a condenser connected to a distillation flask. Mid-boiling point (MBP) is the temperature at which 50 % of the fluid has collected in the cylinder.

Dry point is the temperature at which the last drop of fluid disappears from the bottom of the distillation flask. Final boiling point (FBP) is the highest temperature observed. Front-end volatility and tail-end volatility are the amounts of test sample that evaporate, respectively, at the low and high temperature ranges. If the boiling range is small, the fluid is said to be narrow cut, that is, having components with similar volatilities; if the boiling range is wide, the fluid is termed wide cut.

Distillation may be carried out by several ASTM test methods, including ASTM D 86, D 850, D 1078, and D 1160.

DISTRIBUTIVE MIXING

Distributive mixing or laminar mixing of compatible liquids is usually characterized by the distribution of the droplet or secondary phase within the matrix. This distribution is achieved by imposing large strains on the system such that the interfacial area between the two or more phases increases and the local dimensions, or striation thicknesses, of the secondary phases decrease. This concept is shown schematically in Figure 1. The figure shows a Couette flow device with the secondary component having an initial striation thickness. As the inner cylinder rotates, the secondary component is distributed through the systems with constantly decreasing striation thickness; striation thickness depends on the strain rate of deformation which makes it a function of position. Imposing large strains on the system is not always sufficient to achieve a homogeneous mixture. The type of mixing device, initial orientation and position of the two or more fluid components play a significant role in the quality of the mixture. For example the mixing problem shown below, homogeneously distributes the melt within the region contained by the streamlines cut across by the initial secondary component. The final mixed system will reveal another variation of initial orientation and arrangement of the secondary component. Here, the secondary phase cuts across all streamlines, which leads to a homogeneous mixture throughout the Couette device, under appropriate conditions. (Source: Osswald, T.A. and G. Menges, *Material Science of Polymers for Engineers*, Hanser Publishers, New York, 1996).

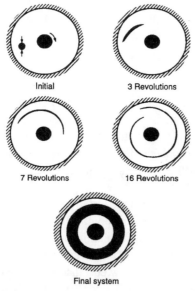

Figure 1. Illustrates distributive mixing in
Couette flow system.

DRAWING COMPOUND

In metal forming, a lubricant for the die or blank used to shape the metal; often contains EP additives to increase die life and to improve the surface finish of the metal being drawn.

DRY SPINNING

Dry spinning is used to form polymeric fibers from solution. The polymer is dissolved in a volatile solvent and the solution is pumped through a spinneret (die) with numerous holes (one to thousands). As the fibers exit the spinneret, air is used to evaporate the solvent so that the fibers solidify and can be collected on a take-up wheel. Stretching of the fibers provides for orientation of the polymer chains along the fiber axis. Cellulose acetate (acetone solvent) is an example of a polymer which is dry spun commercially in large volumes. Due to safety and environmental concerns associated with solvent handling this technique is used only for polymers which cannot be melt spun. Figure 1 is a simplified process diagram. An excellent general reference on fiber spinning is A. Ziabicki, *Fundamentals of Fiber Formation*, Wiley, New York (1976). ISBN 0471982202. A more detailed study of dry spinning is Y. Ohzawa, Y. Nagano, and T. Matsuo, *J. Appl. Polym. Sci.*, **13**, pp. 257-283 (1969). Refer to *Melt Spinning*.

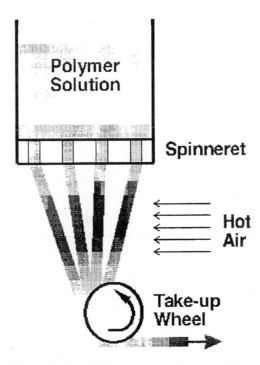

Figure 1. Simplified process diagram of dry spinning of fibers.

DYNAMIC DEMULSIBILITY

Refers to a test of water separation properties of an oil, involving continuous mixture of oil and water at elevated temperatures in an apparatus that simulates a lubricating oil circulating system. Samples are then drawn off both the top and bottom of the test apparatus. Ideally, the top sample should be 100 % oil, and the bottom 100 %water. Because of the severity of the test conditions, separation is virtually never complete.

DYNAMIC MECHANICAL THERMAL ANALYSIS (DMTA)

DMTA is a measurement of the dynamic moduli (in-phase and out-of-phase) in an oscillatory mechanical deformation experiment during a programmed temperature scan at controlled frequency. Thermograms are usually plotted to show elastic modulus, E, and tan δ versus temperature. The peak of the tan δ is a particularly discriminatory measure of Tg, although this is the center of the relaxation whereas in the DSC experiment the onset temperature of the Tg (glass transition temperature) relaxation is usually reported. In such a case the DSC Tg will be lower than that for DMTA by an amount that varies with the specific polymer.

There is, in addition, a frequency effect which puts the mechanical (ca. I Hz) Tg about 17° C higher than that for a DSC measurement (ca. 0.0001Hz) for an assumed activation energy of 400 kJ/mole (typical for polymer Tg). The DMTA has a frequency multiplexing capability which can be used for calculating activation energies using time-temperature superposition software. The temperature range of the DMTA is from -150° C to 300° C and frequencies from 0.033 to 90 Hz. The sample size for the usual flexural test mode is 1 mm. x 10 mm x 40 min; slightly less sample is required in the parallel plate shear mode.

DMTA measurements can be applied to a variety of end-use performance applications testing. In product development on example is in testing or developing polymers for engine mount applications. The following references by the author provide some detailed information on this technique:

1.   *Advanced Polymer Processing Operations*, Noyes Publications, Westwood, New Jersey (1997).

2.   *Handbook of Engineering Polymeric Materials*, Editor, Marcel Dekker Publishers, Inc., N.Y. and Basel (1997).

3.   *Laboratory Guide for Polymer Characterization*, Noyes Publications, Westwood, N.J. (1996).

4.   *An Introduction to Polymer Rheology and Processing*, CRC Press, Boca Raton, FL (1993).

E

ELASTO-HYDRODYNAMIC (EHD) LUBRICATION

Lubrication phenomenon occurring during elastic deformation of two non-conforming surfaces under high load. A high load carried by a small area (as between the ball and race of a rolling contact bearing) causes a temporary increase in lubricant viscosity as the lubricant is momentarily trapped between slightly deformed opposing surfaces.

ELASTICOVISCOUS FLUIDS

A number of materials exist which show properties intermediate to those of classical solids and liquids. In many instances the material may appear to be solid in the everyday meaning of the word but under stress the deformation is not instantaneous and sometimes only partly reversible. Such materials are said to be *viscoelastic*. Theories of viscoelasticity are highly developed and can be applied to the study of the behavior of rubbers in engineering and to the long term behavior under load of all types of polymeric materials. There are also a number of materials which may be outwardly liquid and capable of indefinite deformation but which on release of the deforming stress show some recovery of shape. In industrial processing this type of behavior manifests itself in such phenomena as die swell, calender swell, neck-in, and frozen-in orientations. To the technologist it is these effects rather than a knowledge of shear stress-shear rate data which is often of greatest interest. The term *elasticoviscous* fluid has been coined to describe the behavior. There is no sharp distinction between viscoelastic and elasticoviscous behaviour. A very highly masticated low molecular weight rubber can be almost a simple viscous liquid, a conventionally masticated sample could be considered as elasticoviscous whilst unmasticated rubber and also lightly vulcanized rubber would normally be rated as viscoelastic. Highly cross-linked rubber (ebonite) is a hard solid and is almost (but not quite) Hookean in behavior. It is not normally crucial to have to define the boundaries between such behavior particularly as the mathematical theories of viscoelasticity may be applied to all four types of material (i.e. viscous, elasticoviscous, viscoelastic and elastic). In polymer melts elasticoviscous behavior is generally ascribed to the fact that long chain molecules take up an oriented configuration on application of a stress but on release of the stress bond rotations cause the molecules to take up a random form. This leads to a contraction in (usually) one direction and an expansion in the perpendicular directions. In other cases these effects may be due to structure of particles embedded in a fluid matrix. It has also been found that if droplets of one Newtonian fluid are dispersed in another, elastic recovery occurs on the release of a deforming stress. This has been explained by the fact that the droplets form ellipsoids under shear but surface tension (or strictly speaking interfacial tension) effects encourage the droplets to take up the minimum interfacial surface and a spherical form results on release of the stress.

ELASTOMERS

An elastomer is any rubbery material composed of long, chainlike molecules that are capable of recovering their original shape after being stretched to great extent. Under normal conditions the long molecules making up an elastomeric material are irregularly coiled. With the application of force, however, the molecules straighten out in the direction in which they are being pulled. Upon release, the molecules spontaneously return to their normal compact, random arrangement. The elastomer (from "elastic polymer") with the longest history of use is natural rubber, which is made from the milky latex of various trees, most usually the *Hevea* rubber tree. Natural rubber is still an important industrial polymer, but it now competes with a number of synthetic elastomers, such as styrene-butadiene rubber and polybutadiene, which are derived from by-products of petroleum and natural gas. More than half of all rubber produced goes into automobile tires; the rest goes into mechanical parts such as mountings, gaskets, belts, and hoses, as well as consumer products such as shoes, clothing, furniture, and toys. Compared to thermosets, elastomers are only lightly cross-linked which permits almost full extension of the molecules. However, the links across the molecules hinder them from sliding past each other, making even large deformations reversible. One common characteristic of elastomeric materials is that the glass transition temperature is much lower than room temperature. There are thousands of different grades of polymers available to the design engineer. These materials cover a wide range of properties, from soft to hard, ductile to brittle, and weak to tough. Table 1 lists some common elastomers and typical applications.

Table 1.

Common Elastomers and Their Applications

Polymer	Applications
Polybutadiene	Automotive tires (blended with natural rubber and styrene butadiene rubber), golf ball skin, etc.
Ethylene propylene rubber	Automotive radiator hoses and window seals, roof covering, etc.
Natural rubber (polyisoprene)	Automotive tires, engine mounts, etc.
Polyurethane elastomer	Roller skate wheels, sport arena floors, ski boots, automotive seats (foamed), shoe soles (foamed), etc.
Silicone rubber	Seals, flexible hoses for medical applications, etc.
Styrene Butadiene rubber	Automotive tire treads, etc.

EMBOSSING

Embossing is a special form of calendering. Some rubber products are made from uncured components, which are required to have a surface design that cannot be economically formed by subsequent molding operations. One such example is the cover strip around the sole of canvas shoes. The method employed for producing such strips is similar to that of profiling in that the required design is engraved into the calender roll or shell as a mirror image of the design itself.

EMULSION POLYMERIZATION

This is one of the most widely used methods of manufacturing vinyl polymers. It involves the formation of a stable emulsion (often referred to as a latex) of monomer in water using a soap or detergent as the emulsifying agent. Free-radical initiators, dissolved in the water phase, migrate into the stabilized monomer droplets (known as micelles) to initiate polymerization. The polymerization reaction is not terminated until a second radical diffuses into the swelling micelles, with the result that very high molecular weights are obtained. Reaction heat is effectively dispersed in the water phase. The major disadvantage of emulsion polymerization is that the formulating of the mix is complex compared with the other methods, and purification of the polymer after coagulation is difficult. Purification is not a problem, however, if the finished polymer is to be used in the form of an emulsion, as in latex paints or adhesives.

EMULSION

An emulsion is an intimate mixture of oil and water, generally of a milky or cloudy appearance. Emulsions may be of two types: oil-in-water (where water is the continuous phase) and water-in-oil (where water is the discontinuous phase). Oil-in-water emulsions are used as cutting fluids because of the need for the cooling effect of the water. Water-in-oil emulsions are used where the oil, not the water, must contact a surface--as in rust preventives, non-flammable hydraulic fluids, and compounded steam cylinder oils; such emulsions are sometimes referred to as inverse emulsions. Emulsions are produced by adding an emulsifier. Emulsibility is not a desirable characteristic in certain lubricating oils, such as crankcase or turbine oils, that must separate from water readily. Unwanted emulsification can occur as a result of oxidation products--which are usually polar compounds--or other contaminants in the oil.

EWMA CONTROL CHART

The EWMA Control Chart refers to the *Exponentially Weighted Moving Average* Chart. The common approach is plotting process data as a time series and detecting significant patterns using some statistical rules. An analogy of this approach is studying how the world climate has changed over the past centuries so that we can infer whether the greenhouse effect is taking place. The various ways we use the data to model a process signal, give us the different types of control charts. When we analyze process data, we may consider some of the data points as more important than the others. In the Shewhart Control Chart we put our emphasis entirely on the last data point. This emphasis is illustrated by Figure 1. This chart illustrates that 100% of our information is based on the last observation, observation 16. When we use the 5-point-moving-average method, we put equal emphases on the last 5 data points, while ignoring all the others.

Figure 2 illustrates how 20% of the total weight is placed on each of the last 5 observations, 12 to 16. The CUSUM Chart is based on a very different weighting function: each of the observations carries the same weight. In this example, V16 of the total weight, or 6.25%, is placed on each of the 16 data points.

A comparison between the charts in Figures 3 and 2 show how divergent are the approaches we take when we choose between the Shewhart Control Chart and the CUSUM Chart.

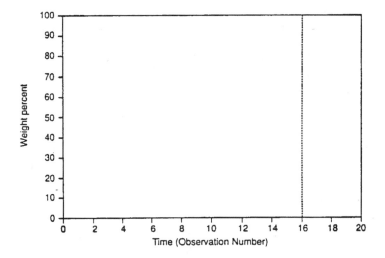

Figure 1. Shows the Shewhart control chart.

The Stepwise SPC Chart serves two individual functions: monitoring process drift and detecting interruptions. To monitor process drift, the Stepwise Chart places total emphasis on the last group of data points and puts equal weight onto each of the data points in that group. In Figure 4, only observations 13 to 16 are active. All the previous observations, 1 to 12, are no longer used. Figure 4 also demonstrates that, within a given segment of the step function, the Stepwise SPC

Chart employs exactly the same weighting function as the CUSUM Chart. To detect outside interruptions the Stepwise SPC Chart places 100% of the weight onto the last data point in the same way as the Shewhart Control Chart. The Stepwise SPC Chart is, therefore, a combination of the two common SPC charts: the Shewhart Control Chart and the CUSUM Chart. Figure 5 illustrates the weighting function of the EWMA Chart. Most emphasis is placed on the last data point and regressively less emphases are placed on the previous data points. In this example, about 20.6% of the total weight is put on observation 16, (20.6 x 0.8)% of the total weight is put on observation 15 and (20.6 x 0.8 x 0.8)% is put on observation 14.

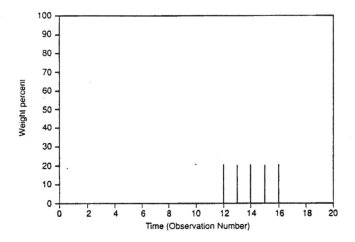

Figure 2. Shows 5-point moving average chart.

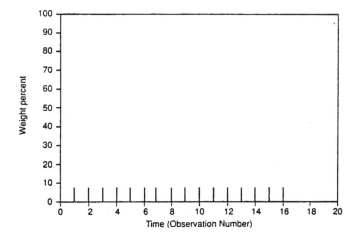

Figure 3. Shows the CUSUM chart.

This pattern, then, is a blend of the weighting functions employed by the Shewhart Control Chart and the CUSUM Chart. We make use of all data points yet more emphasis is placed on the recent ones. Although the Stepwise SPC Chart is a combination of the Shewhart Control Chart and the CUSUM Chart, the Exponentially Weighted Moving Average Chart is a compromise of the two. Figure 6 illustrates the EWMA.

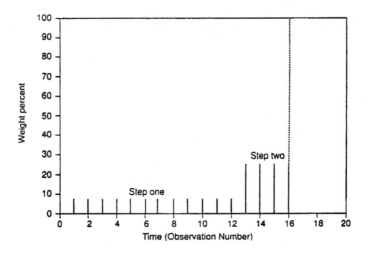

Figure 4. Shows the Stepwise SPC chart.

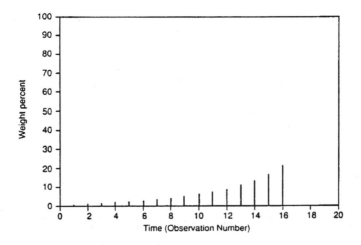

Figure 5. Shows the EWMA chart.

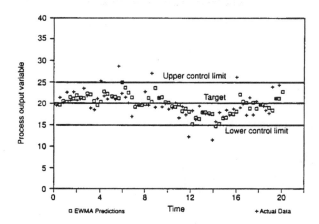

Figure 6. Illustrates the EWMA model.

ENERGY

The capacity to do work. There are many forms of energy, any of which can be converted into any other form of energy. To produce electrical power in a steam turbine-generator system, the chemical energy in coal is converted into heat energy, which (through steam) is converted to the mechanical energy of the turbine, and in turn, converted into electrical energy. Electrical energy may then be converted into the mechanical energy of a vacuum cleaner, the radiant and heat energy of a light bulk, the chemical energy of a charged battery, etc. Conversion from one form of energy to another results in some energy being lost in the process (usually as heat). There are two kinds of mechanical energy: kinetic energy, imparted by virtue of a body's motion, and potential energy, imparted by virtue of a body's position (e.g., a coiled spring, or a stone on the edge of a cliff). Solar (radiant) energy is the basis of all life through the process of photosynthesis, by which green plants convert solar energy into chemical energy. Nuclear energy is the result of the conversion of a small amount of the mass of an unstable (radioactive) atom into energy. The fundamental unit of energy in the Systeme International is the joule. It can be expressed in other energy units, such as the calorie, British thermal unit (Btu), kilowatt-hour, etc. by use of appropriate conversion factors. See *Kinetic Energy* and *Potential Energy*.

EP ADDITIVE

This is a lubricant additive that prevents sliding metal surfaces from seizing under conditions of extreme pressure (EP). At the high local temperatures associated with metal-to-metal contact, an EP additive combines chemically with the metal to form a surface film that prevents the welding of opposing asperities, and the consequent scoring that is destructive to sliding surfaces under high loads. Reactive compounds of sulfur, chlorine, or phosphorus are used to form these inorganic films.

EPOXY RESINS

The epoxy resins were first synthesized in the 1930s. In common with phenolic and polyester resins, the epoxy resins are thermosetting materials. When converted by a curing agent, the thermosetting resins become hard, infusible systems. The system may be visualized as a network crosslinked in all three dimensions.

Thermoplastic resins, on the other hand, such as polyethylene and polyvinyl chloride may be thought of as permanently fusible compounds composed of long linear chains lying together in three dimensions, but not interconnected. In practical effect, the thermoplastic materials will soften progressively with heat or flow with pressure, whereas the thermosetting materials will retain their dimensional stability throughout their design range.

The classification, however, while justified on the basis of structure, is not to be taken as an absolute criterion of performance. Some thermosetting compounds are designed for an extremely limited range and at higher temperatures will distort more readily than the high-heat-resistant thermoplastic compounds. In certain other respects, the two classes of materials also differ. Thermoplastic materials, for instance, are formed under heat and pressure into the desired shapes. This need not be the case with all thermosetting compounds. For example, a low-viscosity epoxy resin and a suitable fluid curing agent can be combined at room temperature and poured into a prepared mold; within a few minutes, the mold can be stripped away to reveal a solid, dimensionally stable block of resin.

Other differences, both in kind and degree, combine to make the thermosetting resins as a class useful in many industrial applications for which the thermoplastics are chemically unsuited. The thermosetting epoxy resins possess a number of unusually valuable properties immediately amenable to use in the formulation of adhesives, sealing liquids, cold solders, castings, laminates, and coatings. The more important of these properties include:

1. *Versatility* - numerous curing agents for the epoxies are available, and the epoxies are compatible with a wide variety of modifiers; hence, the properties of the cured epoxy-resin system can be engineered to widely diverse specifications;

2. *Good handling characteristics* - many epoxy systems can be worked at room temperature, and those which cannot require only moderate heat during mixing. Before the curing agent is incorporated, the resins have indefinite shelf life, provided they are properly made and do not contain any

caustic. The ratio of curing agent to resin is not as critical as with some thermosetting materials. It should be held fairly close to the empirically determined optimum amount if best results are to be obtained, and weighing should be done with care. If too much or too little curing agent is present, the solvent resistance and heat-distortion temperature will be reduced. In general, however, a few percent error in either direction may be tolerated in most applications, and some curing agents permit even wider margins. Cure can be accomplished in almost any specified time period by regulation of cure cycles and proper selection of curing agent. In many cases, pot life, viscosity, and cure schedules can be accommodated to the production situation without seriously influencing the properties of the cured system;

3. *Toughness* - cured epoxy resins are approximately seven times tougher than cured phenolic resins. The relative toughness has been attributed to the distance between crosslinking points and the presence of integral aliphatic chains;

4. *High adhesive properties* - epoxy resins have high adhesive strengths. Epoxy resins have the unique ability to create electromagnetic bonding forces between the epoxy molecule and the adjacent surface.

The epoxy groups, likewise, will react to provide chemical bonds with surfaces, such as metals, where active hydrogens may be found. Since the resin passes relatively undisturbed (i.e., with slight shrinkage) from the liquid to the solid state, the bonds initially established are preserved. In addition, these resins display low shrinkage.

The epoxy resins differ from many thermosetting compounds in that they give off no by-products during cure and, in the liquid state, are highly associated. Cure is by direct addition, and shrinkage is on the order of < 2 percent for an unmodified system, indicating that little internal rearrangement of the molecules is necessary. The condensation and crosslinking of phenolic and polyester resins, on the other hand, yield significantly higher shrinkage values. Cured epoxy resins are very inert chemically.

The ether groups, the benzene rings, and, when present, the aliphatic hydroxyls in the cured epoxy system are virtually invulnerable to caustic attack and extremely resistant to acids. In the case of the phenolic resin, the sodium phenolate formed is readily soluble and when present will cause the ultimate disintegration of the resin chain. In the case of the polyester, the ester linkage is hydrolyzed back to the original alcohol and salt of the carboxyl group. The chemical inertness of the cured epoxy system is enhanced by the dense, closely packed structure of the resinous mass, which is extremely resistant to solvent action.

The epoxy-resin molecule is characterized by the reactive epoxy or ethoxyline groups which serve as terminal linear polymerization points. When crosslinking or cure is accomplished through these groups or through hydroxyls or other groups present, an unusually tough, extremely adhesive, and highly inert solid results. The most widely used liquid epoxy resins have viscosities in the 8,000 to 20,000 centipoise range. Since the commercial resins are not molecularly distilled, they contain some percentages of higher-weight homologs, branched-chain molecules, isomers, and occasionally monoglycidyl ethers in combination with the basic structure. The high-viscosity liquid and the solid commercial resins are predominantly composed of more highly polymerized products considered as homologs of diglycidyl ether of bisphenol A.

EP RUBBERS

Ethylene/propylene copolymers and ethylene/propylene/nonconjugated diene terpolymers are industrially important synthetic elastomers. The outstanding property of the copolymer and the terpolymer are their good weather-resistance compared with polybutadiene (BR), polyisoprene (IR), and styrene/butadiene copolymer (SBR), since they have no double bonds in the backbone of the polymer chains and, thus, are less sensitive to oxygen and ozone. Other excellent properties of these rubbers are high resistance to acid and alkali, electrical properties, and high and low temperature performance. The copolymer and the terpolymer are used in the automotive industry for hoses, gaskets, wipers, bumpers, belts, etc. Furthermore, it is used for cable insulation and for roofing. Production of these olefinic copolymer and terpolymer rubbers have been increased with increasing car production.

The elastomer is produced by vanadium catalysts such as soluble $VOCL_3$ or VCL_4 and $AlEt_{1.5}CL_{1.5}$ or $ALEt_2CL$ as a co-catalyst. It is well known that the catalyst system was first reported by Natta and provided a highly random copolymer of ethylene and propylene. A number of papers have been reported regarding copolymerization of ethylene with propylene by using a hydrocarbon soluble vanadium catalyst (e.g., VCL_4, $VOCL_3$, $V(acac)_3$, $VO(OR)3$, $VOCL(OR)_2$, $VOCL_2,(OR)$, etc.) together with an organoaluminiurn compound. Although the catalysts provide a highly random copolymer of ethylene with propylene, their catalytic activities markedly decrease with temperature. This phenomenon is drastic, particularly at higher temperatures.

With respect to titanium catalysts for polymerization of ethylene or propylene, Ziegler synthesized the first high-density polyethylene and Natta prepared isotactic polypropylene by means of coordination catalysts about 50 years ago. The preparation methods of catalysts have been studied extensively. These $TiCL_3$ catalysts have very high activity for homopolymerization of ethylene and propylene, whereas, they exhibit low activity for random copolymerization of ethylene with propylene when compared to vanadium catalysts. Refer to *Ziegler-Nata Catalyst*, *Vanadium Catalysts*, and *EP Terpolymer*. (Source: *Elastomer Technology Handbook*, N. P. Cheremisinoff - editor, CRC Press, Boca Raton, Florida, 1993).

EP TERPOLYMER

The terpolymers of ethylene and propylene are EPDM (ethylene-propyelene diene monomer). In EPDM, a small amount of a third monomer, called a diene, is introduced. EPDM rubbers are amorphous polymers in which a small amount of a nonconjugated diene is added. The diene introduces unsaturation (i.e., double bonds) which are pendant, or in a side chain. The commonly used dienes are dicyclopentadiene, ethylidene norbornene, and 1,4-hexadiene. The substituent unsaturation imparted by the diene permits sulfur vulcanization. EPDM polymers can also be cross-linked using peroxide systems. The monomer units in EPDM are randomly distributed, meaning that the polymer does not follow a regular alteration sequence of ethylene and propylene units, but rather contains short runs of both polyethylene and polypropylene interspersed among longer segments of random copolymers. The three commercial comonomers used to introduce unsaturation are illustrated in Figure 1.

dicyclopentadiene (DCPD)

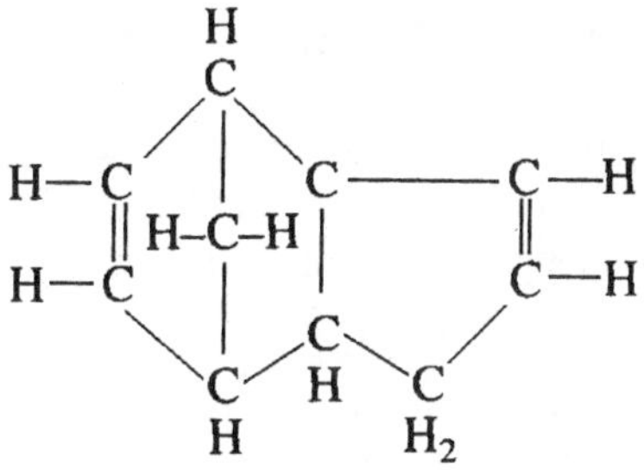

ethylidene norbornene (ENB)

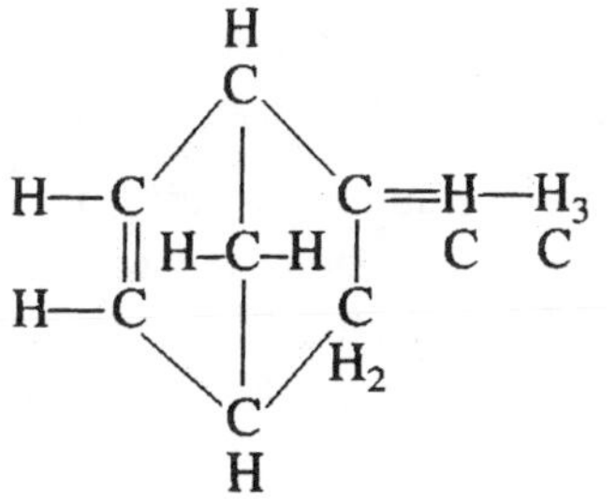

and 1,4-hexadiene (1,4-HD)

$$CH_2=CH-CH_2-CH=CH-CH_3$$

Figure 1. Three common commercial dienes.

In all three cases the polymer's double bond is the one illustrated on the left side of the molecule. The most common termonomer is ethylidene norbornene because of its ease of incorporation and its greater reactivity in sulfur vulcanization. Refer to *Sulfur Vulcanization*. (Source: *Handbook of Polymer Science and Technology: Volume 2 - Performance Properties of Plastics and Elastomers*, N. P. Cheremisinoff - editor, Marcel Dekker Inc., New York, 1989).

ESTER

A chemical compound formed by the reaction of an organic or inorganic acid with an alcohol or with another organic compound containing the hydroxyl (-OH) radical. The reaction involves replacement of the hydrogen of the acid with a hydrocarbon group. The name of an ester indicates its derivation: e.g., the ester resulting from the reaction of ethyl alcohol and acetic acid is called ethyl acetate. Esters have important uses in the formulation of some petroleum additives and synthetic lubricants.

EVA (ETHYLENE VINYL ACETATE)

EVA copolymers are used in a wide variety of binder applications. They display low odor, contain non-APE surfactant stabilization, and generally tend to have low levels of VOC and formaldehyde. They provide excellent adhesion of natural and synthetic fibers. Some typical binder applications are as follows:

1. Wipes: Features include high wet and solvent strength, absorbency, good adhesion to polyester and cellulose. The polymers can be used to produce durable and disposable wipes for a variety of end uses, both industrial and consumers. They are considered excellent for use in air-laid applications.

2. Lightweight Disposable: Features include synthetic fiber adhesion, resiliency, solvent resistance, absorbency, fire retardant, and salt stable. There are a broad range of polymers used in this application.

3. Interlinings: Features include durability, resiliency, high solvent resistance, fast curing, polyester adhesion, good color, heat sealable, and pigment compatible. They provide strong, tough products with good aesthetic appeal of finished garments. They also provide good color for white products, and have wash and dry cleanability.

4. Industrial Durables: Features include high-temperature heat resistance, controlled migration, excellent color, solvent resistance, durability and a wide glass transition temperature range of grades. Their benefits include low elongation in end use, flexibility, non-brittleness, they have good aesthetic appeal, and toughness.

5. Medical and Hygiene: Features include repellent compatibility, high tear

strength, sterilizability, excellent adhesion to pulp, rayon, cotton and polyester. Grades are available as either hydrophylic or hydrophobic. Benefits include toughness and the ability to produce strong products that are soft and drapable. They are excellent binders for acquisition layers and to serve as absorbent core materials.

Table 1 provides some average properties of this polymer group. In reviewing the general properties of this polymer, note the use of the following legend: A = amorphous - Cr = crystalline - C = clear - E = excellent - G = good - P = poor - O = opaque - T = translucent- R = Rockwell - S = Shore.

Table 1

Average Properties of EVAs

STRUCTURE: A	FABRICATION		
	Bonding	Ultrasonic	Machining
	G	P	P
SPECIFIC DENSITY: 0.93	DEFLECTION TEMPERATURE (° F) @ 66 psi: 144 @ 264 psi: 95		
WATER ABSORPTION RATE (%): 0.07	UTILIZATION TEMPERATURE (° F) min: -76 max: 131		
ELONGATION (%): 800	MELTING POINT (° F): 176		
TENSILE STRENGTH (psi): 2000	COEFFICIENT OF EXPANSION: 0.00009		
COMPRESSION STRENGTH (psi): 1450	TRANSPARENCY: O		
FLEXURAL STRENGTH (psi): 1400	UV RESISTANCE: G		
FLEXURAL MODULUS (psi): 8700	CHEMICAL RESISTANCE		
IMPACT (IZOD ft. lbs/in): NB	Acids	Alkalis	Solvents
HARDNESS: R40	P	G	G

EVAPORATION

Evaporation is the gradual change of a liquid into a gas without boiling. The molecules of any liquid are constantly moving. The average molecular speed depends on the temperature, but individual molecules may be moving much faster or slower than the average. At temperatures below the boiling point, faster molecules approaching the liquid's surface may have enough energy to escape as gas molecules. Because only the faster molecules escape, the average speed of the remaining molecules decreases, lowering the liquid's temperature, which depends on the average speed of the molecules.

EXTENSIONAL RHEOMETRY

Extensional rheometry is the least understood field of rheology. The simplest way to measure extensional viscosities is to stretch a polymer rod held at elevated temperatures at a speed that maintains a constant strain rate as the rod reduces its cross-sectional area. The viscosity can be computed as the ratio of instantaneous axial stress to elongational strain rate. The biggest problem when trying to perform this measurement is to grab the rod at its ends as it is pulled apart. The most common way to grab the specimen is with toothed rotary clamps to maintain a constant specimen length. A schematic of Meissner's extensional rheoMeter incorporating rotary clamps is shown in Figure 1.

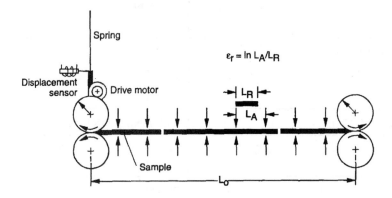

Figure 1. Meissner extensional rheometer.

EXTRUSION

The extrusion process is basically designed to continuously convert a soft material into a particular form. The heart of this processing/fabrication machine is a screw conveyer. It carries the cold plastic material (in granular or powdered form) forward by the action of the screw and squeezes it, and, with heat from external heaters and the friction of viscous flow, changes it to a molten stream (refer to Figure 1). As it does this, it develops pressure on the material, which is highest right before the molten plastic enters the die. The screen pack, consisting of a number of fine or coarse mesh gauzes supported on a breaker plate and placed between the screw and the die, filter out dirt and unfused polymer lumps. The pressure on the molten plastic forces it through an adapter and into the die, which dictates the shape of the final extrudate.

A die with a round opening produces pipe; a square die opening produces a square profile, etc. Other continuous shapes, such as the film, sheet, rods, tubing, and filaments, can be produced with appropriate dies. Extruders are also used to apply insulation and jacketing to wire and cable and to coat substrates such as paper, cloth, and foil.

When thermoplastic polymers are extruded, it is necessary to cool the extrudate below Tm (melting temperature) or Tg (glass transition temperature) to impart dimensional stability. This cooling can often be done simply by running the product through a tank of water, by spraying cold water, or, even more simply, by air cooling. When rubber is extruded, dimensional stability results from cross-linking (vulcanization).

Extruders have several other applications in polymer processing: in the blow-molding process they are used to make hollow objects such as bottles; in the blown-film process they are used for making wide films; they are also used for compounding plastics (i.e., adding various ingredients to a resin mix) and for converting plastics into the pellet shape commonly used in processing.

Refer to the following subject entries for additional information: *Screw Configurations*, *Screw Transfer Molding*, *Preplasticating Equipment*, *Injection Molding*, and *Sheet Extrusion*.

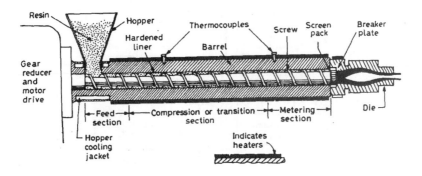

Figure 1. Features of a single-screw extruder.

F

FABRICATION PROCESSES

Although the family of polymers is extraordinarily large and varied, there are fairly broad and basic approaches that are used when fabricating a product out of polymers or polymers compounded with other ingredients. The type of fabrication process to be used depends on the properties and characteristics of the polymer and on the shape and form of the final product. It is widespread practice in industry to distinguish between thermoplastic and thermosetting resins. Compression and transfer molding are the most common methods of processing thermosetting plastics. For thermoplastics, the more important processing techniques are extrusion, injection, blow molding, and calendering; other processes are thermoforming, slush molding, and spinning.

FATIGUE TESTING

The standard fatigue tests for polymeric materials are the ASTM-D 671 test and the DIN 50100 test. In the ASTM test, a cantilever beam, shown in Figure 1, is held in a vise and bent at the other end by a yoke which is attached to a rotating variably eccentric shaft.

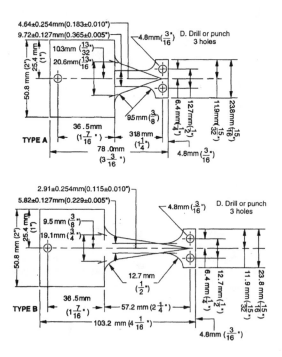

Figure 1. ASTM constant force fatigue test specimens.

A constant stress throughout the test region in the specimen is achieved by its triangular shape. Fatigue testing results are plotted as stress amplitude versus number of cycles to failure. The graphs are usually called S-N curves a term inherited from metal fatigue testing.

Fatigue in plastics is strongly dependent on the environment, the temperature, the frequency of loading, surface, etc. For example, due to surface irregularities and scratches, initiation at the surface is more likely in a polymer component that has been machined one that was injection molded.

As an example, an injection molded piece is formed by several layers of different orientation. In such parts the outer layers serve as a protective skin that inhibits crack initiation. In an injection molded article, cracks are likely to initiate inside the component by defects such as weld lines and filler particles. The gate region is also a prime initiator of fatigue cracks. Corrosive environments accelerate crack initiation and failure via fatigue.

At a low frequency and low stress level, the temperature inside the polymer specimen will rise and eventually reach thermal equilibrium when the heat generated by hysteretic heating equals the heat removed from the specimen by conduction. As the frequency is increased, viscous heat is generated faster, causing the temperature to rise even further. After thermal equilibrium has been reached, a specimen eventually fails by conventional brittle fatigue, assuming the stress is above the endurance limit. However, if the frequency or stress level is increased even further, the temperature will rise to the point that the test specimen softens and ruptures before reaching thermal equilibrium.

This mode of failure is usually referred to as *thermal fatigue*. The temperature rise in the component depends on the geometry and size of test specimen. For example, thicker specimens will cool slower and are less likely to reach thermal equilibrium. Similarly, material around a stress concentrator will be subjected to higher stresses which will result in temperatures that are higher than the rest of the specimen leading to crack initiation caused by localized thermal fatigue. To neglect the effect of thermal fatigue, cyclic tests with polymers must be performed at very low frequencies that make them much lengthier than those performed with metals and other materials which have high thermal conductivity. Stress concentrations have a great impact on the fatigue life of a component. Material irregularities caused by filler particles or by weld fines also affect the fatigue of a component. Up to this point, we assumed a zero mean stress. However, a great part of polymer components that are subjected to cyclic loading have other loads and stresses applied to them, leading to non-zero mean stress values. This superposition of two types of loading will lead to a combination of creep, caused by the mean stress, and fatigue, caused by the cyclic stress.

Test results from experiments with cyclic loading and non-zero mean stresses are complicated by the fact that some specimens fail due to creep and others due to conventional brittle fatigue. (Source: Osswald, T.A. and G. Menges, *Material Science of Polymers for Engineers*, Hanser Publishers, New York, 1996).

FDA (FOOD AND DRUG ADMINISTRATION)

Agency administered under the U.S. Department of Health and Human Services (formerly Health, Education and Welfare) "to enforce the Federal Food, Drug, and Cosmetic Act and thereby carry out the purpose of Congress to insure that foods are safe, pure, and wholesome, and made under sanitary conditions; drugs and therapeutic devices are safe and effective for their intended uses; cosmetics are safe and prepared from appropriate ingredients; and that all of these products are honestly and informatively labeled and packaged."

FIBERGLASS

Fibrous reinforcement in popular usage is almost synonymous with fibreglass, although other fibrous materials (carbon, boron, metals, aramid polymers) are also used. Glass fiber is supplied as mats of randomly oriented microfibrils, as woven cloth, and as continuous or discontinuous filaments. Hand lay-up is a versatile method employed in the construction of large structures such as tanks, pools, and boat hulls.

In hand lay-up mats of glass fibers are arranged over a mold and sprayed with a matrix-forming resin, such as a solution of unsaturated polyester (60 parts) in styrene monomer (40 parts) together with free-radical polymerization initiators. The mat can be supplied already impregnated with resin. Polymerization and network formation may require heating, although free-radical "redox" systems can initiate polymerization at room temperature. The molding may be compacted by covering the mold with a blanket and applying a vacuum between the blanket and the surface or, when the volume of production justifies it, by use of a matching metal mold.

Continuous multi-filament yarns consist of strands with several hundred filaments, each of which is 5 to 20 micrometers in diameter. These are incorporated into a plastic matrix through a process known as filament winding, in which resin-impregnated strands are wound around a form called a mandrel and then coated with the matrix resin. When the matrix resin is converted into a network, the strength in the hoop direction is very great (being essentially that of the glass fibers).

Epoxies are most often used as matrix resins, because of their good adhesion to glass fibers, although water resistance may not be as good as with the unsaturated polyesters. A method for producing profiles (cross-sectional shapes) with continuous fiber reinforcement is pultrusion. As the name suggests, pultrusion resembles extrusion, except that the impregnated fibers are pulled through a die that defines the profile while being heated to form a dimensionally stable network. Source: N. P. Cheremisinoff, *Guide to Fiberglass Reinforced Plastics*, Noyes Publishers, Westwood, N.J., 1996.

FIBER ORIENTATION

Thermosetting molding compounds contain fillers or fibrous reinforcements, often as high in content as 50-75% of the total compound weight. These materials range from wood flour, talc, and cotton fibers to glass and organic fibers of various lengths. During the molding process, these reinforcements will tend to "orient" themselves in various ways - largely dependent on such factors as the type and size of the reinforcement, the molded part configuration as it relates to the flow pattern of the compound within the confines of the mold, the rheology of the compound, the molding process - compression/transfer/injection, and the gate location and configuration in the case of either transfer or injection molding. The main role of reinforcements in a compound is to impart strength - primarily impact and flexural to the molded article.

The ultimate properties of fibrous reinforcements are directional so that their orientation within the molded article during molding will greatly affect the article's final physical property profile. If the fibers are unidirectional, tensile and modulus will be maximized, while a transversal alignment of the fibers tends to reduce those values.

On the other hand, a random orientation will result in property values uniform in all directions but not as high as is achieved with a unidirectional orientation. Fiber orientation can also affect molded dimensions because the alignment of the fibers within a molded part will influence thermal expansion, a principal cause of molding shrinkage. It is primarily the resin that shrinks, but its direction and amount of shrinkage are highly dependent on the reinforcement and its final orientation within the molded part.

FIBER REINFORCEMENT

The term polymer-matrix composite is applied to a number of plastic-based materials in which several phases are present. It is often used to describe systems in which a continuous phase (the matrix) is polymeric and another phase (the reinforcement) has at least one long dimension. The major classes of composites include those made up of discrete layers (sandwich laminates) and those reinforced by fibrous mats, woven cloth, or long, continuous filaments of glass or other materials. Plywood is a form of sandwich construction of natural wood fibres with plastics. The layers are easily distinguished and are both held together and impregnated with a thermosetting resin, usually urea formaldehyde. A decorative laminate can consist of a half-dozen layers of fibrous kraft paper (similar to paper used for grocery bags) together with a surface layer of paper with a printed design-- the entire assembly being impregnated with a melamine-formaldehyde resin. For both plywood and the paper laminate, the cross-linking reaction is carried out with sheets of the material pressed and heated in large laminating presses. Refer to *Fiberglass*.

FIRE POINT

Temperature at which the vapor concentration of a combustible liquid is sufficient to sustain combustion, as determined by test method ASTM D 92, Cleveland Open Cup. Refer to *Flash Point*.

FLAME IONIZATION

Flame atomization is an analytical technique that produces ions as well as atoms. Since only atoms are detected, it is important that the ratio of atoms to ions remain constant for the element being analyzed. This ratio is affected by the presence of other elements in the sample matrix. The addition of large amounts of an easily ionized element such as potassium to both the sample and standards helps mask the ionization interference. The capabilities of Flame AA can be extended by employing the following modifications: a cold quartz tube for containing mercury vapor (for mercury determination); a heated quartz tube for decomposing metallic hydride vapors for As, Se, Sb, Pb, Te, Sn, and Bi determination; a graphite quartz tube for decomposing involatile compounds of metals, with extremely high sensitivity. Refer to *Atomic Absorption*.

FLAMMABLE LIMITS

The percent concentration in air - (by volume) is given for the lower (LFL) and upper (UFL) limit. The values, along with those for flash point and ignition temperature, give an indication of the relative flammability of the chemical. The limits are sometimes referred to as "lower explosive limit" (LEL) and "upper explosive limit" (UEL). Refer to *Flash Point*.

FLAMMABILITY RANGE

Defined as the difference between the UFL and LFL. This difference provides an indication of how wide the flammability limits of a chemical are. Generally, the wider the range, the more hazardous the chemical may be considered from a fire standpoint.

FLASH POINT

The lowest temperature at which the vapor of a combustible liquid can be made to ignite momentarily in air, as distinct from fire point. Flash point is an important indicator of the fire and explosion hazards associated with a petroleum product. There are a number of ASTM tests for flash point, e.g., Cleveland open cup, Pensky-Martens closed tester, Tag closed tester, Tag open cup.

FOAMING

Foams, also called expanded plastics, possess inherent features that make them suitable for certain applications. For instance, the thermal conductivity of a foam is lower than that of the solid polymer. Also, a foamed polymer is more rigid than the solid polymer for any given weight of the material. Finally, compressive stresses usually cause foams to collapse while absorbing much energy, an obvious advantage in protective packaging. Properties such as these can be tailored to fit various applications by the choice of polymer and by the manner of foam formation or fabrication.

The largest markets for foamed plastics are in home insulation (polystyrene, polyurethane, phenol formaldehyde) and in packaging, including various disposable food and drink containers. Also refers to the occurrence of frothy mixture of air and a slurry that can reduce the effectiveness of the product, and cause sluggish hydraulic operation, air binding of oil pumps, and overflow of tanks or sumps.

Foaming can result from excessive agitation, improper fluid levels, air leaks, cavitation, or contamination with water or other foreign materials. Foaming can be inhibited with an anti-foam agent. The foaming characteristics of a lubricating oil can be determined by blowing air through a sample at a specified temperature and measuring the volume of foam, as described in test method ASTM D 892. Refer to *Foamed Thermoplastics* and *Foamed Thermosets*.

FOAMED THERMOPLASTICS

Polystyrene pellets can be impregnated with isopentane at room temperature and modest pressure. When the pellets are heated, they can be made to fuse together at the same time that the isopentane evaporates, foaming the polystyrene and cooling the assembly at the same time. Usually the pellets are prefoamed to some extent before being put into a mold to form a cup or some form of rigid packaging.

The isopentane-impregnated pellets may also be heated under pressure and extruded, in which case a continuous sheet of foamed polystyrene is obtained that can be shaped into packaging, dishes, or egg cartons while it is still warm. Structural foams can also be produced by injecting nitrogen or some other gas into a molten thermoplastic such as polystyrene or polypropylene under pressure in an extruder.

Foams produced in this manner are more dense than the ones described above, but they have excellent strength and rigidity, making them suitable for furniture and other architectural uses. One method of making foams of a variety of thermoplastics is to incorporate a material that will decompose to generate a gas when heated. This is known as a blowing agent.

To be an effective blowing agent, the material should decompose at about the molding temperature of the plastic, decompose over a narrow temperature range, evolve a large volume of gas, and, of course, be safe to use. One commercial agent is azodicarbonamide, usually compounded with some other ingredients in order to modify the decomposition temperature and to aid in dispersion of the agent in the resin. One mole of azodicarbonamide generates about 39,000 cc of nitrogen and other gases at 200° C. Thus 1 gram added to 100 grams of polyethylene can result in foam with a volume of more than 800 cc. Polymers that can be foamed with blowing agents include polyethylene, polypropylene, polystyrene, polyamides, and plasticized PVC.

FOAMED THERMOSETS

The rapid reaction of isocyanates with hydroxyl-bearing prepolymers to make polyurethanes is typical of the RIM (Reaction Injection Molding) process. These materials also can be foamed by incorporating a volatile liquid, which evaporates under the heat of reaction and foams the reactive mixture to a high degree. The rigidity of the network depends on the components chosen, especially the prepolymer. Hydroxyl-terminated polyethers are often used to prepare flexible foams, which are used in furniture cushioning. Hydroxyl-terminated polyesters, on the other hand, are popular for making rigid foams such as those used in custom packaging of appliances. The good adhesion of polyurethanes to metallic surfaces has brought about some novel uses, such as filling and making rigid certain aircraft components (rudders and elevators, for example). Another rigid thermoset that can be foamed in place is based on phenol-formaldehyde resins. The final stage of network formation is brought about by addition of an acid catalyst in the presence of a volatile liquid. Refer to *Reaction Injection Molding*.

FOOD ADDITIVE

Non-nutritional substance added directly or indirectly to food during processing or packaging. Polymer food additives are usually refined waxes or white oils that meet applicable FDA standards. Applications include direct additives, such as coatings for fresh fruits and vegetables, and indirect additives, such as impregnating oils for fruit and vegetable wrappers, dough divider oils, defoamers for yeast and beet sugar manufacture, release and polishing agents in confectionery manufacture, and rust preventives for meat processing equipment.

FOUR-BALL METHOD

Either of two lubricant test procedures, the FourBall Wear Method (ASTM D 2266) and Four-Ball EP (extreme pressure) Method (ASTM D 2596), based on the same principle. Three steel balls are clamped together to form a cradle upon which a fourth ball rotates on a vertical axis. The balls are immersed in the lubricant under investigation. The Four-Ball Wear Method is used to determine the anti-wear properties of lubricants operating under boundary lubrication conditions. The test is carried out at a specified speed, temperature, and load. At the end of a specified test time, the average diameter of the wear scars on the three lower balls is reported. The Four-Ball EP Method is designed to evaluate performance under much higher unit loads. The loading is increased at specified intervals until the rotating ball seizes and welds to the other balls. At the end of each interval the average scar diameter is recorded. Two values are generally reported-load wear index (formerly mean Hertz load) and weld point.

FREEZING POINT

A specific temperature that can be defined in two ways, depending on the ASTM test used. In ASTM D 1015, which measures the freezing point of high purity products. Freezing point is the temperature at which a liquid solidifies. In ASTM D 2386, which measures the freezing point of aviation fuel, freezing point is that temperature at which hydrocarbon crystals formed on cooling disappear when the temperature of the fuel is allowed to rise.

FRETTING

A form of wear resulting from small-amplitude oscillations or vibrations that cause the removal of very finely divided particles from rubbing surfaces (e.g., the vibrations imposed on the wheel bearings of an automobile when transported by rail car). With ferrous metals the wear particles oxidize to a reddish, abrasive iron oxide, which has the appearance of rust or corrosion, and is therefore sometimes called fretting corrosion; other terms applied to this phenomenon are false Brinelling (localized fretting involving the rolling elements of a bearing) and friction oxidation.

FRICTION

The resistance to the motion of one surface over another. The amount of friction is dependent on the smoothness of the contacting surfaces, as well as the force with which they are pressed together. Friction between unlubricated solid bodies is independent of speed and area. The coefficient of friction is obtained by dividing the force required to move one body over a horizontal surface at constant speed by the weight of the body; e.g., if a force of 4 kilograms is required to move a body weighing 10 kilograms, the coefficient of friction is 4/10, or 0.4. Coefficients of rolling friction (e.g., the motion of a tire or ball bearing) are much less than coefficients of sliding friction (back and forth) motion over two flat surfaces. Sliding friction is thus more wasteful of energy and can cause more wear. Fluid friction occurs between the molecules of a gas or liquid in motion, and is expressed as shear stress. Unlike solid friction, fluid friction varies with speed and area. In general, lubrication is the substitution of low fluid friction in place of high solid-to-solid friction. In dealing with granular or powdery solids, friction is important in the design and selection of storage and feed bins. The *effective angle of friction*, δ, is used to define the strength that the solids develop under flowing conditions; that is under continuous deformation and pressure. Angle δ is a property of the material, and should not be confused with the angle of internal friction, ϕ, which is used in solid mechanics. Experiments on bulk solids have shown that steady flow occurs only under certain stress conditions in solids. For a given material, angle δ has been found to vary within a few degrees for a range of pressures which occur in gravity flow bins. Hence, for pelletized resin feeding or general bulk solids handling, δ can be important. The reader can refer to the following reference for more details: N. P. Cheremisinoff, *Chemical Engineer's Condensed Encyclopedia of Process Equipment*, Gulf Publishing Co., Houston, TX, 2000.

G

GAS CHROMATOGRAPHY

In gas chromatography the sample is usually injected at high temperature to ensure vaporization. Obviously, only materials volatile at this temperature can be analyzed. If the stationary phase is a solid, the technique is referred to as *gas-solid chromatography*.

The separation mechanism is principally one of adsorption. Those components more strongly adsorbed are held up longer than those which are not. If the stationary phase is a liquid, the technique is referred to as *gas-liquid chromatography* and the separation mechanism is principally one of partition (solubilization of the liquid phase). Gas chromatography has developed into one of the most powerful analytical tools available to the organic chemist. The technique allows separation of extremely small quantities of material. The characterization and quantitation of complex mixtures can be accomplished with this process. The introduction of long columns, both megabore and capillary, produces a greater number of theoretical plates increasing the efficiency of separation beyond that of any other available technique.

The technique is applicable over a wide range of temperatures (-40 to 350° C) making it possible to chromatograph materials covering a wide range of volatiles. The laboratory uses packed columns along with megabore and capillary. In this way the broadest range of chromatographic problems can be addressed. The detector used to sense and quantify the effluent provides the specificity and sensitivity for the analytical procedure. Refer to *Chromatography*. (Source: Cheremisinoff, N.P. *Polymer Characterization: Laboratory Techniques and Analysis*, Noyes Publishers, New Jersey, 1996).

GAS-PHASE POLYMERIZATION

This polymerization process is used with gaseous monomers such as ethylene, tetrafluoroethylene, and vinyl chloride. The monomer is introduced under pressure into a reaction vessel containing a polymerization initiator. Once polymerization begins, monomer molecules diffuse to the growing polymer chains. The resulting polymer is obtained as a granular solid.

GEL PERMEATION CHROMATOGRAPHY

Gel Permeation Chromatography (GPC), also known as Size Exclusion Chromatography (SEC), is a technique used to determine the average molecular weight distribution of a polymer sample. Using the appropriate detectors and analysis procedure it is also possible to obtain qualitative information on long chain branching or determine the composition distribution of copolymers. As the name implies, GPC or SEC separates the polymer according to size or hydrodynamic radius. This is accomplished by injecting a small amount of polymer solution (0.01 to 0.6 %) into a set of columns that are packed with porous beads. Smaller molecules can penetrate the pores, and are, therefore, retained to a greater extent than the larger molecules, which continue down the columns and elute faster. This process is illustrated in Figure 1.

One or more detectors is attached to the output of the columns. For routine analysis of linear homopolymers, this is most often a Differential Refractive Index (DRI) or a UV detector. For branched or copolymers, however, it is necessary to have at least two sequential detectors to determine molecular weight accurately. Branched polymers can be analyzed using a DRI detector coupled with a "molecular weight sensitive" detector such as an on-line viscometer (VIS) or a low-angle laser light scattering (LALLS) detector.

The compositional distribution of copolymers, i.e., average composition as a function of molecular size, can be determined using a DRI detector coupled with a selective detector such as UV or FTIR. It is important to consider the type of polymer and information that is desired before submitting a sample. The following outline describes each instrument that is currently available.

GPC/DRI/LALLS - One can use two instruments with sequential LALLS and DRI detectors. The unit is operated using TCB at 135° C and is used to analyze PE, EP, and PP samples. The other, operating at 60° C, is for butyl type polymers which dissolve in TCB at lower temperatures. The data consists of two chromatograms, plots of detector mV signal (LALLS and DRI) versus retention time. The DRI trace corresponds to the concentration profile whereas the LALLS signal is proportional to concentration *M, resulting in more sensitivity at the high molecular weight end.

An example of the output is shown in Figure 2 for polyethylene NBS 1476. The LALLS trace shows a peak at the high molecular weight end (low retention time) which is barely noticeable on the DRI trace. This suggests a very small amount of high molecular weight, highly branched material. This type of bimodal peak in the DRI trace is often seen in branched EP (ethylene-propylene polymers) or LDPE (low density polyethylene) samples.

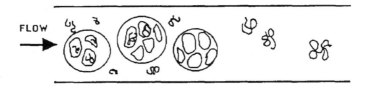

Figure 1. Polymer solution flow through a GPC column.

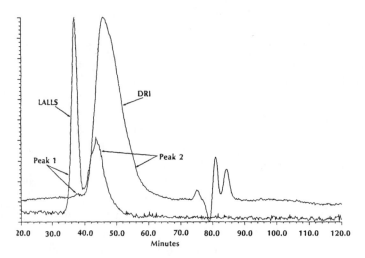

Figure 2. Sample output from LALLS detector.

For a linear polymer (if all the calibration constants are known), the molecular weights from both pages should agree within 10%. A LALLS report that gives higher molecular weights than the DRI suggests that the sample is branched, and the values from the LALLS report should be used (again, assuming that the calibration constants are correct). Occasionally, some of the sample, gel or insolubles, is filtered out during the sample preparation and analysis. The percentage should be indicated on the report.

GPC/DRI/VIS - GPC with an on-line viscometer can be used instead of a LALLS detector to analyze branched polymers. In this case the intrinsic viscosity is measured so that the Mark-Houwink parameters are not needed. It is complementary to the LALLS instrument in intrinsic viscosities.

GPC/DRI/UV - The UV detector is used to analyze chromophores. Its most common use is for graft or block copolymers containing PS or PMS. The data from this instrument consists of two chromatograms, the UV and DRI traces. An example is shown Figure 3 for an EP-g-PS coploymer (peak 1) with risidual PS homopolymer (peak 2). The UV absorption relative to the DRI signal corresponds to the copolymer composition, which is why the relative UV absorption is higher for the pure PS in peak 2. The results report consists of two pages. One is the molecular weight report from the DRI calibration curve as described above. Note that the molecular weights are reported as if the sample is a homopolymer not copolymer. The other page using the UV data gives an effective extinction coefficient E' which is the UV/DRI ratio. A higher E' indicates a higher composition of the UV active chromophore (for example, more PS in the graft copolymer).

This technique is also used to determine the compositional distribution of

ENB (ethylidene norbonene) in EPDM, i.e., whether the ENB is evenly distributed across the molecular weight distribution or concentrated in the low or high molecular weight end. The GPC/DRI/UV instrument can be used to analyze samples that dissolve in THF at 30-45° C.

GPC/DRI/FrIR - The GPC/DRI/FTIR instrument is complementary to the UV detector for compositional distribution. It runs at 135° C in TCB and can be used for EP analysis. Typical applications include ethylene content as a function of molecular weight, maleic anhydride content in maleated EP, or PCL content in caprolactone-g-EP copolymers. The FTIR detector is off-line so that 5-10 fractions of the eluant are collected on KBr plates and analyzed. This procedure gives calibration of IR absorption bands. This method is much more labor intensive than the other techniques and should be used with discretion. (Source: Cheremisinoff, N.P. *Polymer Characterization: Laboratory Techniques and Analysis*, Noyes Publishers, New Jersey, 1996).

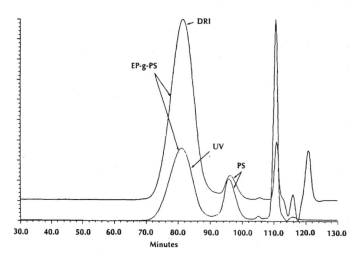

Figure 3. Sample UV detector output.

GRAFTING (RADIATION TYPE)

Radiation grafting is a versatile technique by which surface properties of a polymer can be tailored through the choice of different monomers. The most common radiation sources are high-energy electrons, γ radiation, and ultraviolet and visible light. Grafting is usually performed by irradiating the polymer in the presence of a solvent containing a monomer. Alternatively, grafting can be initiated thermally by contacting the polymer, which has been pre-irradiated in air to produce reactive groups on the surface, with a monomer. Systems are available for both batch and continuous modes of operation. Refer to *Surface Grafting*.

H

HALF-LIFE

The term "half-life" is generally interpreted as meaning a "rate constant" in describing cross-linking reactions or catalyst activity in a polymerization reaction. With respect to vulcanization, half-life is the time it takes one-half of any quantity of curing agent present to thermally decompose (as in the case of peroxides). The time is independent of the quantity of curing agent present and is primarily dependent on temperature. In the first half-life, 50% of the added agent decomposes. During the second half-life, 50% of the remaining agent decomposes. This is 25% of the original quantity, and the total amount decomposed is now 75%. The process continues and theoretically never reaches 100% decomposed. Practically, the level is so low after 7 half-lives that the effect of additional cure time is insignificant. Table 1 shows the relationship between number of half-lives and percent decomposed for a peroxide cure.

Table 1
Half Life Relation to Cure

Number of Half-Lives	% of Original Peroxide Decomposed or % of Ultimate Cure State
1	50
2	75
3	87.5
4	93.75
5	96.9
6	98.4
7	99.2
8	99.6
9	99.8
10	99.9

Time/Temperature as a Function of the Half-Life: The peroxide thermal decomposition reaction is the first one in the curing or cross-linking sequence, and determines the rate. This reaction follows first-order kinetics. The kinetic order was determined by partially curing compounds for varying cure times and then analyzing unreacted or unused peroxide. The rate of peroxide thermal decomposition thus obtained coincided with the decomposition rate as determined on a Monsanto oscillating disc rheometer (ODR). Figure 1 shows a peroxide decomposition curve and the cross-linking curve. These plots show the direct dependence of cross-linking on peroxide decomposition. The plot shows that a peroxide cure reaches a plateau when all of the peroxide has decomposed. Changing the peroxide level in the compound will change the state of cure, but will not change the time necessary to reach a plateau. Cure time can be altered only when cure temperatures are changed.

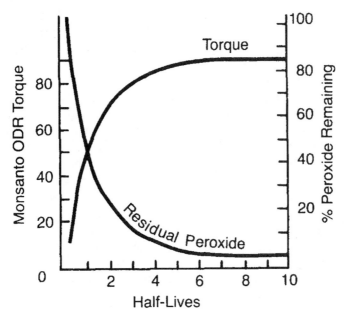

Figure 1. ODR curve.

HALOGEN

Any of a group of five chemically related nonmetallic elements: chlorine, bromine, fluorine, iodine, and astatine. Chlorine compounds are used as EP additives in certain lubricating oils, and as constituents of certain petrochemicals (e.g., vinyl chloride, chlorinated waxes). Chlorine and fluorine compounds are also used in some synthetic lubricants.

HALOGENATED HYDROCARBONS

A halogenated hydrocarbon is defined as a derivative of a hydrocarbon in which a hydrogen atom is replaced by a halogen atom. Since all of the halogens react similarly, and the number of hydrocarbons (including all saturated hydrocarbons, unsaturated hydrocarbons, aromatic hydrocarbons, other cyclical hydrocarbons, and all the isomers of these hydrocarbons) is large, the number of halogenated hydrocarbons can also be very large. The most common hydrocarbon derivatives are those of the first four alkanes and the first three alkenes (and, of course, the isomers of these hydrocarbons). There are some aromatic hydrocarbon derivatives, but, again, they are of the simplest structure. Whatever the hydrocarbon backbone is, it is represented in the general formula by its formula, which is R-. Therefore, the halogenated hydrocarbons will have formulas such as R-F, R-Cl, R-Br, and R-I for the respective substitution of fluorine, chlorine, bromine, and iodine on to the hydrocarbon backbone. As a rule, the general formula can be written R-X, with the R as the hydrocarbon backbone, the X standing for the halide (any of the halogens), and the ''-'' the covalent bond between the hydrocarbon backbone and the halogen. R-X is read as "alkyl halide".

Radicals of the alkanes are referred to as alkyl radicals. There are two other important radicals; they are the vinyl radical, which is produced when a hydrogen atom is removed from ethylene, and the phenyl radical, which results when a hydrogen atom is removed from benzene. The term halogenated means that a halogen atom has been substituted for a hydrogen atom in a hydrocarbon molecule. The most common halogenated hydrocarbons are the chlorinated hydrocarbons. The simplest chlorinated hydrocarbon is methyl chloride, whose molecular formula is CH_3Cl. The structural formula for methyl chloride shows that one chlorine atom is substituted for one hydrogen atom. Methyl chloride has many uses, such as a herbicide, as a topical anesthetic, extractant, and low-temperature solvent, and as a catalyst carrier in low-temperature polymerization. It is a colorless gas that is easily liquified and is flammable; it is also toxic in high concentrations. Methyl chloride is the common name for this compound, while its proper name is chloromethane. Proper names are determined by the longest carbon chain in the molecule, and the corresponding hydrocarbon's name is used as the last name of the compound. Any substituted groups are named first, and a number is used to designate the carbon atom that the functional group is attached to, if applicable.

It is possible to substitute more than one chlorine atom for a hydrogen atom on a hydrocarbon molecule; such substitution is done only when the resulting compound is commercially valuable or is valuable in another chemical process. An example is methylene chloride (the common name for dichloromethane), which is made by substituting two chlorine atoms for two hydrogen atoms on the methane molecule. Its molecular formula is CH_2Cl_2. Methylene chloride is a colorless, volatile liquid with a sharp, ether-like odor. It is listed as a non-flammable liquid, but it will ignite at 1,224° F; it is narcotic at high concentrations. It is most commonly used as a stripper of paints and other finishes. It is also a good degreaser and solvent extractor and is used in some plastics processing applications.

Substituting a third chlorine on the methane molecule results in the compound whose proper name is trichloromethane (tri- for three; chloro- for chlorine; and methane, the hydrocarbon's name for the one-carbon chain). It is more commonly known as chloroform. Its molecular formula is $CHCl_3$. Chloroform is a heavy, colorless, volatile liquid with a sweet taste and characteristic odor. It is classified as non-flammable, but it will burn if exposed to high temperatures for long periods of time. It is narcotic by inhalation and toxic in high concentrations. It is an insecticide and a fumigant and is very useful in the manufacture of refrigerants. The total chlorination of methane results in a compound whose proper name is tetrachloromethane (tetra- for four), but its common name is carbon tetrachloride (or carbon tet). This is a fire-extinguishing agent that is no longer used since it has been classified as a carcinogen. It is still present, though, and its uses include refrigerants, metal degreasing, and chlorination of organic compounds. Its molecular formula is CCl_4. It is possible to form analogues of methyl chloride, (methyl fluoride, methyl bromide, methyl iodide), methylene chloride (substitute fluoride, bromide, and iodide in this name also), chloroform (fluoroform, bromoform, and iodoform), and carbon tetrachloride (tetrafluoride, tetrabromide, and tetraiodide). Each of these halogenated hydrocarbons has some commercial value.

What was true for one hydrocarbon compound is true for most hydrocarbon compounds, particularly straight-chain hydrocarbons; that is, you may substitute a functional group at each of the bonds where a hydrogen atom is now connected to the carbon atom. Where four hydrogen atoms exist in methane, there are six hydrogen atoms in ethane; you recall that the difference in make-up from one compound to the next in an analogous series is the "unit" made up of one carbon and two hydrogens. Therefore, it is possible to substitute six functional groups onto the ethane molecule. You should also be aware that the functional groups that would be substituted for the hydrogens need not be the same, that is, you may substitute chlorine at one bond, fluorine at another, the hydroxyl radical at a third, an amine radical at a fourth, and so on.

Substituting one chlorine atom for a hydrogen atom in ethane produces ethyl chloride, a colorless, easily liquifiable gas with an ether-like odor and a burning taste, which is highly flammable and moderately toxic in high concentrations. It is used to make tetraethyl lead and other organic chemicals. Ethyl chloride is an excellent solvent and analytical reagent, as well as an anesthetic. Its molecular formula is C_2H_5Cl.

Although we are using chlorine as the functional group, it may be any of the other halogens. In addition, we are giving the common names, while the proper names may be used on the labels and shipping papers. Ethyl chloride's proper name is chloroethane.

Substituting another chlorine produces ethylene dichloride (proper name 1,2-dichloroethane). In this case, an isomer is possible, which would be the chlorinated hydrocarbon where both chlorines attached themselves to the same carbon atom, whereby 1,1-dichloroethane is formed. These compounds have slightly different properties and different demands in the marketplace. As further chlorination of ethane occurs, we would have to use the proper name to designate which compound is being made. One of the analogues of ethylene dichloride is ethylene dibromide, a toxic material that is most efficient and popular as a grain fumigant, but it is known to be a carcinogen in test animals.

There are many uses for the halogenated hydrocarbons. Many of them are flammable; most are combustible. Some halogenated hydrocarbons are classified as neither, and a few are excellent fire-extinguishing agents (the Halons®), but they will all decompose into smaller, more harmful molecular fragments when exposed to high temperatures for long periods of time.

HAND LAY-UP TECHNIQUE

This refers to an open mold technique used for the production of many articles where the resin and reinforcement systems can be tailored to meet specific end use requirements. Typical products produced by this method are skis, boat hulls, tubs, and shower stalls. The molded cavity is sprayed with a gel coat to a desired thickness to produce a desired surface finish or color. The reinforcement, which is usually glass fibers, mats, or sheets, is then hand rolled to secure proper wetting of the fibers as well as the required wall thickness. The final curing takes place through the use of a catalyst or hardener reaction. Matched metal molds are often used with hand lay-up formulations in order to achieve high quality surface finishes on both sides of the molded article.

HDPE (HIGH DENSITY POLYETHYLENE)

High-density polyethylene (HDPE) is a major thermoplastic used for a multitude of applications. Polyethylene has been manufactured since the early 1950s, but because of its industrial importance, the production of HDPE continues to be very active. New improved catalysts and processes have been developed to produce HDPE with tailored properties at lower manufacturing costs.

Three types of polymerization processes are used today for low pressure ethylene polymerization; these are (a) liquid slurry polymerization, (b) solution polymerization, and (c) gas phase polymerization. A brief description of each follows.

Liquid Slurry Polymerization

The liquid slurry polymerization process encompasses by far the largest group of HDPE technologies. In most cases this process utilizes a catalyst of activity such that catalyst deashing is not required. Excellent temperature control is a major attraction of the liquid slurry process. However, when linear low-density polyethylene (LLDPE) is made by copolymerizing ethylene with a higher α-olefin comonomer in a liquid slurry process, the swelling of the polymer in the slurry medium is a major problem. The swelling severely lowers the polymer production rate and polymer density to a minimum of about 0.925-0.935 g/cc. Currently, long-jacketed loop reactors and continuous stirred tank reactors (CSTRs) are most widely used for slurry polymerization.

The loop reactors, which are recycled tubular reactors, are used by the Phillips Petroleum Co. and Solvay et Cie. The Phillips process is characterized by the use of a light hydrocarbon diluent such as isopentane or isobutane in loop reactors which consist of four jacketed vertical pipes. Figure 1 shows the schematic flow diagram for the loop reactor polyethylene process. The use of high-activity supported chromium oxide catalyst eliminates the need to deash the product. This reactor is operated at about 35 atm and 85-110° C with an average polymer residence time of 1.5 hr. Solid concentrations in the reactor and effluent are reported as 18 and 50 wt %, respectively. The reactor diameter is 30 in. (O.D.) and the length of the reactor loop is about 450 ft.

Although reactor fouling was more severe at high temperatures with increased solubility of the polymer in the diluent, higher operating temperatures have become possible in the latest Phillips process because of reduced reactor fouling. This is accomplished by injecting a small quantity of additives such as sodium dioctyl sulfosuccinate and aluminum mono- and dihexadecyl-salicylic acid. The contents of the reactor are discharged continuously to a flash tank, utilizing the sensible heat to evaporate most of the diluent and the small amount of unreacted ethylene. The dry polyethylene powder is discharged from the flash tank through a series of licks to a drier to remove additional diluent dissolved in the polymer particles.

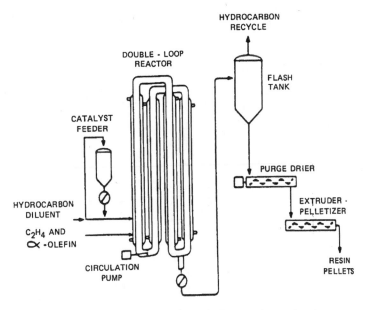

Figure 1. Loop reactor system for ethylene polymerization.

Then the powder is purged with nitrogen and transported pneumatically to the finishing area for stabilization and pelletizing. It has been reported that ethylene conversion in the reactor is around 98%. This process relies on polymerization temperature for average molecular weight control, while the MWD is controlled by the type of catalyst employed and certain proprietary operational adjustments which alter MWD.

Continuous stirred tank reactors are also widely used for hexane slurry ethylene polymerization by many manufacturers. In the Hoechst process, the reaction is carried out in four CSTRs arranged in series such that the slurry phase and the vapor phase move in concurrent flow. Polymerization occurs at 100 psig and 85° C with 98% conversion of ethylene. The residence time in the reactor is about 2.7 hr. The product slurry is pumped into centrifuges, which separate the bulk of the hydrocarbon diluent liquid from the polymer fluff.

Solution Polymerization

Solution processes have some unique advantages over slurry processes in that the MWD can be controlled better, and the process variables are also more easily controlled because the polymerization occurs in a homogeneous phase. The high polymerization temperature (130-150° C) also leads to high reaction rates and high polymer throughputs from the reactor. However, very high molecular weight polymers cannot be produced easily at these high temperatures, and since the solid content is relatively low compared with the slurry process, greater diluent recovery may be required. Figure 2 shows the DuPont solution polymerization process. The catalyst components, cyclohexane, ethylene, octene-1, and hydrogen, are charged continuously to a CSTR operating at a temperature in excess of 150° C and a pressure of about 80 atm.

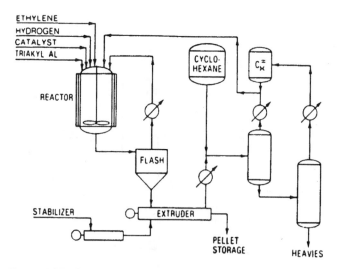

Figure 2. DuPont solution polymerization process.

Because of the short residence time (5-10 min) and high polymer concentration (35%), relatively small reactors may be used either in series or parallel to alter the MWD (molecular weight distribution) of the product.

Gas Phase Polymerization

Polymer separation, diluent recovery, and polymer drying steps may be completely eliminated in gas phase polymerization. The first commercial gas phase HDPE plant using a fluidized bed reactor was constructed by Union Carbide in 1968. Other gas phase polymerization processes using different types of reactors have also been developed by BASF, Naphthachimie, and Amoco. In the fluidized bed polymerization process shown in Figure 3, ethylene and butene-1 are copolymerized over a chromium- or titanium-based high-activity catalyst. Dry catalyst particles less than 250 μm are injected to the fluidized bed reactor, which is maintained by a large volume of circulating gas. Reaction occurs at 300 psig and 85-100° C for HDPE or 75-100° C for LLDPE. The heat of polymerization is adequately removed by the circulating cooled monomer gas at high velocity. The catalyst activity in this process is more than 600 kg/g Cr (or Ti).

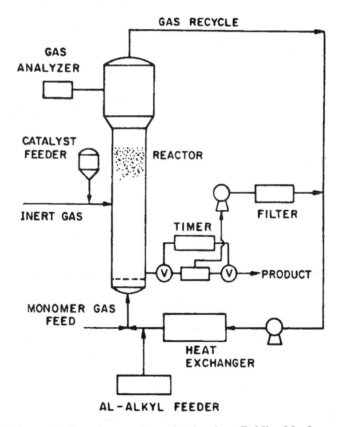

Figure 3. Gas phase polymerization in a fluidized bed.

The polymerization rate is controlled by the rate of catalyst injection to the bed. The gas velocity in the fluidized bed is maintained at about 5-10 times the weight of the fresh feed. The polymer particles are withdrawn intermittently or continuously near the distributor plate through a gas lock chamber into discharge tanks, where the ethylene gas is separated by adjusting the product withdrawal rate.

In a newer version of the Union Carbide fluidized bed polymerization process, a smaller reactor containing an internal cooler is used for more efficient temperature control. Although the conversion per pass is low (2-5%), the overall conversion of ethylene and butene to polymer is about 97%. Polymer particles grow to an average size of about 1000 μm in diameter during their 3-5 hr residence in the fluidized bed reactor. When high-activity catalyst is used, the average particle size of the polymer becomes about 15-20 times larger than the size of the original catalyst particle. Average molecular weight is controlled principally by manipulating the chain transfer agent concentration (hydrogen) and the polymerization temperature, while the MWD is strongly affected by the specific catalyst type and, to a much lower extent, by reactor operating conditions. The gas phase LLDPE reactor is operated at a temperature close to the polymer's softening point.

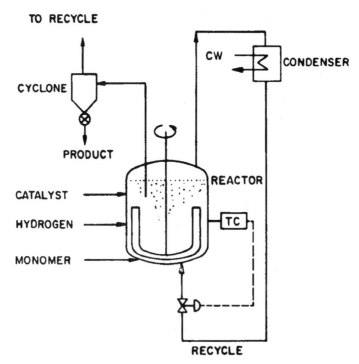

Figure 4. Continuous stirred bed reactor for gas phase olefin polymerization.

Two potential difficulties in operating the gas phase fluidized bed reactors are (a) the possibility of sintering and agglomeration of the polymer in the bed due to poor fluidization and/or poor heat removal, and (b) the possible inability to achieve control of polymer particle size and shape under a broad range of operating conditions. In Naphthachimie's gas phase process, highly active Ziegler-type catalysts supported on magnesium compound are used to polymerize ethylene in a similar fluidized bed reactor. BASF uses a continuous stirred bed reactor (CSBR) for gas phase ethylene polymerization as shown in Figure 4.

The reactor is operated at a higher pressure and temperature (500 psig, 100-110° C) than employed in Union Carbide's fluidized bed process. The makeup feed and ethylene recycle enter at the bottom of the reactor and rise through the bed at a very slow velocity of about 0.1 ft/sec. The polymer bed is agitated rather than fluidized. A uniform temperature of about 110° C is maintained in the bed, and the unreacted ethylene gas exits the top of the reactor at 105° C. About 9 % of the ethylene recycle leaves the reactor with the effluent polyethylene powder and is then separated from the product at 40 psig. The recycle ethylene is compressed to 1500 psig and then cooled to 32° C before being reintroduced to the reactor.

Properties

Table 1 provides average properties of HDPE. In reviewing the general properties of this polymer, note the use of the following legend: A = amorphous - Cr = crystalline - C = clear - E = excellent - G = good - P = poor - O = opaque - T = translucent - R = Rockwell - S = Shore.

Table 1

Average Properties of HDPE.

STRUCTURE: Cr	FABRICATION		
	Bonding	Ultrasonic	Machining
	P	P	G
Specific Density: 0.95	Deflection Temperature (° F)		
	@ 66 psi	@ 264 psi	
	176	131	
Water Absorbtion Rate (%): 0.01	Utilzation Temperature (° F)		
	min	max	
	-180	248	

Table 1 continued

Elongation (%): 100	Melting Point (° F): 266		
Tensile Strength (psi): 4550	Coefficient of Expansion: 0.00007		
Compression Strength (psi): 2900	Arc Resistance: 180		
Flexural Strength (psi): 5800	Dielectric Strength (kV/mm): 22		
Flexural Modulus (psi): 120000	Transparency: T		
Impact (Izod ft. lbs/in): NB	UV Resistance: P		
Hardness: SD65	**CHEMICAL RESISTANCE**		
	Acids	**Alkalis**	**Solvents**
	G	E	E

Refer to *LDPE*, *LLDPE* and *HMWDPE*. (Source: *Handbook of Polymer Science and Technology: Volume 1 - Performance Synthesis and Properties*, N. P. Cheremisinoff - editor, Marcel Dekker Inc., New York, 1989).

HMWDPE (HIGH MOLECULAR WEIGHT POLYETHYLENE)

Table 1 provides average properties of HMWDPE. In reviewing the general properties of this polymer, note the use of the following legend: A = amorphous - Cr = crystalline - C = clear - E = excellent - G = good - P = poor - O = opaque - T = translucent - R = Rockwell - S = Shore.

Table 1. Average Properties of HMWDPE.

STRUCTURE: Cr	**FABRICATION**		
	Bonding	**Ultrasonic**	**Machining**
	P	P	G
Specific Density: 0.95	**Deflection Temperature (° F)**		
	@ 66 psi	**@ 264 psi**	
	176	131	
Water Absorption Rate (%): 0.01	**Utilization Temperature (° F)**		
	min.	**max.**	
	-180	250	

Table 1 continued

Elongation (%): 700	**Melting Point (° F)**: 266
Tensile Strength (Psi): 3700	**Coefficient of Expansion**: 0.00007
Compression Strength (psi): 2500	**Arc Resistance**: 180
Flexural Strength (psi): 1500	**Dielectric Strength (kV/mm)**: 22
Flexural Modulus (psi): 175000	**Transparency**: T
Impact (Izod ft. lbs/in): NB	**UV Resistance**: P
Hardness: SD68	**CHEMICAL RESISTANCE**

	Acids	**Alkalis**	**Solvents**
	G	E	E

HEAT OF COMBUSTION

A measure of the available energy content of a fuel, under controlled conditions specified by test method ASTM D 240 or D 2382. Heat of combustion is determined by burning a small quantity of a fuel in an oxygen bomb calorimeter and measuring the heat absorbed by a specified quantity of water within the calorimeter. Heat of combustion is expressed either as calories per gram or British thermal units per pound. Also called thermal value, heating value, calorific value.

HEAT OF DECOMPOSITION

The value is the amount of heat liberated when the specified weight decomposes to more stable substances. Most chemicals are stable and do not decompose under the conditions of temperature and pressure encountered during shipment. A negative sign before the value simply indicates that heat is given off during the decomposition. The value does not include heat given off when the chemical burns. The units typically used are Btu per pound, calories per gram, and joules per kilogram.

HEAT OF SOLUTION

The value represents the heat liberated when the specified weight of chemical is dissolved in a relatively large amount of water at 25° C ("infinite dilution"). A negative sign before the value indicates that heat is given off, causing a rise in temperature. (A few chemicals absorb heat when they dissolve, causing the temperature to fall.) The units used are Btu per pound, calories per gram, and joules per kilogram. In those few cases where the chemical reacts with water and the reaction products dissolve, the heat given off during the reaction is included in the heat of solution.

HEAT OF POLYMERIZATION

The value is the heat liberated when the specified weight of the compound (usually called the monomer) polymerizes to form the polymer. In some cases the heat liberated is so great that the temperature rises significantly, and the material may burst its container or catch fire. The negative sign before the value indicates that heat is given off during the polymerization reaction. The units used are Btu per pound, calories per gram, and joules per kilogram.

HEPTANE

Liquid paraffinic hydrocarbon containing seven carbon atoms in the molecule, which may be straight-chain (normal) or branched-chain (iso). Heptane can be used in place of hexane where a less volatile solvent is desired, as in the manufacture of certain adhesives and lacquers, and in extraction of edible and commercial oils. Heptane is blended with isooctane to create a standard reference fuel in laboratory determinations of octane number.

HEXANE

A highly volatile paraffinic hydrocarbon containing six carbon atoms in the molecule; it may also contain six-carbon isoparaffins. Widely used as a solvent in adhesive and rubber solvent formulations, and in the extraction of a variety of edible and commercial oils. Hexane is a neurotoxin and must be handled with adequate precautions.

HIGH-PRESSURE POLYMERIZATION

These processes are based on the use of high pressures from 1.0×10^8 N/M^2 up to 3.0×10^8 N/M^2 and at temperatures ranging up to 350 °C. The polymerization reaction is initiated either by oxygen or more commonly by organic peroxides. The polymerization reaction is typical free-radical polymerization involving free-radical initiation, polymer chain propagation, and radical recombination.

Several important side reactions involving chain transfer occur due to the highly reactive nature of the polyethylene free radical and the high temperatures normally employed for manufacture. The molecular weight is usually determined by kinetic chain transfer either by hydrogen abstraction from an added chain transfer agent or by an internal rearrangement forming a double bond and a small radical. Chain transfer agents which are used include alkanes, olefins, ketones, aldehydes, and hydrogen.

Radical recombination generally makes only a small contribution to molecular weight control and since chain transfer has a higher activation energy than the propagation reaction, the molecular weight decreases with increasing temperature.

Short branches, mainly butyl and ethyl groups, are produced by intramolecular chain transfer, probably via six-membered ring transition states. Long-chain branches are produced by intramolecular chain transfer between a dead polymer molecule and a growing polymeric radical.

The principal problems of the high-pressure ethylene polymerization process are those concerned with the compression and handling of high-pressure gases and with the control of the highly exothermic reaction. The heat of polymerization of ethylene in the gas phase is about 22-23 kcal/mol or about 800 cal/g. At 1.4×10^8 N/M^2 the specific heat of ethylene in the range 150-300° C is 0.60-0.68 cal/(g)(°C). Unless heat is removed during the reaction the temperature will consequently rise about 12-13° C for each 1% of ethylene converted during polymerization.

If the reaction temperature becomes too high, alternative decomposition reactions of ethylene to give a mixture of carbon, methane, and hydrogen may occur. These are also strongly exothermic, for example, the decomposition of ethylene to carbon and methane alone evolves 34 kcal/mol, and once initiated such reactions may cause pressure increases of explosive violence. Thus efficient control of reaction temperature is of paramount importance.

Pressure exerts a marked effect on the polymerization reaction rate constant and can be used to control the reaction rate and molecular weight in addition to the more usual variables of initiator concentration and temperature. Since the number of short branches and the molecular weight are determined by chain transfer reactions which are more influenced by temperature and less by pressure than the polymerization reaction, it follows that the molecular weight decreases and the degree of short branching increases with increasing temperature (and vice versa with pressure).

A simplified schematic of a high-pressure polyethylene synthesis plant is illustrated in Figure 1. This scheme produces LDPE of density between 0.915 to 0.940 g/cc. The dashed lines on the diagram indicate heating control zones in the process. (Source: *Handbook of Polymer Science and Technology: Volume 2 -*

Performance Properties of Plastics and Elastomers, N. P. Cheremisinoff - editor, Marcel Dekker Inc., New York, 1989).

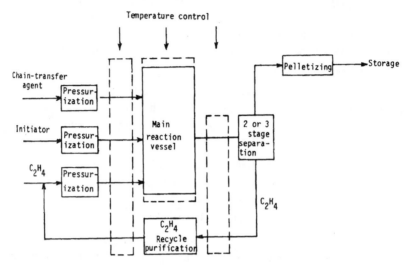

Figure 1. Schematic of high-pressure LDPE process.

HIGH IMPACT POLYSTYRENE (HIPS)

Isotactic polystyrene is a highly crystalline material with a melting point of around 240° C, whereas atactic polystyrene (the polymer that is widely used in numerous applications) is a completely amorphous, glassy polymer with a softening point typically as low as 80° C. Table 1 provides average properties of commercial grade high impact polystyrene. In reviewing the general properties of this polymer, note the use of the following legend: A = amorphous - Cr = crystalline - C = clear - E = excellent - G = good - P = poor - O = opaque - T = translucent- R = Rockwell - S = Shore.

Table 1

Average Properties of HIP

STRUCTURE: A	FABRICATION		
	Bonding	Ultrasonic	Machining
	E	G	G
Specific Density: 1.04	Deflection Temperature (° F)		
	@ 66 psi	@ 264 psi	
	195	180	

Table 1 continued

Water Absorption Rate (%): 0.1	Utilization Temperature (° F)		
	min.	max.	
	-22	140	
Elongation (%): 55	Melting Point (° F): 185		
Tensile Strength (psi): 4000	Coefficient of Expansion: 0.000042		
Compression Strength (psi): 7500	Arc Resistance: 100		
Flexural Strength (psi): 8700	Dielectric Strength (kV/mm): 18		
Flexural Modulus (psi): 280000	Transparency: T		
Impact (Izod ft. lbs/in): 2	UV Resistance: P		
Hardness: R65	**CHEMICAL RESISTANCE**		
	Acids	Alkalis	Solvents
	G	E	P

HOMOGENIZATION

Refers to the intimate mixing of an emulsion by intensive shearing action to obtain more uniform dispersion of the components. Refer to *Mixing*.

HUMIDITY

Water vapor in the atmosphere. Absolute humidity is the amount of water vapor in a given quantity of air; it is not a function of temperature. Relative humidity is a ratio of actual atmospheric moisture to the maximum amount of moisture that could be carried at a given temperature, assuming constant atmospheric pressure.

HYDROCARBON

Any chemical compound of hydrogen and carbon; also called an *organic compound*. Hydrogen and carbon atoms can be combined in virtually countless ways to make a diversity of products. Carbon atoms form the skeleton of the hydrocarbon molecule, and may be arranged in chains (aliphatic) or rings (cyclic). There are three principal types of hydrocarbons that occur naturally in petroleum: *paraffins, naphthenes,* and *aromatics,* each with distinctive properties. Paraffins are aliphatic, the others cyclic. Paraffins and naphthenes are saturated; that is, they have a full complement of hydrogen atoms and, thus, only single bonds between carbon atoms.

Aromatics are unsaturated, and have as part of their molecular structure at least one *benzene* ring, i.e., six carbon atoms in a ring configuration with alternating single and double bonds. Because of these double bonds, aromatics are usually more reactive than paraffins and naphthenes, and are thus prime starting materials for chemical synthesis.

Other types of hydrocarbons are formed during the petroleum refining process. Important among these are *olefins* and *acetylenes*. Olefins are unsaturated hydrocarbons with at least one double bond in the molecular structure, which may be in either an open chain or ring configuration; olefins are highly reactive. Acetylenes are also unsaturated and contain at least one triple bond in the molecule.

HYDROCARBON WAXES

Waxes are derived from petroleum and are of two common types, *paraffin* and *microcrystalline*. Paraffin waxes have lower carbon numbers, a higher proportion of straight-chain hydrocarbons, and lower melting points compared to the microcrystallines. Typical carbon numbers are n = 20 - 50 for paraffin waxes and n = 30 - 70 for microcrystallines.

If a wax is present in a vulcanizate at a level exceeding its solubility, some of it will migrate to the rubber surface where it can form a physical barrier to prevent the penetration of ozone.

Waxes are essentially unreactive to ozone, so that there is no appreciable element of chemical protection. Commercial waxes are usually blends of paraffin and microcrystalline waxes. The solubility and mobility of various waxes are affected by the polymer type, the filler types and loading, the state of cure, and the time and temperature of storage. Each type of commercial wax has particular solubility and migration characteristics that can be matched to the expected operating environment of the vulcanizate.

The various components of typical blends will migrate efficiently over a temperature range of - 10 to 60° C (refer to Figure 1). Blended waxes have the advantage that the various oligomers (ultra-low molecular weight components of the polymer or blend) will provide an effective *bloom* over a wider temperature range than will a paraffin or microcrystalline wax alone. At higher temperatures waxes become more soluble in the rubber so that less bloom formation occurs.

Some advantages of waxes are:

1.   they are of relatively low cost,

2.   they are nonstaining, and

3.   they generally have no adverse effects on rubber processing/ vulcanization.

Waxes act as internal lubricants in rubber and increase the scorch safety somewhat. Unfortunately, waxes have a number of shortcomings. First, they are ineffective under dynamic stress conditions. This is likely due to a lack of adhesion between the wax film and the rubber and to the inextensibility of the wax bloom.

Second, their protection capability is highly dependent on exposure temperature. Protection is difficult to achieve at both very low ($<10°$ C) and very high ($>50°$ C) temperatures. Third, waxes may easily be lost during storage or use by embrittlement, scuffing, or delamination. Because of these limitations, chemical antiozonants must be used (often in combination with waxes) in most diene rubber applications.

Refer to *Ozone Cracking*. (Source: *Handbook of Polymer Science and Technology, Volume 2: Performance Properties of Plastics and Elastomers*, N. P. Cheremisinoff, Macrel Dekker Publishers, New York, 1989).

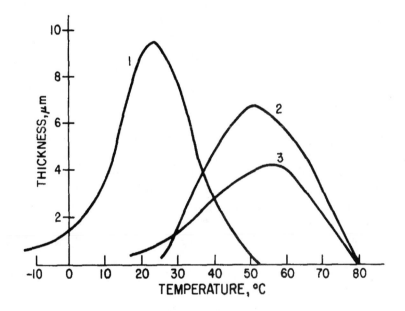

Figure 1. Shows thickness of wax bloom as a function of temperature. Rubber is butadiene/α-methyl-styrene copolymer. Curve 1, paraffin wax; Curve 2, microcrystalline wax; Curve 3, blended wax.

I

IGNITION TEMPERATURE

This is the minimum temperature at which the material will ignite without a spark or flame being present. Along with the values of flash point and flammable limits in air, it gives an indication of the relative flammability of the chemical. It is sometimes called the *autoignition temperature*. The method of measurement is given in ASTM A2155.

ICP ATOMIC EMISSION SPECTROSCOPY

In *inductively coupled plasma atomic emission spectroscopy* (ICP), the sample is vaporized and the element of interest atomized in an extremely high temperature (7000° C) argon plasma, generated and maintained by radio frequency coupling. The atoms collide with energetically excited argon species and emit characteristic atomic and ionic spectra that are detected with a photomultiplier tube. Separation of spectral lines can be accomplished in two ways. In a sequential or scanning ICP, a scanning monochromator with a movable grating is used to being the light from the wavelength of interest to a single detector. In a simultaneous or direct reader ICP, a polychromator with a diffraction grating is used to disperse the light into its component wavelength. Detectors for the elements of interest are set by the vendor during manufacture. Occasionally a scanning channel is added to a direct reader to allow measurement of an element not included in the main polychromator. ICP is used for the determination of ppm levels of metals in liquid samples. It is not suitable for the noble gases, halogens, or light elements such as H, C, N, and O. Sulfur requires a vacuum monochromator.

A direct reader ICP excels at the rapid analysis of multi-element samples. Common sample types analyzed by ICP include trace elements in polymers, wear metals in oils, and numerous one-of-a-kind catalysts. ICP instruments are limited to the analysis of liquids only. Solid samples require some sort of dissolution procedure prior to analysis. The final volume of solution should be at least 25 mL. The solvent can be either water, usually containing 10% acid, or a suitable organic solvent such as xylene. ICP offers good detection limits and a wide linear range for most elements. With a direct reading instrument multi-element analysis is extremely fast.

Chemical and ionization interferences frequently found in atomic absorption spectroscopy are suppressed in ICP analysis. Since all samples are converted to simple aqueous or organic matrices prior to analysis, the need for standards matched to the matrix of the original sample is eliminated. The requirement that the sample presented to the instrument must be a solution necessitates extensive sample preparation facilities and methods. More than one sample preparation method may be necessary per sample depending on the range of elements requested. Spectral interferences can complicate the determination of trace elements in the presence of other major metals. ICP instruments are not rugged. Constant attention by a trained operator, especially to the sample introduction and torch systems, is essential. Spectral interferences, such as line overlaps, are prevalent and must be corrected for accurate quantitative analysis. With a scanning instrument it may be possible to move to an interference free line. With a direct reader, sophisticated computer programs apply mathematical corrections based on factors previously determined on multi-element standards. (Source: Cheremisinoff, N.P. *Polymer Characterization: Laboratory Techniques and Analysis*, Noyes Publishers, New Jersey, 1996).

IMPACT STRENGTH

In practice, nearly all polymer components are subjected to impact loads. Since many polymers are tough and ductile, they are often well suited for this type of loading. However, under specific conditions even the most ductile materials, such as polypropylene, can fail in a brittle manner at very low strains. These types of failure are prone to occur at low temperature and at very high deformation rates.

A significantly high rate of deformation leads to complete embrittlement of polymers which results in a lower threshold of elongation at break. On the other hand, the stiffness and the stress at break of the material under consideration increases with the rate of deformation. Figure 1 shows the stress-strain and fracture behavior of a polymer tested at various rates of deformation. The area under the stress-strain curves represent the volume-specific energy to fracture.

For impact, the elongation at break of 2.2% and the stress at break of 135 MPa represent a minimum of volume-specific energy because the stress increases with higher rates of deformation, but the elongation at break remains constant. If the stress-strain distribution in the polymer component is known, one can estimate the minimum energy absorption capacity.

The impact strength of a copolymer and polymer blend of the same materials can be quite different. It should be pointed out here that elastomers usually fail by ripping. The ripping or tear strength of elastomers can be tested using the ASTM D1004, ASTM D1938, or DIN 53507 test methods. The latter two methods make use of rectangular test specimens with clean slits cut along the center. The tear strength of elastomers can be increased by introducing certain types of particulate fillers. For example, a well dispersed carbon black filler can double the ripping strength of a typical elastomer.

In general, one can say if the filler particles are well dispersed and have diameters between 20 nm and 80 nm, they will reinforce the matrix. Larger particles will act as microscopic stress concentrators and will lower the strength of the polymer component.

Impact tests are widely used to evaluate a material's capability to withstand high velocity impact loadings. The most common impact tests are the izod and the Charpy tests.

The izod test evaluates the impact resistance of a cantilevered notched bending specimen as it is struck by a swinging hammer. Figure 2 shows a typical izod-type impact machine. The pendulum or hammer is released from a predetermined height and after striking the specimen, travels to a recorded height. The energy absorbed by the breaking specimen is determined from the difference between the two heights. The standard test method that describes the izod impact test is ASTM-D 256.

(Source: Osswald, T.A. and G. Menges, *Material Science of Polymers for Engineers*, Hanser Publishers, New York, 1996).

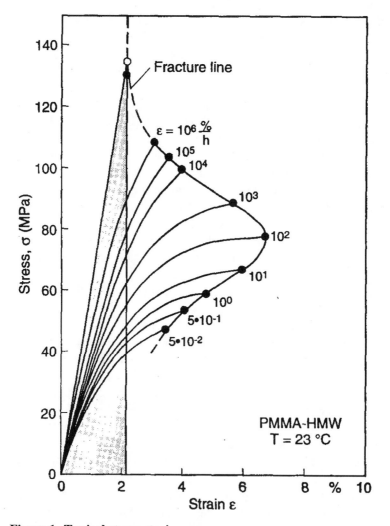

Figure 1. Typical stress-strain curve.

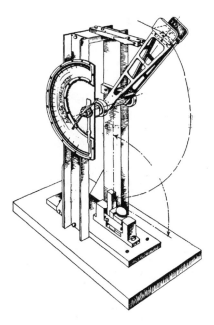

**Figure 2. Cantilever beam izod
impact machine.**

INJECTION MOLDING

Injection molding is the most important molding method for thermoplastics. It is based on the ability of thermoplastic materials to be softened by heat and to harden when cooled. The process consists of softening the material in a heated cylinder and injecting it under pressure into the mold cavity, where it hardens by cooling. Each step is carried out in a separate zone of the same apparatus in a cyclic operation. Granular material (the plastic resin) falls from the hopper into the barrel when the plunger is withdrawn. The plunger then pushes the material into the heating zone, where it is heated and softened (plasticized or plasticated). Rapid heating takes place due to spreading of the polymer into a thin film around a torpedo. The already molten polymer displaced by this new material is pushed forward through the nozzle, which is in intimate contact with the mold. The molten polymer flows through the sprue opening in the die, down the runner, past the gate, and into the mold cavity.

The mold is held tightly closed by the clamping action of the press platen. The molten polymer is thus forced into all parts of the mold cavities, giving a perfect reproduction of the mold. The material in the mold must be cooled under pressure before the mold is opened and the molded part is ejected. The plunger is then withdrawn, a fresh charge of material drops down, the mold is closed under a locking force, and the entire cycle is repeated. Mold pressures of 8000 to 40,000 psi and cycle times as low as 15 sec are achieved on some machines. The feed mechanism of the injection molding machine is activated by the plunger stroke. The function of the torpedo in the heating zone is to spread the polymer melt into thin film in close contact with the heated cylinder walls. The fins, which keep the torpedo centered, also conduct heat from the cylinder walls to the torpedo, although in some machines the torpedo is heated separately. Injection-molding machines are rated by their capacity to mold polystyrene in a single shot. A 2-oz machine can melt and push 2 oz of general-purpose polystyrene into a mold in one shot. This capacity is determined by a number of factors such as plunger diameter, plunger travel, and heating capacity.

The main components of an injection -molding machine are:

1.   the injection unit which melts the molding material and forces it into the mold;
2.   the clamping unit which opens the mold and closes it under pressure;
3.   the mold used; and
4.   the machine controls.

Injection-molding machines are known by the type of injection unit used in them. The more common injection molding machine in use today is screw-transfer injection. An example of this type of machine is shown in Figure 1. The machine has two sections mounted on a common base. One section constitutes the plasticizing and injection unit, which includes the feed hopper, the heated barrel that encloses the screw, the hydraulic cylinder which pushes the screw forward to inject the plasticized material into the mold, and a motor to rotate the screw. The other section clamps and holds the mold halves together under pressure during the injection of the hot plastic melt into the mold. The thermosetting material (in granular or pellet form) suitable for injection molding is fed from the hopper into the barrel and is then moved forward by the rotation of the screw. The material receives conductive heat from the wall of the heated barrel and frictional heat from the rotation of the screw.

For thermosetting materials, the screw used is a zero compression-ratio screw (i.e., the depths of flights of the screw at the feed-zone end and at the nozzle end are the same). By comparison, the screws used in thermoplastic molding machines have compression ratios such that the depth of flight at the feed end is one and one-half to five times that at the nozzle end. This difference in screw configuration is a major difference between thermoplastic and thermosetting-molding machines. As the material moves forward in the barrel due to rotation of the screw, it changes in consistency from a solid to a semifluid, and as it starts accumulating at the nozzle end, it exerts a backward pressure on the screw. This back pressure is thus used as a processing variable. The screw stops turning when the required amount of material-the charge-has accumulated at the nozzle end of the barrel, as sensed by a limit switch. (The charge is the exact volume of material required to fill the sprue, runners, and cavities of the mold.) The screw is then moved forward like a plunger by hydraulic pressure (up to 20,000 psi) to force the hot plastic melt through the sprue of the mold and into the runner system, gates, and mold cavities.

The nozzle temperature is controlled to maintain a proper balance between a hot mold (350-400° F), and a relatively cool barrel (150-200° F). Molded-in inserts are commonly used with thermosetting materials. However, since the screw-injection process is automatic, it is desirable to use post-assembled inserts rather than molded-in inserts because molded-in inserts require that the mold be held open each cycle to place the inserts. A delay in the manual placement disrupts an automatic cyclic operation, affecting both the production rate and the product quality. Thermosetting materials used in screw-injection molding are modified from conventional thermosetting compounds.

These modifications are necessary to provide the working time-temperature relationship required for screw plasticating. The most commonly used injection-molding thermosetting materials are the phenolics. Other thermosetting materials often molded by the screw-injection process include melamine, urea, polyester, alkyd, and diallyl phthalate (DAP). (Source: Chanda, M. and S. K. Roy, *Plastics Technology Handbook*, Marcel Dekker, Inc., New York, NY, 1987).

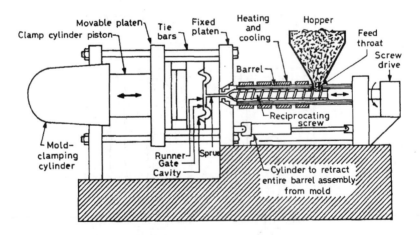

Figure 1. Direct screw transfer molding machine.

INJECTION STRETCH BLOW MOLDING

The process steps involved in injection stretch blow molding are illustrated by the series of drawings in Figure 1. The advantages of this process are:

- ability to fabricate high clarity, high drop impact containers;
- ability to fabricate a precision formed neck (thread) finish;

- it is a scrapless process (i.e., no regrind);
- container sizes up to 2 liters can be fabricated with this process;
- container thicknesses that are possible range from 0.25 to 2 mm;
- articles produced have high barrier properties due to braxially stretching.

The preferred resins used with this fabrication process include PET, PS, PP, PC, and nylon. Container applications derived from this process include soft drinks, vegetable oil bottles, peanut butter jars, mouth wash bottles, cough syrup bottles, non dairy creamer containers, over the counter drug containers (e.g., aspirin bottles and juice bottles.

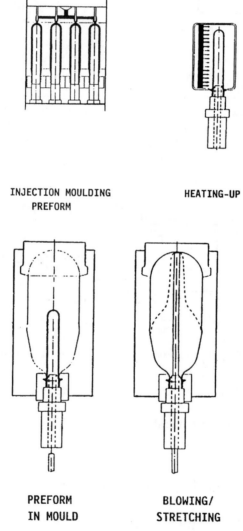

INJECTION MOULDING
PREFORM

HEATING-UP

PREFORM
IN MOULD

BLOWING/
STRETCHING

Figure 1. Injection blow molding steps.

IN-LINE SCREW INJECTION MOLDS

Mold designs employed for screw injection molding of thermosets resemble those used for transfer or screw transfer molding. The molds are in the closed position when the heated compound is forced, under pressure, into the mold cavity. The centrally located pot used in transfer molds is replaced in this case with a "sprue bushing". The bushing is contracted by the molding machine nozzle and the compound flows through it into the runner system. Molds are constructed with clamping plates, retainer plates, backing plates, support pillars. All molding components such as cavities, runner blocks, sprue bushing, and slide plates are constructed from high grade tool steel, hardened, highly polished, and hard chrome plated for wear resistance, good mold release and to provide superior molded part finish.

INTERNAL BATCH MIXERS

Powdered or particulate rubber can be processed in internal batch mixers. The two systems most often used are the Bayer Sikoplast Screw/Hopper and the Farrel M.V.X. These are shown schematically in Figures 1 and 2. There are several stages in the conversion of elastomers and other compounding ingredients into rubber compounds. These stages are receipt of raw materials, testing, storage, weighing, feeding, mixing, batch-off, cooling, testing, storage, and dispatch. These stages can be divided into three groups:
1.   material flow to the mixer;
2.   mixing; and
3.   material flow away from the mixer.

The main raw materials classes handled in rubber compounding operations are as follows:

Elastomers. Elastomers are normally supplied in bale form. This necessitates handling with pallets, forklifts and muscle as far as the feed conveyor. From here on, there should be no need for manual effort. There are systems available that can handle four different grades or types of elastomer in bale form, weighing them simultaneously to 1 percent accuracy, and collecting them on a conveyor in batches up to 300 kg (660 lbs). The operation is controlled by a minicomputer.

Carbon Black. There are several justifications for bulk handling of carbon black. It is cheaper in bulk and picks up less moisture. Handling is cleaner and more efficient.

Nonblack Fillers. Because these are usually liquids, they are the one class of raw materials that most mixing shops today handle at least semiautomatically. For those plasticizers used in large quantities, bulk handling is advantageous. Preheating is usually necessary for some of these materials in order to reduce their viscosity for proper processing.

The mixing operation consists of three processes: simple mixing, laminar mixing, and dispersive mixing. The relative importance of each depends on the particular compound formulation (in terms of the attraction between the particles of solid additives and the flow properties of the rubber), the geometry of the mixer, and the operating conditions. In a specific case, any one of the three may be the efficiency-determining process. Simple mixing or homogenization involves the moving of a particle from one point to another, without changing its physical shape. This increase in randomness or entropy is also called extensive mixing. If the shear forces are sufficiently large, particles may fracture (dispersive mixing), and the polymer may flow (laminar mixing). In addition, if the deformation of the elastomer exceeds its breaking strain, then it will break into super-molecular flow units. There are four physical changes which take place during the mixing cycle:

Incorporation. At the beginning of a mixing cycle, the rubber is forced into the working area between the pairs of rotors and between the rotors and chamber wall. The identity of the original bales or particles is destroyed. The incorporation state is when the initially free ingredients become attached to the rubber. This is referred to as the wetting stage. The elastomer first undergoes a large deformation, increasing the surface area for accepting filler agglomerates, and then sealing them inside. Second, the elastomer breaks down into small pieces and mixes with the filler and once again seals the filler inside. The former mechanism is easily observed in an open mill.

Dispersion. The filler agglomerates are gradually broken down, distributed through the rubber (by simple mixing), and are then dispersed (that is, broken down to the ultimate size) giving a fine scale of mixing. This is important in the case of carbon black because at this stage an intimate contact between the surface of the carbon black and the elastomer develops, resulting in bound rubber. Both the disruption of the filler aggregates and the forcing of the elastomer onto the filler surface require high-shear stress. However, the shear stresses do not all result from the imposed shear field because microscopic shear fields are generated from elongational deformations also.

Distribution. This process of increasing homogenization takes place throughout the mixing cycle.

Plasticization. In this stage of mixing, the rheological properties are modified to suit subsequent operations.

Overmixing wastes machine time and energy and exposing the material to shearing and high temperatures can result in excessive carbon-black interaction, viscosity increases, and in some cases reversion of the rubber. Continuous compounding of elastomers is being practiced in several applications in the cable, shoe, and insulation or electrical cable industry. These applications generally involve highly filled formulations based on EPDM, EVA, or thermoplastic rubber. (Source: N. P. Cheremisinoff, *Guidebook to Extrusion Technology*, PTR Prentice Hall, New Jersey, 1993).

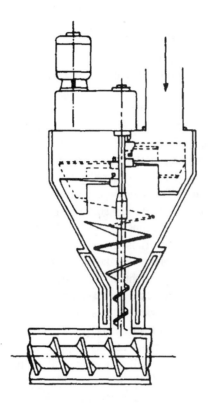

Figure 1. Sikoplast screw and hopper configuration.

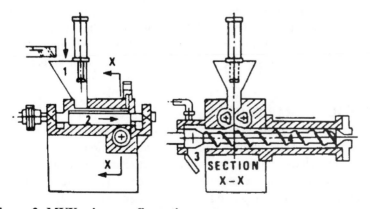

Figure 2. MVX mixer configuration.

INTERNAL MIXING PROCEDURES

There are three basic procedures for mixing rubber compounds in an internal mixer; namely, the conventional method, the rapid oil addition method, and the upside-down mix method. Many variations of these three methods are also used to suit the special characteristics of individual formulations and equipment. It is, in general, necessary to add particulate fillers early in the mixing cycle, so that good dispersion is achieved as a result of the high shear stress and high viscosity at the lower temperatures then prevailing. Similarly, the oils and plasticizers which reduce viscosity should be added later. Upside-down procedures and variants of it are attempts to implement these ideas in practice.

The **conventional mixing method** consists of adding the elastomer first, then the dry ingredients, then the liquid ingredients after the dry materials are well dispersed in the elastomer. This method can achieve a homogenous dispersion of all ingredients, including filler of very small particle size. However, the mixing time required is usually long because it is more difficult to incorporate the liquid ingredients once the dry materials are dispersed. With fillers which are of lowbulk density or with fillers which cake when dry, a variation of this technique is to add part of the liquid ingredients at the same time as the dry ingredients.

The **rapid oil-addition method** involves adding the elastomer first and the dry ingredients as soon afterward as possible. After about 1 to 2 minutes of mixing, all liquids are added together. The proper time for the addition of liquids needs to be determined. Dispersion usually improves if the addition of liquids is delayed slightly; however, this will extend the mixing time. Use of this method can give very good dispersion if liquids are added at the proper time. It is often used for compounds containing a large volume of liquid plasticizers. However, it can lead to an extended mixing cycle due to the lubricating effect of the liquids between the rubber and the metal parts of the mixer.

The **upside-down mix** method is the fastest and the simplest way of mixing. It is especially efficient and effective for those compounds containing a large volume of liquid plasticizers and large particle-size fillers. This method involves adding all ingredients into the mixer before lowering the ram and commencement of mixing. All dry ingredients are added to the mixer first, then all liquids, and finally all elastomers on top. This method is not suitable for those compounds containing low structure, small particle-size carbon black, or compounds having high loadings of both soft mineral filler and oil, together with an elastomer of high Mooney viscosity.

ION CHROMATOGRAPHY (IC)

Commercial ion chromatograph instruments have become available since early 1976. Ion chromatography (IC) is a combination of ion exchange chromatography, eluent suppression and conductimetric detection. For anion analysis, a low capacity anion exchange resin is used in the separator column and a strong cation exchange resin in the H+ form is used in the suppressor column.

A dilute mixture of $NaCO_3$ is used as the eluent, because carbonate and bicarbonate are conveniently neutralized to low conductivity species and the different combinations of carbonate-bicarbonate give variable buffered pH values. This allows the ions of interest in a large range of affinity to be separated.

The anions are eluted through the separating column in the background of carbonatebicarbonate and conveniently detected based on electrical conductivity. As a result of these reactions in the suppressor column, the sample ions are presented to the conductivity detector as H'X-, not in the highly conducting background of carbonate-bicarbonate, but in the low conducting background of H_2CO_3.

In the system a dilute aqueous sample is injected at the head of the separator column. The anion exchange resin selectively causes the various sample anions of different types to migrate through the bed at different respective rates, thus effecting the separation. The effluent from the separator column then passes to the suppressor column where the H^+ form cation exchange resin absorbs the cations in the eluent stream. Finally, the suppressor column effluent passes through a conductivity cell.

The highly conductive anions in a low background conductance of H_2CO_3 are detected at high sensitivity by the conductivity detector. The nonspecific nature of the conductimetric detection allows several ions to be sequentially determined in the same sample. The conductimetric detection is highly specific and relatively free from interferences.

Different stable valance states of the same element can be determined. On the other hand, because of the nonspecific nature of the conductivity detector, the chromatograph peaks are identified only by their retention times. Thus, the two ions having the same or close retention times will be detected as one broad peak giving erroneous results. A typical chromatogram for the standard common anions F^-, Cl^-, NO_2^-, PO_4^{-3}, Br^-, NO_3 and SO_4^{-2} is shown in Figure 1.

Numerous applications of ion chromatography have been illustrated in the literature for a variety of complex matrices. The advantages of ion chromatography are:

1.   Sequential multi-anion capability; eliminates individual determinations of anion by diverse technique;
2.   Small sample size required ($<$ 1 mL);
3.   Rapid analysis (- 10 minutes for about 7 anions);
4.   Large dynamic range over four decades of concentration;
5.   Speciation can be determined.

The principle disadvantages of IC are:

1.   Interferences possible if two anions have similar retention times;
2.   Determination difficult in the presence of an ion present in very large excess over others;

3.   Sample has to be in aqueous solution;
4.   Method not suitable for anions with pKa of < 7.

In addition to the common inorganic anions analyzed by IC, a number of other species can also be determined by using appropriate accessories. Some of these applications include: Ion chromatography; Chemistry - IC; Mobile Phase IC; Electrochemical Detection. Species determinations include: Carboxylic Acids; Formaldehyde; Borate; Ammonia; Fatty Acids; Ethanol-Amines; Phenols, CN-, Br-, I-, S-2, etc. (Source: Cheremisinoff, N.P. *Polymer Characterization: Laboratory Techniques and Analysis*, Noyes Publishers, New Jersey, 1996).

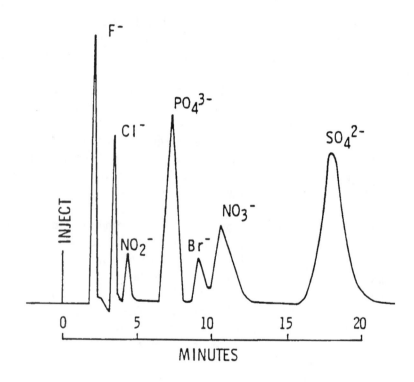

Figure 1. Ion chromatograph for inorganic ions analysis.

IONOMERS (SURLYN)

Table 1 provides average properties of this polymer. In reviewing the general properties of this polymer, note the use of the following legend: A = amorphous - Cr = crystalline - C = clear - E = excellent - G = good - P = poor - O = opaque - T = translucent- R = Rockwell - S = Shore.

Table 1

Average Properties of Ionomers (Surlyn)

STRUCTURE: A	FABRICATION		
	Bonding	Ultrasonic	Machining
	P	P	P
Specific Density: 0.94	Deflection Temperature (° F)		
	@ 66 psi	@ 264 psi	
	122	102	
Water Absorption Rate (%): 0.08	Utilization Temperature (° F)		
	min	max	
	-148	167	
Elongation (%): 400	Melting Point (° F): 194		
Tensile Strength (psi): 4000	Coefficient of Expansion: 0.000067		
Compression Strength (psi): 1400	Arc Resistance: 90		
Flexural Strength (psi): 1000	Transparency: C		
Flexural Modulus (psi): 36000	UV Resistance: P		
Impact (Izod ft. lbs/in): 6	CHEMICAL RESISTANCE		
Hardness: R40	Acids	Alkalis	Solvents
	P	E	E

ION SELECTIVE ELECTRODES (ISE)

ISE measures the ion activities or the thermodynamically effective free ion concentrations. ISE has a membrane construction that serves to block the interfering ions and only permit the passage of ions for which it was designed. However, this rejection is not perfect, and hence some interferences from other ions occur. The electrode calibration curves are good over 4 to 6 decades of concentration. The typical time per analysis is about a minute, though some electrodes need 15 minutes for adequate response. The response time is faster as more concentrated solutions are analyzed. Although a single element technique, many elements can be determined sequentially by changing electrodes, provided calibration curves are prepared for all ions. Also, the instrument is portable and is thus useful for field studies. Sample volumes needed are typically about 5 ml, although 300 ml or less can be measured with special modifications. An accuracy of 2-5 % is achieved.

The ISE measures the activity of the ions in solution. This activity is related to concentration and thus, in effect, measures the concentration. However, if an ion such as fluoride, which complexes with some metals - Fe or Al - is to be measured, it must be decomplexed from these cations by the addition of a reagent such as citric acid or EDTA. ISEs for at least 22 ionic species are commercially available. An example is described here for the measurement of fluoride ions in solution. The fluoride electrode uses a LaF_3 single crystal membrane and an internal reference, bonded into an epoxy body. The crystal is an ionic conductor in which only fluoride ions are mobile. When the membrane is in contact with a fluoride solution, an electrode potential develops across the membrane. This potential, which depends on the level of free fluoride ions in solution, is measured against an external constant reference potential with a digital pH/mv meter or specific ion meter.

The measured potential corresponding to the level of fluoride ions in solution is described by the Nernst equation. The level of fluoride, A, is the activity or *effective concentration* of free fluoride ions in solution. The total fluoride concentration, C, may include some bound or complexed ions as well as free ions. Ionic activity coefficients are variable and largely depend on total ionic strength. (Source: Cheremisinoff, N.P. *Polymer Characterization: Laboratory Techniques and Analysis*, Noyes Publishers, New Jersey, 1996).

ISO (INTERNATIONAL ORGANIZATION FOR STANDARDIZATION)

The International Organization for Standardization (ISO) is a worldwide federation of national standards bodies from some 130 countries, one from each country. ISO is a non-governmental organization established in 1947. The mission of ISO is to promote the development of standardization and related activities in the world with a view to facilitating the international exchange of goods and services, and to developing cooperation in the spheres of intellectual, scientific, technological and economic activity.

ISO's work results in international agreements which are published as International Standards. Many people will have noticed a seeming lack of correspondence between the official title when used in full, International Organization for Standardization, and the short form, ISO. Shouldn't the acronym be 'IOS'? Yes, if it were an acronym – which it is not. In fact, "ISO" is a word, derived from the Greek isos, meaning *equal*, which is the root of the prefix "iso-" that occurs in a host of terms, such as "isometric" (of equal measure or dimensions) and "isonomy" (equality of laws, or of people before the law). From "equal" to "standard", the line of thinking that led to the choice of

"ISO" as the name of the organization is easy to follow. In addition, the name ISO is used around the world to denote the organization, thus avoiding the plethora of acronyms resulting from the translation of "International Organization for Standardization" into the different national languages of members, e.g. IOS in English, OIN in French (from Organisation internationale de normalisation). Whatever the country, the short form of the Organization's name is always ISO.

The existence of non-harmonized standards for similar technologies in different countries or regions can contribute to so-called "technical barriers to trade". Export-minded industries have long sensed the need to agree on world standards to help rationalize the international trading process. This was the origin of the establishment of ISO. International standardization is well-established for many technologies in such diverse fields as information processing and communications, textiles, packaging, distribution of goods, energy production and utilization, shipbuilding, banking and financial services. It will continue to grow in importance for all sectors of industrial activity for the foreseeable future. The main reasons are:

1. *Worldwide progress in trade liberalization* - Today's free-market economies increasingly encourage diverse sources of supply and provide opportunities for expanding markets. On the technology side, fair competition needs to be based on identifiable, clearly defined common references that are recognized from one country to the next, and from one region to the other. An industry-wide standard, internationally recognized, developed by consensus among trading partners, serves as the language of trade;

2. *Interpenetration of sectors* - No industry in today's world can truly claim to be completely independent of components, products, rules of application, etc., that have been developed in other sectors. Bolts are used in aviation and for agricultural machinery; welding plays a role in mechanical and nuclear engineering, and electronic data processing has penetrated all industries. Environmentally friendly products and processes, and recyclable or biodegradable packaging are pervasive concerns;

3. *Worldwide communications systems* - The computer industry offers a good example of technology that needs quickly and progressively to be standardized at a global level. Full compatibility among open systems fosters healthy competition among producers, and offers real options to users since it is a powerful catalyst for innovation, improved productivity and cost-cutting;

4. *Global standards for emerging technologies* - Standardization programs in completely new fields are now being developed and implemented. Such fields include advanced materials, the environment, life sciences,

urbanization and construction. In the very early stages of new technology development, applications can be imagined but functional prototypes do not exist. Here, the need for standardization is in defining terminology and accumulating databases of quantitative information;

5.   *Developing countries* - Development agencies are increasingly recognizing that a standardization infrastructure is a basic condition for the success of economic policies aimed at achieving sustainable development. Creating such an infrastructure in developing countries is essential for improving productivity, market competitiveness, and export capability.

Industry-wide standardization is a condition existing within a particular industrial sector when the large majority of products or services conform to the same standards. It results from consensus agreements reached between all economic players in that industrial sector - suppliers, users, and often governments. They agree on specifications and criteria to be applied consistently in the choice and classification of materials, the manufacture of products, and the provision of services. The aim is to facilitate trade, exchange and technology transfer through:

1.   enhanced product quality and reliability at a reasonable price;

2.   improved health, safety and environmental protection, and reduction of waste;

3.   greater compatibility and interoperability of goods and services;

4.   simplification for improved usability;

5.   reduction in the number of models, and thus reduction in costs;

6.   increased distribution efficiency, and ease of maintenance. Users have more confidence in products and services that conform to International Standards.

Assurance of conformity can be provided by manufacturers' declarations, or by audits carried out by independent bodies. All aspects of the rubber and plastics industry in North America follow one form or another.

ISO STANDARDS

The two major standards are ISO 9000 and ISO 14000. Both "ISO 9000" and "ISO 14000" are actually families of standards which are referred to under these generic titles for convenience. Both families consist of standards and guidelines relating to management systems, and related supporting standards on terminology and specific tools, such as auditing (the process of checking that the management system conforms to the standard). *ISO 9000* is primarily concerned with "quality management". Like 'beauty', everyone may have his or her idea of what "quality" is. In plain language, the standardized definition of "quality" in ISO 9000 refers to all those features of a product (or service) which are required by the customer.

"Quality management" means what the organization does to ensure that its products conform to the customer's requirements. *ISO 14000* is primarily concerned with "environmental management". In plain language, this means what the organization does to minimize harmful effects on the environment caused by its activities. Both ISO 9000 and ISO 14000 concern the way an organization goes about its work, and not directly the result of this work. In other words, they both concern processes, and not products – at least, not directly. Nevertheless, the way in which the organization manages its processes is obviously going to affect its final product. In the case of ISO 9000, it is going to affect whether or not everything has been done to ensure that the product meets the customer's requirements.

In the case of ISO 14000, it is going to affect whether or not everything has been done to ensure a product will have the least harmful impact on the environment, either during production or disposal, either by pollution or by depleting natural resources. However, neither ISO 9000 nor ISO 14000 are product standards.

The management system standards in these families state requirements for what the organization must do to manage processes influencing quality (ISO 9000) or the processes influencing the impact of the organization's activities on the environment (ISO 14000). In both cases, the philosophy is that these requirements are generic.

No matter what the organization is or does, if it wants to establish a quality management system and an environmental management system (EMS), then both systems have a number of essential features which are spelled out in ISO 9000 and ISO 14000.

What is an EMS?

An EMS, an environmental management system, is an approach, a tool, a set of procedures, a planned and organized way of doing things - i.e., a *system*. It is any planning and implementation system that a company employs to manage the way it interacts with the natural environment. An EMS is built around the way a company operates, around its production processes and general management system, not around its emissions, effluents and solid wastes the way environmental regulations are. There are many types of EMS (not just ISO 14000). Two well-known ones are the British Standard Institution's BS 7750 and the European Union's EMAS (Eco-Management and Auditing Scheme, which allows ISO 14001 to serve as its core component. (Source: N. P. Cheremisinoff and A. Bendavid-Val, *Green Profits: A Managers Guide to Pollution Prevention and ISO 14000*, Butterworth-Heinemann, U.K., 2001).

J

JELLY, WHITE PETROLEUM

Also referred to as Vaseline USP/BP Grade. Table 1 provides typical commercial specifications.

Table 1
Typical Specifications for White Petroleum Jelly

Grade	USP/BP & pharmaceutical grade [medical]
Specifications	Tasteless, odorless, semi-solid, white ASH...........................not more than 0.08% Melting point......................48-52%
Uses	Using as raw material for preparing of all kinds of ointments, cosmetic for making creams, face cream, hair pomade/hair wax, etc.
Packing	In 170 kgs net in new iron drums [blue color] in 1x20' FCL = 98 drums = 16.660 kgs. net.

JETNESS

In plastics applications, medium and high-color carbon blacks are employed for tinting and to provide the property of *jetness* (an aesthetic property). Carbon black is also used as anti-photo and as thermal oxidation agents in the compounding of polyoelfins. As an example, clear polyethylene cable coating is susceptible to crazing or cracking, accompanied by a rapid loss of physical properties and dielectric strength when exposed to sunlight for extended periods of time. The carbon black essentially serves as a "black body" which absorbs ultraviolet light and infrared radiation. It also serves to terminate free radical chains and hence, provides good protection against thermal degradation. This however, is not a property of jetness. The term generally refers to the aesthetic appearance of smooth, black, glossy surface textures.

JOULE

The unit of energy called the joule is equal to 1 watt-second, or 10 million ergs, or about 0.000948 British thermal unit. The British Thermal Unit, in science and engineering, is a unit measurement of heat or energy, usually abbreviated as Btu or BTU. One Btu was originally defined as the quantity of heat required to raise the temperature of 1 lb (0.45 kg) of water from 59.5° F (15.3° C) to 60.5° F (15.8° C) at constant pressure of 1 atmosphere; for very accurate scientific or engineering measurements, however, this value was not precise enough. The Btu has now been redefined in terms of the joule as equal to 1055 joules; in engineering, a Btu is equivalent to approximately 0.293 watt-hour. Refer to *Units of Measurement*.

JOULE'S LAW

Joule's law of electric heating states that the amount of heat produced each second in a conductor by a current of electricity is proportional to the resistance of the conductor and to the square of the current. Joule experimentally verified the law of conservation of energy in his study of the transfer of mechanical energy into heat energy.

JOULE-THOMSON EFFECT

Using many independent methods, Joule determined the numerical relation between heat and mechanical energy, or the mechanical equivalent of heat. Together with the physicist William Thomson (later Baron Kelvin), Joule found that the temperature of a gas falls when it expands without doing any work. This principle, which became known as the Joule-Thomson effect, underlies the operation of common refrigeration and air conditioning systems.

K

KINEMATIC VISCOSITY

The absolute viscosity of a fluid divided by its density at the same temperature of measurement. It is the measure of a fluid's resistance to flow under gravity, as determined by test method ASTM D 445. To determine kinematic viscosity, a fixed volume of the test fluid is allowed to flow through a calibrated capillary tube *(viscometer)* that is held at a closely controlled temperature. The kinematic viscosity, in centistokes (cSt), is the product of the measured flow time in seconds and the calibration constant of the viscometer.

KINETIC ENERGY

Kinetic energy is the energy possessed by an object, resulting from the motion of that object. The magnitude of the kinetic energy depends on both the mass and the speed of the object according to the equation

$$E = \tfrac{1}{2}mv^2$$

where m is the mass of the object and v^2 is its speed multiplied by itself. The value of E can also be derived from the following equation:

$$E = (ma)d$$

where a is the acceleration applied to the mass, m, and d is the distance through which a acts. The relationships between kinetic and potential energy and among the concepts of force, distance, acceleration, and energy can be illustrated by the lifting and dropping of an object.

When the object is lifted from a surface a vertical force is applied to the object. As this force acts through a distance, energy is transferred to the object. The energy associated with an object held above a surface is termed potential energy. If the object is dropped, the potential energy is converted to kinetic energy. Refer to *Potential Energy*.

KNIT LINE

In the extrusion of hollow products, such as tubing and pipe, the plastic melt has to flow around a central core or mandrel. The mandrel is usually held in place by a number of spider supports, as shown in Figure 1. As the plastic melt flows past the spider supports it splits and flows together again after the spider. When the melt stream recombines, it takes a certain amount of time before full, intimate contact is established. This is because the long plastic molecules take some time to re-entangle. If there is insufficient time available for the molecules to re-entangle, a weak spot will form along the length of the extruded product. This is called the weld line or knit line. Such a knit line can cause the premature failure of a product, particularly if the product is internally pressurized (which is often the case as in tubing or with pipe).

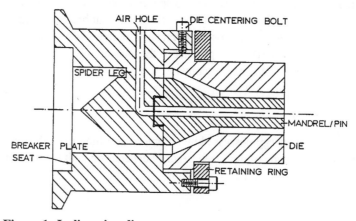

Figure 1. In-line pipe die.

L

LAMINATION

Lamination is a fabrication process in which various materials such as paper, aluminum foil and polymer film are joined together. Contrary to the coextrusion technique, the materials are joined in the solid phase. The multilayer structure combines the properties of the components. Adhesion between the different substrates is accomplished by means of solvent-based or water-based glues. Lamination lines are operated at high line speeds. Typical applications include foodpackaging, pharmaceuticals and flexible packaging of industrial products.

Typical contributions of single layers are:

- Gas barrier   nylon, polyester
- Rigidity   paper, nylon
- Gloss, transparency   polyester, polypropylene
- Sealability   PE, EVA, EAA
- Thermoformability   PP, PVC

A schematic of a glue lamination process scheme is illustrated below.

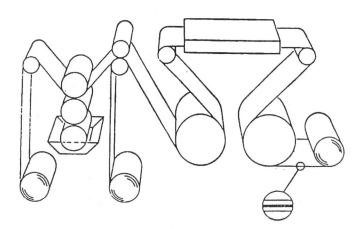

Figure 1. General layout of a glue lamination process.

Extrusion Coating

Paper, paperboard, cellulose film, fiberboard, metal foils, or transparent films are coated with resins by direct extrusion. The resins most commonly used are the polyolefins, such as polyethylene, Polypropylene, ionomer, and ethylene-vinyl acetate copolymers. Nylon, PVC, and polyester are used to a lesser extent. Combinations of these resins and substrates are used to provide a multilayer structure. A related technique, called extrusion laminating, involves two or more substrates, such as paper and aluminum foil, combined by using a plastic film, (e.g., polyethylene) as the adhesive and as a moisture barrier. The equipment employed for extrusion coatings is similar to that used for the extrusion of flat film. A thin molten film from the extruder is drawn down into the nip between a chill roll and a pressure roll positioned below the die. The pressure between these two rolls forces the film onto the substrate. The substrate moves at a speed faster than the extruded film, and hence draws the film to a desired thickness. The molten film is cooled with a chill roll.

LATENT HEAT OF VAPORIZATION

The value is the heat that must be added to the specified weight of a liquid before it can change to vapor (gas). It varies with temperature; the value given is that at the boiling point at 1 atm. The units used are Btu per pound, calories per gram, and joules per kilogram. No value is given for chemicals with very high boiling points at 1 atm, because such substances are considered essentially nonvolatile.

LATE TOXICITY

Where there is evidence that the chemical can cause cancer, mutagenic effects, teratogenic effects, or a delayed injury to vital organs such as the liver or kidney, a qualitative description of the effect is often given on a material safety data sheet. The term can be interpreted as implying long term or chronic effects due to exposure to the chemical. In this respect, a distinction must be made between acute and chronic effects. An acute effect is one in which there is a short term or immediate response, usually due to exposure of the chemical at a high concentration. A chronic effect implies a long term exposure to small doses, with symptoms sometimes taking years to materialize. Refer to *Lethal Concentration* and *Lethal Dose*.

LETHAL CONCENTRATION

Lethal concentration or LC_{50} refers to the toxicity of a material where 50 % mortality is observed in test animals. It is the concentration in air of a volatile chemical compound at which half the test population of an animal species dies when exposed to the compound. It is expressed as parts per million by volume of the toxicant per million parts of air for a given exposure period. (Source: N. P. Cheremisinoff, *Handbook of Industrial Toxicology and Hazardous Materials*, Marcel Dekker Publishers, New York, 1999).

LETHAL DOSE

Lethal dose or LD_{50} refers to 50% mortality, a general measurement of toxicity. It is the dose of a chemical compound that, when administered to laboratory animals, causes death in one-half the test population. It is expressed in milligrams (mg) of toxicant per kilogram of animal weight. The route of administration may be oral, epidermal, or intraperitoneal. (Source: N. P. Cheremisinoff, *Handbook of Industrial Toxicology and Hazardous Materials*, Marcel Dekker Publishers, New York, 1999).

LOW DENSITY POLYETHYLENE (LDPE)

Polyethylene of density ranging from 0.91 to 0.94 g/cc is classified broadly as low-density polyethylene (LDPE), and this polymer can be manufactured by high- or low-pressure processes. After a half century of LDPE production with high-pressure tubular or autoclave reactors, radical new technology capable of operating at less than 300 psi and near 100° C emerged in the mid-1970s. This low-pressure LDPE technology has rapidly established itself as a low-cost route to polyethylene having many process and property advantages over conventional high-pressure LDPE.

A key to the new low-pressure LDPE technology is the family of transition metal catalysts that triggers the polymerization reaction at very low pressure and temperature. Moreover, most of these new low-pressure processes utilize higher α-olefins such as butene-1 or hexene-1 as a comonomer to regulate the density of polyethylene. The α-olefins comonomers add side groups which spread the plates of the polyethylene crystal apart sufficiently to reduce the density of the polyethylene to 0.920 g/cc or even lower.

The LDPE manufactured by copolymerizing ethylene with butene-1 has a substantial degree of ethyl group branching and vinyl unsaturation, whereas the LDPE prepared by conventional high-pressure technology has longer chain branches and essentially no vinyl unsaturation.

The observable differences in polymer properties between high-pressure LDPEs and low-pressure LDPEs are caused by the linearity of the main polymer chains, the molecular weight distribution, and the type of chain branching. Figure 1 illustrates schematically the polymer chain structure of various polyethylenes.

Three types of polymerization processes are used today for low-pressure ethylene polymerization: (a) liquid slurry polymerization, (b) solution polymerization, and (c) gas phase polymerization.

Table 1 provides some average properties of LDPE. In reviewing the general properties of this polymer, note the use of the following legend: A = amorphous - Cr = crystalline - C = clear - E = excellent - G = good - P = poor - O = opaque - T = translucent- R = Rockwell - S = Shore.

Refer also to *Polyethylene, LDPE (Low Density Polyethylene)*, *HMWDPE (High Molecular Weight Polyethylene)*, and *HDPE (High Density Polyethylene)*. (Source: *Handbook of Polymer Science and Technology: Volume 1 - Synthesis and Properties*, N. P. Cheremisinoff - editor, Marcel Dekker Inc, New York, 1989.)

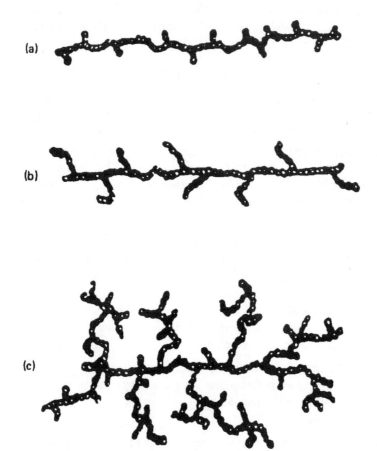

Figure 1. Chain structures of polyethylene: (a) HDPE (by low-pressure process); (b) LLDPE (by low-pressure process); (c) LDPE (by high-pressure process).

Table 1

Average Properties of LDPE

STRUCTURE: Cr	FABRICATION		
	Bonding	Ultrasonic	Machining
	P	P	P
Specific Density: 0.92	Deflection Temperature (°F)		
Water Absorption Rate (%): 0.01	@ 66 psi	@ 264 psi	
	113	95	

Table 1 continued

Elongation (%): 500	Utilization Temperature (° F)		
	min		max
	-76		210
	Melting Point (° F): 248		
Tensile Strength (psi): 1700	Coefficient of Expansion: 0.00009		
Compression Strength (psi): 1400	Transparency: T		
Flexural Strength (psi): 1400	UV Resistance: P		
Flexural Modulus (psi): 22000	CHEMICAL RESISTANCE		
Impact (Izod ft. lbs/in): NB	Acids	Alkalis	Solvents
Hardness: SD55	G	E	E

LIQUID CHROMATOGRAPHY

In liquid chromatography (LC) the sample is first dissolved in the moving phase and injected at ambient temperature. Thus there is no volatility requirement for samples. However, the sample must dissolve in the moving phase. Note that LC has an important advantage over GC; namely, that the solubility requirement can usually be met by changing the moving phase. The volatility requirement is not so easily overcome. There are four kinds of liquid chromatography, depending on the nature of the stationary phase and the separation mechanism. These are as follows:

Liquid/Liquid Chromatography (LLC) -- is partition chromatography or solution chromatography. The sample is retained by partitioning between mobile liquid and stationary liquid. The mobile liquid cannot be a solvent for the stationary liquid. As a subgroup of liquid/liquid chromatography there is paper chromatography.

Liquid/Solid Chromatography (LSC)-- is adsorption chromatography. Adsorbents such as alumina and silica gel are packed in a column and the sample components are displaced by a mobile phase. Thin layer chromatography and most open column chromatography are considered liquid/solid chromatography.

Chromatography--employs zeolites and synthetic organic and inorganic resins to perform chromatographic separation by an exchange of ions between the sample and the resins. Compounds which have ions with different affinities for the resin can be separated.

Exclusion Chromatography--is another form of liquid chromatography. In the process a uniform nonionic gel is used to separate materials according to their molecular size. The small molecules get into the polymer network and are retarded, whereas larger molecules cannot enter the polymer network and will be swept our of the column.

The elution order is the largest molecules first, medium next and the smallest sized molecules last. The term "gel permeation chromatography" has been coined for separation polymers which swell in organic solvent. The trend in liquid chromatography has tended to move away from open column toward what is called high pressure liquid chromatography (HPLC) for analytical as well as preparative work. The change in technique is due to the development of high sensitivity, low dead volume detectors. The result is high resolution, high speed, and better sensitivity liquid chromatography.

The output of a chromatographic instrument can be of two types: (1) A plot of area retention time versus detector response. The peak areas represent the amount of each component present in the mixture. (2) A computer printout giving names of components and the concentration of each in the sample. The units of concentration are reported in several ways: as weight percent or ppm by weight, as volume percent or ppm by volume, or as mole percent. Refer to *Chromatography*. (Source: Cheremisinoff, N.P. *Polymer Characterization: Laboratory Techniques and Analysis*, Noyes Publishers, New Jersey, 1996).

LIQUID CRYSTAL POLYMERS

Liquid crystal polymers (LCPs) were introduced over the last three decades. In the liquid state, either as a solution (lyotropic) or a melt (thermotropic), they lie between the boundaries of solid crystals and isotropic liquids. This polymeric state is also referred to as a mesomorphic structure, or a mesophase, a combined term adopted from the Greek language (mesos = intermediate; morphe = form). This state does not meet all the criteria of a true solid or a true liquid, but it has characteristics similar to both a solid and a liquid. For instance, the anisotropic optical properties of LC polymeric fluids are like those of crystalline solids, but their molecules are free to move as in liquids.

The main difference between these polymers and the conventional liquid crystals used in electrical display devices is the molecular weight. LCPs have a much higher molecular weight. This difference provides unique features. Especially important is the existence of a glass transition point which permits freezing of an LC phase for use over a wide range of temperatures. In the 1940s and 1950s polymeric liquid crystallinity was recognized. The systems studied were mostly natural and synthetic lyotropic biopolymers such as tobacco mosaic virus, collagen, and poly(γ- benzyl-l-glutamate).

The LC state is a unique condition in which long-range molecular orientational order persists in the absence of various types of short-range translational order.

This class of ordering is a direct result of specific types of inter-molecular interactions and manifests itself in unique macroscopic physical properties. This theoretical prediction was well demonstrated when, in 1965, Stephanie Kwolek of the DuPont Company discovered that certain rigid wholly aromatic polyamides gave anisotropic solutions in alkylamide and alkylurea solvents, an observation that led ultimately to the development of Kevlar ararnid fiber. This was a milestone as it raised the question in the fibers industry:

Could a similar phenomenon occur in the melt? Were thermotropic polymers possible? They were possible, and indeed the number reported to date by academic and industrial researchers is nearly beyond count. In addition, as it later turned out, aromatic thermotropic polymers were found to offer a great many more useful properties than just their now well-known tensile capabilities. These polymers are injection moldable, albeit at temperatures in the vicinity of 400° C, a temperature not compatible with common melt-spinning equipment. The rate of thermal degradation of such a polyester at 400°C makes stable fiber production particularly difficult. Moreover, most conventional injection-molding equipment requires modification to operate at the high temperatures needed to ensure reasonable processing of this polymer.

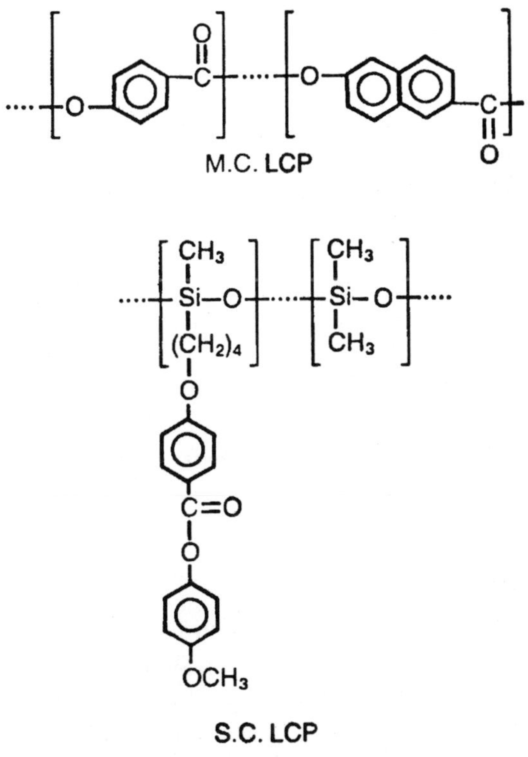

Figure 1. Shows main-chain and side-chain LCPs.

There are two types of chemical structure for LC polymers: main-chain LCPs (MCLCPs) and side-chain LCPs (SCLCPs). In the MCLCPs the mesogenic groups are a part of the backbone of the molecular chains. In the SCLCPs the mesogenic units are linked as pendant side chains to a polymer background. Both types can form thermotropic or lyotropic LC states. Examples of both types are shown in Figure 1. The reader may refer to Chapter 19 of *Handbook of Polymer Science and Technology: Volume 2 - Performance Properties of Plastics and Elastomers*, N. P. Cheremisinoff - editor, Marcel Dekker Inc, New York, 1989. The authors of this chapter provide a detailed overview of the synthesis and structure of these polymers. The following are some average properties of LCP with 20 % glass fiber.

Specific Gravity	1.52
Shrinkage,in/in,1/8 in. thick	0.0003
Shrinkage, in/in,1/4 in. thick	NA
Water Absorption,% 24hrs	NA
Mechanical Properties	
Impact,Izod,Notched (Ft-Lb/in)	4.00
Impact, Izod, Unnotched (Ft-Lb/in)	12.00
Tensile Strength (Psi)	19,000
Tensile Elongation (%	1.800
Tensile Modulus (Psi x 10^6)	2.20
Flexual Strength (Psi)	25,000
Flexural Modulus (Psi x 10^6)	1.80
Compressive Strength (Psi)	NA
Hardness (Rockwell R)	105.0
Electrical Properties	
Dielectric Strength (V/Mil)	550
Dielectric Constant (@ 1 MC dry)	3.60
Dissipation Factor (@ 1 MC dry)	0.030
Arc Resistance (sec)	NA
Thermal Properties	
Heat Deflection Temperature @ 264 psi (°F)	520
Heat Deflection Temperature @ 66 psi (°F)	NA

Note - NA refers to not applicable.

LIQUID HEAT CAPACITY

The value is the heat (in Btu) required to raise the temperature of one pound of the liquid one degree Fahrenheit at constant pressure. For example, it requires almost 1 Btu to raise the temperature of 1 pound of water from 68° F to 69° F. The value is useful in calculating the increase in temperature of a liquid when it is heated, as in a fire. The value increases slightly with an increase in temperature. Refer to *Joule* and *Units of Measurement*.

LIQUID VISCOSITY

The value (in centipoise) is a measure of the ability of a liquid to flow through a pipe or a hole; higher values indicate that the liquid flows less readily under a fixed pressure head. For example, heavy oils have higher viscosities (i.e., are more viscous) than gasoline. Liquid viscosities decrease rapidly with an increase in temperature. A basic law of fluid mechanics states that the force per unit area needed to shear a fluid is proportional to the velocity gradient. The constant of proportionality is the viscosity.

LLDPE (LOW LINEAR DENSITY POLYETHYLENE)

Table 1 provides average properties of LLDPE. In reviewing the general properties of this polymer, note the use of the following legend: A = amorphous - Cr = crystalline - C = clear - E = excellent - G = good - P = poor - O = opaque - T = translucent- R = Rockwell - S = Shore.

Table 1
Average Properties of LLDPE

STRUCTURE: Cr	FABRICATION		
	Bonding	**Ultrasonic**	**Machining**
	P	P	P
Specific Density: 0.92	Deflection Temperature (° F)		
	@ 66 psi	**@ 264 psi**	
	113	95	
Water Absorption Rate (%): 0.01	Utilization Temperature (° F) min: -55 max: 170		
Elongation (%): 700	MELTING POINT (° F): 250		
Tensile Strength (psi): 2000	Coefficient of Expansion: 0.00014		
Compression Strength (psi): 1600	Transparency: T		
Flexural Strength (psi): 1900	UV Resistance: P		
Flexural Modulus (psi): 70000	CHEMICAL RESISTANCE		
Impact (Izod ft. lbs/in): NB	**Acids**	**Alkalis**	**Solvents**
Hardness: SD52	G	E	E

Refer also to *Polyethylene*, *LDPE* (*Low Density Polyethylene*), *HMWDPE* (*High Molecular Weight Polyethylene*), and *HDPE* (*High Density Polyethylene*).

LONG CHAIN BRANCHING

Polymers are manufactured from low-molecular-weight compounds called monomers by polymerization reactions, in which large numbers of monomer molecules are linked together. Depending on the structure of the monomer or monomers and on the polymerization method employed, polymer molecules may exhibit a variety of architectures. Most common are the linear, branched, and network structures.

The linear structure is a chainlike molecule made from the polymerization of ethylene. With the chemical formula $CH_2=CH_2$, ethylene is essentially a pair of double-bonded carbon atoms (C), each with two attached hydrogen atoms (H).

A polyethylene chain from which other ethylene repeating units branch off is known as low-density polyethylene (LDPE); this polymer demonstrates the branched structure.

An example of a network structure is that of phenol-formaldehyde (PF) resin. PF resin is formed when molecules of phenol (C_6H_5OH) are linked by formaldehyde (CH_2O) to form a complex network of interconnected branches. The PF repeating unit is represented by phenol rings with attached hydroxyl (OH) groups and connected by methylene groups (CH_2).

Branched polymer molecules cannot pack together as closely as linear molecules can; hence, the intermolecular forces binding these polymers together tend to be much weaker. This is the reason why the highly branched LDPE is very flexible and finds use as packaging film, while the linear HDPE is tough enough to be shaped into such objects as bottles or toys.

The properties of network polymers depend on the density of the network. Polymers having a dense network, such as PF resin, are very rigid--even brittle--whereas network polymers containing long, flexible branches connected at only a few sites along the chains exhibit elastic properties.

Long-chain branching has the distinct advantage of providing good processability. In other words, a polymer with a degree of branching tends to extrude and mix more easily than a linear version of the grade. Additionally, in polymer finishing operations such as drying, bailing and pelletizing, polymers that are branched tend to demonstrate more robust finishing characteristics. The term finishing refers to polymer drying and bailing or pelletizing.

On the other hand, branching tends to retard cure performance. Branched polymers are more prone to slower cure, which may impact on other performance properties gained in post-curing operations. As such, lower compression sets are often observed. Polymer designers are therefore faced with tradeoffs in developing tailored grades that are good processing yet must meet demanding cure performance by the compounder. However, this is not necessarily a disadvantage, since there are a number of ways to incorporate branchiness, thereby providing an additional degree of freedom in polymer product design.

M

MACHINING

Rigid thermoplastics and thermosets can be machined by conventional processes such as drilling, sawing, turning on a lathe, sanding, and other operations. Glass-reinforced thermosets are machined into gears, pulleys, and other shapes, especially when the number of parts does not justify construction of a metal mold. Various forms can be stamped out (die-cut) from sheets of thermoplastics and thermosets. The cups made by vacuum forming, for instance, are cut out of the mother sheet using a sharp die. In the case of a thermoplastic such as polystyrene, the scrap sheet left over can be reground and remolded.

MASTIC

Refers to any of various semi-solid substances, usually formulated with rubber, other polymers, commonly used as a tile adhesive caulking, and a sound reducing treatment on various surfaces.

MASS SPECTROMETRY (MS and GC/MS)

Most of the spectroscopic and physical methods employed by the chemist in structure determination are concerned only with the physics of molecules. Mass spectroscopy deals with both the chemistry and the physics of molecules, particularly with gaseous ions. In conventional mass spectrometry, the ions of interest are positively charged ions. The mass spectrometer has three functions:

1. To produce ions from the molecules under investigation;
2. To separate these ions according to their mass to charge ratio;
3. To measure the relative abundances of each ion.

As early as the 1950s the chemistry of functional groups in directing fragmentation, and the power of mass spectrometry for organic structure determination began to develop. Today, mass spectrometry has achieved status as one of the primary spectroscopic methods to which a chemist faced with a structural problem turns.

The great advantage of the method is found in the extensive structural information which can be obtained from sub-microgram quantities of material. The methodology of mass separation is governed by both the kinetic energy of the ion and the ion's trajectory in an electromagnetic field. There exists a balance between the centripetal and centrifugal forces which the ion experiences. Centripetal forces are caused by the kinetic energy and centrifugal forces by the electromagnetic field.

Mass spectrometers provide a wealth of information concerning the structure of organic compounds, their elemental composition and compound types in complex mixtures. A detailed interpretation of the mass spectrum frequently allows the positions of the functional groups to be determined. Moreover, mass spectrometry is used to investigate reaction mechanisms, kinetics, and is also used in tracer work. Analyses are calculated to give mole %, weight %, or volume %. Either individual components, compound types by carbon number, or total compound type are reported. This is determined by the nature of the sample and the requirements of the submitter.

A wide variety of materials from gases to solids and from simple to complex mixtures can be analyzed. The molecular weight and atomic composition are generally determined. Only a very small amount of sample is required. Most calibration coefficients can be used for long periods of time. Some compounds such as long chain esters and polyethers decompose in the inlet system, and the spectrum obtained is not that of the initial substance. Calibration coefficients are required for quantitative analyses. The sample introduced to the instrument cannot usually be recovered. Some classes of compounds, such as olefins and naphthenes, give very similar spectra and cannot be distinguished except by analysis before and after hydrogenation or dehydrogenation. Figure 1 shows the basic components of the instrument.

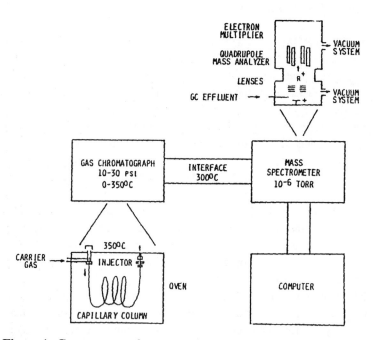

Figure 1. Components of mass spectroscopy.

MATCHED MOLD FORMING

Matched mold forming is a thermoforming process in which matched molds are used to assure a contained, defined part. Foamed PS egg cartons are a product example. See *Thermoforming*.

MBS (POLYMETHACRYLATE BUTADIENE STYRENE)

Table 1 provides average properties of this polymer. In reviewing the general properties of this polymer, note the use of the following legend: A = amorphous - Cr = crystalline - C = clear - E = excellent - G = good - P = poor - O = opaque - T = translucent- R = Rockwell - S = Shore.

Table 1
Average Properties of MBS

STRUCTURE: A	FABRICATION		
	Bonding	**Ultrasonic**	**Machining**
	E	E	P
Specific Density: 1.10	**Deflection Temperature (° F)** @ 66 psi: 194 @ 264 psi: 185		
Water Absorption Rate (%): 0.2	**Utilization Temperature (° F)** min: -40 max: 180		
Elongation (%): 20	**Melting Point (° F)**: 212		
Tensile Strength (psi): 7200	**Coefficient of Expansion**: 0.000045		
Compression Strength (psi): 8700	**Arc Resistance**: 80		
Flexural Strength (psi): 10000	**Dielectric Strength (kV/mm)**: 18		
Flexural Modulus (psi): 290000	**Transparency**: C		
Impact (Izod ft. lbs/in): 0.8	**UV Resistance**: P		
Hardness: R110	**CHEMICAL RESISTANCE**		
	Acids	**Alkalis**	**Solvents**
	G	E	P

MELAMINES

Melamine molding compounds have a wide range of grades, utilizing wood flour, minerals, and/or α-cellulose fibers as their reinforcements. These formulations result in free-flowing compounds suitable for all premolding applications as well as molding in all the conventional thermosetting molding processes. Additionally, melamine compounds are also available in the higher strength, higher bulk factor compounds using glass fibers and/or chopped cotton flock as reinforcement. Compounds can be molded in compression and transfer molding but require the use of auxiliary equipment in order to overcome their poor porability when employed in the screw injection process.

MELTING POINT OF WAX

The temperature at which a sample of wax either melts or solidifies from the solid or liquid state, respectively, depending on the ASTM test used. Low melting point generally indicates low viscosity, low blocking point, and relative softness.

MELT PHASE THERMOFORMING

This term refers to the normal method of most thermoforming processes in which a material is formed at a temperature above the melting point in semi-crystalline polymers or above the softening point in amorphous polymers. See *Pressure Forming*, *Vacuum Forming* and *Thermoforming*.

MELT FLOW INDEXER

The melt flow indexer is often used in industry to characterize a polymer melt and is a simple and quick means of quality control. It takes a single point measurement using standard testing conditions specific to each polymer class on a ram type extruder known as a extrusion plastometer. The standard procedure for testing the flow rate of thermoplastics using a extrusion plastometer is described in the ASTM D 1238 test. During the test, a sample is heated in the barrel and extruded from a short cylindrical die using a piston actuated by a weight. The weight of the polymer in grams extruded during the 10-minute test is the melt flow index (MFI) of the polymer.

MELT FRACTURE

The occurrence of irregularly shaped extrudates has been a matter of common knowledge to rubber technologists for a very long time and has been associated with situations where the processing temperatures were too low, where the rubber was insufficiently masticated (i.e., of too high a molecular weight) and where there was little filler or processing aid present. Figure 1 illustrates the different forms of melt fracture that are commonly observed during extrusion. The ways of overcoming this fault were well understood and few, if any, formal investigations were ever carried out. Whilset smooth extrudates may be obtained at comparatively low shear rates, above a critical shear rate (and corresponding critical shear stress) the extrudate becomes uneven. Some workers report that they have observed a point of inflection in the flow curve at this point but others have not been able to detect such a change. The form of the distortion varies widely. In some cases the extrudate has the form of a screwed thread, in others the extrudate has a rod-like cross section which is twisted into the form of a spiral. The determination of melt fracture can be done in laboratory-scale extrusion tests, and is an important phenomenon to assess when polymer suppliers are attempting to provide or introduce new polymer grades to the marketplace. Customer processing conditions can often be specific, and many applications for end-user part requirements generally tend to have high aesthetic appearance specifications; as such melt fracture is unacceptable. Although these are generalizations, melt fracture can be related to some degree to excessive crystallinity, lack of branchiness, and high scorchiness. Hence, the polymer designer may "tweak" such polymer product design parameters as broadening MWD (molecular weight distribution), or decreasing diene levels (e.g., in EPDM) in order to make a product less *scorchi*. There are also compounding practices and ingredients that can aid in minimizing melt fracture.

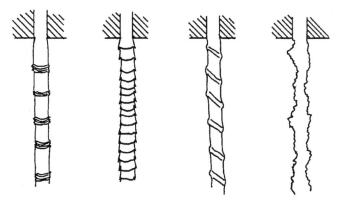

Figure 1. Illustrates common forms of melt fracture.

MELT SPINNING

Melt spinning is the preferred method of manufacture for polymeric fibers. The polymer is melted and pumped through a spinneret (die) with numerous holes (one to thousands). The molten fibers are cooled, solidified, and collected on a take-up wheel. Stretching of the fibers in both the molten and solid states provides for orientation of the polymer chains along the fiber axis. Polymers such as poly(ethylene terephthalate) and nylon 6,6 are melt spun in high volumes.

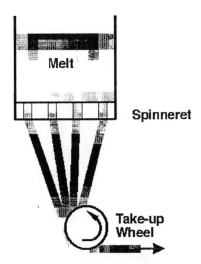

Figure 1. Melt spinning process.

Figure 1 provides a simplified process schematic. An excellent general reference on fiber spinning is A. Ziabicki, *Fundamentals of Fiber Formation*, Wiley, New York (1976). ISBN 0471982202. A classic article which emphasizes structure development during melt spinning is: J.R. Dees and J.E. Spruiell, *J. Appl. Polym. Sci.*, 18, pp. 1053-1078 (1974).

MICROSCOPY

Microscopic techniques have been applied for decades to polymer characterization. For polymer blends a minimum domain size of 1 μm can be examined in the optical microscope using one or more of the following techniques. A schematic of a typical optical microscope is shown in Figure 1.

Phase contrast - thin sections (100-200 μm) in thickness (and having refractive indices which differ by approximately .005) are supported on glass slides and examined "as is" or with oil to remove microtoming artifacts, e.g., determination of the number of layers in coextruded films, dispersion of fillers, and polymer domain size.

Polarized light - is used if one of the polymer phases is crystalline or for agglomeration of inorganic filters, e.g., nylon/EP blends and fillers such as talc.

Incident - is used to examine surfaces of bulk samples, e.g., carbon black dispersion in rubber compounds.

Bright field - Mainly used to examine thin sections of carbon black loaded samples, e.g., carbon black dispersion in thin films of rubber compounds. Refer to *Scanning Electron Microscopy*.

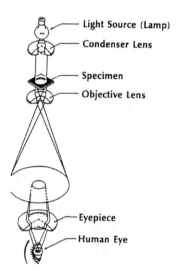

Figure 1. Essential elements of an optical microscope.

MILL MIXING

Mill mixing is the original method used to mix rubber stock and is still used for specialty compounds, small batches, and in small operations. Temperature control is very important. Cooling is usually accomplished by flooding or spraying the inside of the mill rolls with water, or circulating water through channels drilled in the roll walls. Chilled water is used where it is available. The compound temperature is adjusted by regulating the rate of the flow of the water through the rolls. Steam heating is used where a temperature increase is required. The roll temperature depends on the nature of the elastomer and such factors as the types and quantities of fillers and plasticizers to be incorporated. Nitrile and lightly loaded EPDM compounds are usually milled with roll temperatures in the range of $10\text{-}30^\circ$ C, chlorprene compounds at $20\text{-}40^\circ$ C, and butyl and highly loaded EPDM compounds at $60\text{-}80^\circ$ C.

For butyl and EPDM compounds, the front roll should be approximately 10° C cooler than the back roll because these elastomers tend to release from the hotter roll. The elastomer is first added to the nip, about 6 mm (0.25 in.) wide, at the top of rolls. A band of elastomer then comes through the nip and is formed preferably around the front roll. Depending on the elastomer used, varying degrees of difficulty may be encountered in forming the band at the beginning. After a few passes a band will form and be fed back into the nip continuously. The elastomer is then cut back and forth twice to assure proper blending and to allow the elastomer in the bank to go through the nip. It is important for efficient mixing to maintain a rolling bank on the mill during the incorporation of ingredients.

All dry ingredients except the fillers and the cure system are then added into the nip and the compound is crosscut back and forth twice to assure good dispersion of these dry ingredients throughout the batch. Next, the mill is opened slightly to add the fillers slowly to the batch. In order to prevent excessive loading of fillers at the center of the mill, strips of compound are cut from the end of the rolls several times during this operation and thrown back into the nip. When most of the fillers have dispersed in the compound, the liquid and the remaining fillers are added slowly and alternately to the batch.

When no loose filler is visible, the batch is crosscut back and forth twice more to assure good dispersion of fillers and plasticizers. It is often a useful procedure to "pig" roll the batch and feed the "pigs" back into the nip at right angles as a part of the crossblending process. The next step is to open the mill more and add vulcanizing agents to the batch.

When the vulcanizing agents are well dispersed, the entire batch is cut back and forth at least five times to assure thorough crossblending before being sheeted off the mill. This is a slow labor-intensive operation (about 30 minutes/batch versus 8 minutes or less in an internal mixer). It is not only time consuming but difficult to control because it so heavily depends on the skill of the operator. The internal mixer is much more widely used in the industry because of its versatility, rapid mixing, and large throughput.

MINERAL SPIRITS

Naphthas of mixed hydrocarbon composition and intermediate volatility, within the boiling range of 149° C to 204° C and with a flash point greater than 38° C; widely used as solvents or thinners in the manufacture of cleaning products, paints, lacquers, inks, and rubber. Also used uncompounded for cleaning metal and fabrics.

MIXING PRACTICES

The quality of the finished product in almost all polymer processes depends in part to how well the material was mixed. Mixing occurs inside internal mixers and similarly as an element of the processing step (e.g., inside single and twin screw extruders used in the fabrication of polymer parts). Both the material properties and the formability of the compound into shaped parts are highly influenced by the quality of the mixing. Hence, a better understanding of the mixing process will help to achieve optimum processing conditions and increase the quality of the final part.

The process of polymer blending or mixing is accomplished by distributing and/or dispersing a minor or secondary component within a major component which serves as a matrix.

The major component can be thought of as the continuous phase, and the minor components can be thought of as distributed or dispersed phases in the form of droplets, filaments or agglomerates. There are three general categories of mixtures that can be created; namely:
1.   Homogeneous mixtures of compatible polymers,
2.   Single phase mixtures of partly incompatible polymers, and
3.   Multi-phase mixtures of incompatible polymers.

Table 1 lists examples of compatible, partially incompatible and incompatible polymer blends. When creating a polymer blend, one must always keep in mind that the blend will most probably be remelted in subsequent processing or shaping processes. For example, a rapidly cooled system that was frozen as a homogenous mixture can split into separate phases, due to coalescence, when reheated. For all practical purposes, such a blend would not be acceptable for processing. To avoid this problem, compatibilizers which are macromolecules used to ensure compatibility in the boundary layers between the two phases, are common.

The mixing action that takes place during blending of these three general types of polymer blends and the physical phenomena that dominates each one of them can be broken down into two major categories - *distributive* mixing and *dispersive* mixing.

The morphology development of polymer blends is determined by competing distributive mixing, dispersive mixing and coalescence mechanisms. Figure 1 presents a model that helps visualize these mechanisms which govern morphology development in polymer blends.

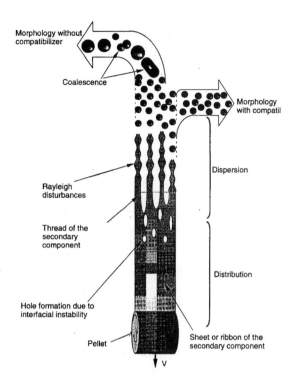

Figure 1. Shows mechanisms for morphology development in polymer blends.

Table 1
List of Common Polymer Blends

Compatible polymer blends

Natural rubber and polybutadiene

Polyamides (e.g., PA 6 and PA 66)

Polyphenylene ether (PPE) and polystyrene

Partially incompatible polymer blends

Polyethylene and polyisobutylene

Polyethylene and polypropylene (5% PE in PP)

Polycarbonate and polybutylene terephthalate

Table 1 continued

Incompatible polymer blends

Polystyrene/polyethylene blends

Polyamide/polyethylene blends

Polypropylenelpolystyrene blends

Refer to *Distributive Mixing*. (Source: Osswald, T.A. and G. Menges, *Material Science of Polymers for Engineers*, Hanser Publishers, New York, 1996).

MOLDS

The purpose of a mold is to determine the shape of the molded part. It conducts the hot plasticized material from the heating cylinder to the cavity, vents off the entrapped air or gas, cools the part until it is rigid, and ejects the part without leaving marks or causing damage. The mold design, construction, and craftsmanship largely determine the quality of the part and its manufacturing cost. The injection mold is normally described by a variety of criteria, including:

1.   number of cavities in the mold;
2.   material of construction, e.g., steel, stainless steel, hardened steel, beryllium copper, chrome-plated aluminum, and epoxy steel;
3.   parting line, e.g., regular, irregular, two-plate mold, and three-plate mold;
4.   method of manufacture, e.g., machining, hobbing, casting, pressure casting, electroplating, and spark erosion;
5.   runner system, e.g., hot runner and insulated runner;
6.   gating type, e.g., edge, restricted (pinpoint), submarine, sprue, ring, diaphragm, tab, flash, fan, and multiple; and
7.   method of ejection, e.g., knockout pins, stripper ring, stripper plate, unscrewing cam, removable insert, hydraulic core pull, and pneumatic core pull.

Molds used for injection molding of thermoplastic resins are usually flash molds, because in injection molding, as in transfer molding, no extra loading space is needed. However, there are many variations of this basic type of mold design. The design most commonly used for all types of materials is the two-plate design.

The cavities are set in one plate, the plungers in the second plate. The sprue bushing is incorporated in that plate mounted to the stationary half of the mold. With this arrangement it is possible to use a direct center gate that leads either into a single-cavity mold or into a runner system for a multicavity mold. The plungers and ejector assembly and, in most cases, the runner system, belong to the moving half of the mold. This is the basic design of an injection mold, though many variations have been developed to meet specific requirements. Use of multiple mold cavities permits greater increase in output speeds. However, the greater complexity of the mold also increases significantly the manufacturing cost. In a single-cavity mold the limiting factor is the cooling time of the molding, but with more cavities in the mold the plasticizing capacity of the machine tends to be the limiting factor. Cycle times therefore do not increase prorate with the number of cavities. The optimum number of mold cavities depends on factors such as the complexity of the molding, the size and type of the machine, cycle time, and the number of moldings required. Figure 1 provides an example of a rubber mold.

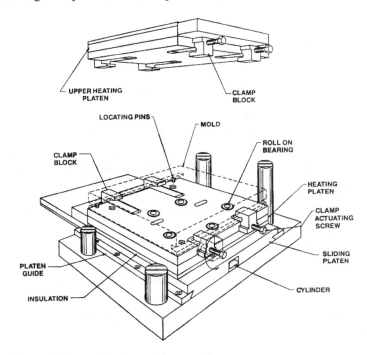

Figure 1. Example of a rubber mold.

MOLECULAR WEIGHT

Polymer properties are strongly linked to the molecular weight and the molecular weight distribution. As an example, Figure 1 shows the effect on molecular weight on physical properties. A polymer such as polystyrene is stiff and brittle at room temperature with a degree of polymerization of 1,000. However, at a degree of polymerization of 10, polystyrene is sticky and soft at room temperature. The stiffness properties reach an asymptotic maximum, whereas the flow temperature increases with molecular weight. In contrast, degradation temperature steadily decreases with increasing molecular weight. Hence, it is necessary to find the molecular weight that renders ideal material properties for the finished product, while having flow properties that make it easy to shape the material during the manufacturing process.

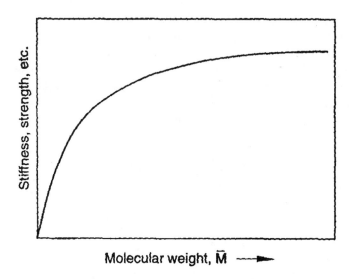

Figure 1. Shows effect of molecular weight on physical properties.

With the exception of maybe some naturally occurring polymers, most polymers have a molecular weight distribution such as shown by Figure 2. For such a molecular weight distribution function, we can define a number average, weight average, and viscosity average. The number average is the first moment and the weight average the second moment of the distribution function. In terms of mechanics this is equivalent to the center of gravity and the radius of gyration as first and second moments, respectively. The number average is defined by:

$$\bar{M}_n = \frac{\sum m_i}{\sum n_i} = \frac{\sum n_i M_i}{\sum n_i}$$

where m_i is the weight, M_i the molecular weight and n_i the number of molecules with i repeat units. The weight average is calculated using:

$$\bar{M}_w = \frac{\sum m_i M_i}{\sum m_i} = \frac{\sum n_i M_i^2}{\sum n_i M_i}$$

Another form of molecular weight average is the viscosity *average* which is calculated using:

$$\bar{M}_v = \left(\frac{\sum m_i \, M_i^{\alpha+1}}{\sum m_i} \right)^{1/\alpha}$$

where α is a material dependent parameter which relates the *intrinsic* viscosity to the molecular weight of the polymer. This relation is sometimes referred to as *Mark-Houwink* relation.

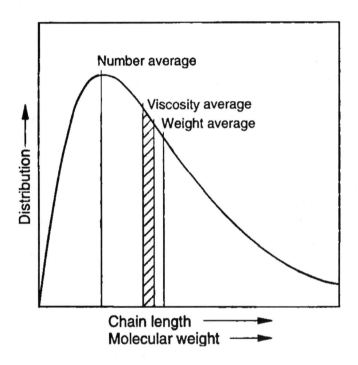

Figure 2. Molecular weight distribution.

MOONEY VISCOSITY

A measure of the resistance of raw or unvulcanized rubber to deformation, as measured in a Mooney viscometer. A steel disc is embedded in a heated rubber specimen and slowly rotated. The resistance to the shearing action of the disc is measured and expressed as a Mooney viscosity value. Viscosity increases with continued rotation, and the time required to produce a specified rise in Mooney viscosity is known as the *Mooney scorch value*, which is an indication of the tendency of a rubber mixture to cure, or vulcanize, prematurely during processing.

MULTIPLE-SCREW EXTRUDERS
General Considerations

Multiple-screw extruders (that is, extruders with more than a single screw) were developed largely as a compounding device for uniformly blending plasticizers, fillers, pigments, stabilizers, etc., into the polymer. Subsequently, the multiple-screw extruders also found use in the processing of plastics. Multiple-screw extruders differ significantly from single-screw extruders in mode of operation.

In a single-screw machine, friction between the resin and the rotating screw makes the resin rotate with the screw, and the friction between the rotating resin and the barrel pushes the material forward, and this also generates heat. Increasing the screw speed and/or screw diameter to achieve a higher output rate in a single-screw extruder will therefore result in a higher buildup of frictional heat and higher temperatures.

In contrast, in twin-screw extruders with intermeshing screws the relative motion of the flight of one screw inside the channel of the other pushes the material forward almost as if the machine were a positive displacement gear pump which conveys the material with very low friction. In twin-screw extruders, heat is therefore controlled independently from an outside source and is not influenced by screw speed. This fact becomes especially important when processing a heat-sensitive plastic like PVC.

Multiple-screw extruders are therefore gaining wide acceptance for processing vinyls, although they are more expensive than single-screw machines. For the same reason, multiple-screw extruders have found a major use in the production of high-quality rigid PVC pipe of large diameter. Several types of multiple-screw machines are available, including intermeshing corotating screws (in which the screws rotate in the same direction, and the flight of one screw moves inside the channel of the other), intermeshing counter-rotating screws (in which the screws rotate in opposite directions), and nonintermeshing counterrotating screws. Multiple-screw extruders can involve either two screws (twin-screw design) or four screws. A typical four-screw extruder is a two-stage machine, in which a twin-screw plasticating section feeds into a twin-screw discharge section located directly below it. The multiple screws are generally sized on output rates (lb/hr) rather than on L/D ratios or barrel diameters.

Twin Screw Extruders

There are a variety of twin-screw extruder designs employed throughout the polymers industry, with each type having distinct operating principles and applications in processing. Designs can be generally categorized as corotating and counterrotating twin-screw extruders. Different arrangements of twin-screw systems are illustrated in Figure 1. The figure illustrates differences between corotating and counterrotating as well as between intermeshing and nonintermeshing screw arrangements. Note the difference between fully intermeshing and partially intermeshing systems and between open- and closed-chamber types. For example, a counterrotating screw shown in Figure 1 (type 1) is an axially closed, single-chamber pumping system, whereas a corotating screw (type 4) is an axially open mixing system relying on drag forces.

A screw system open in axial direction is actually open in the longitudinal direction of the screw channel since it has a passage from the inlet to the outlet of the apparatus. This means that material exchange can take place lengthwise along the channel. In a closed arrangement the screw flights in the longitudinal direction are closed at intervals. It is important that the cross section of the screw channel be open in order for material exchange to take place from one flight to the other in a direction normal to the screw channel. There is usually some leakage over the screw crests and through the areas required for mechanical clearances. Whether the screws are open lengthwise or crosswise, or have a closed geometry, has a direct effect on conveying conditions, mixing action, and the pressure buildup capacity of the system.

Nonintermeshing systems, for example, are open lengthwise and crosswise (refer to Figure 1). Fully intermeshing, counterrotating systems can be closed lengthwise and crosswise, and regardless of mechanical clearances they can develop closed chambers. This is the case with screw pumps. Lengthwise and crosswise closed, fully intermeshing corotating systems (Figure 1, type 2) are theoretically impossible, as is a lengthwise open and crosswise closed counterrotating system. Fully intermeshing, corotating screws are open lengthwise. When normal screw flights are employed they are closed crosswise, and when staggered screw discs are used they are open crosswise. This design is illustrated by types 4 and 6 in Figure 1.

There are several types of partially intermeshing screws, e.g., lengthwise open and crosswise closed systems and lengthwise and crosswise open systems. Most commercial intermeshing counterrotating screw extruders are combinations of Systems 1, 9a and 9b, shown in Figure 1. Commercial nonintermeshing, counterrotating screws correspond to System 11, and are used along with a single-screw discharge extruder.

Counterrotating screws operate with both shafts turning with equal angular velocity in a counteracting rotation. Figure 2 shows a vector diagram illustrating the velocity fields and forces of a counterrotating screw operation. In contrast, with corotating screws as shown in Figure 1, when screw 1 is held stationary, screw 2 will not roll off, but will make a circular translation motion around screw 1. The center point M2 moves in this fashion on a circular path around M1, with the radius M1MPV2; screw 2 itself produces no rotation of its own. Each point describes a circular path with the same radius. If the arc P1-P2 on the circumference of disc 2 is a screw crest, the points of this arc describe (as shown with two intermediary points P3 and P4, Figure 1), a family of arcs which cut out a halfmoon-shaped surface area F from screw 1.

Each point along the axial screw surface is tangentially wiped by the other screw with a constant relative velocity. For simplicity, consider the screw simply as a flat disc as shown in Figure 2.

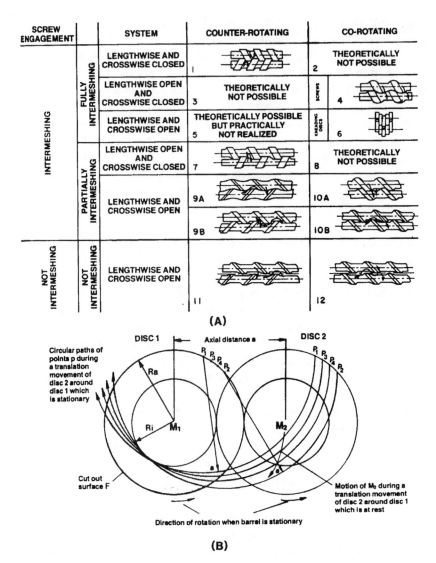

Figure 1. (A) Twin screw configurations. (B) Movement principle of corotating screw.

If disc 1 is held stationary, disc 2 with rolling circle radius AM2 rolls off the base circle with the radius M1A of disc 1. Each point of disc 2 describes an elongated or abbreviated epicycloid depending on its distance from M2. Let the arc P1-P2 of the circumference of disc 2 define a screw crest; then the points of this arc represented with two intermediary points (P3 and P4) describe a family of cycloids which "cut out" the surface area F from disc 1. The shape of the cutout area F corresponds to the crosscuts of the screw channel perpendicular to the turning axis. Figure 2B illustrates two examples of these screw crosscuts. The forward rotation of the thin discs in a direction lengthwise to the shaft axis results in screws with positive and negative pitch. Each point of the surface area will be wiped by the other screw. The screw flanks are developed in the crosscut from a cycloidal family and both screws roll off one another in a fashion resembling that of gear wheels. In the wiping points, the relative velocity depends on the distance of these points from the rotating axis and is in the area of $-2\pi n(R_a - R_1) < V_{rel} < +2\pi n(R_a - R_1)$.

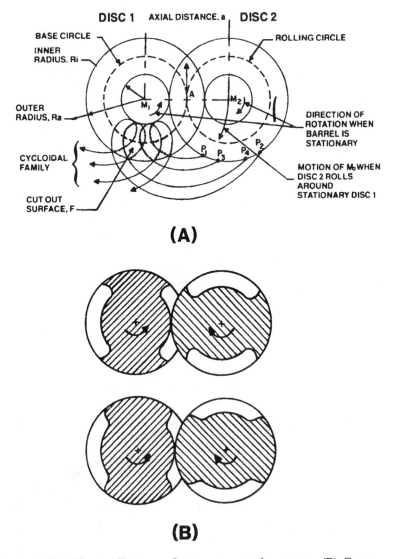

Figure 2. (A) Vector diagram of a counterrotating screw; (B) Cross section of counterrotating screws.

In contrast, with counterrotating arrangements, the screws in the wedge area roll off one another in a rolling motion with locally different gliding velocities. This gliding velocity is small compared to the circumferential speed and is typically $+2\pi n(R_a - R_1)$. Corotating screws do not roll off one another at any point, but instead have a translation movement, where one crest edge wipes a screw flank tangentially with equally high relative velocity. The relative velocity is high compared to the circumferential speed and is $V_{rel} = +2\pi n(R_a - R_1)$. In the wedge area the screws have equal pitch and equal pitch direction are intermeshing.

The open area in the wedge part between the two screws is different. In counterrotating designs, there are closed C-shaped chambers with a very small volume such that no material transfer from one screw to the other takes place. In this type of design the distributive mixing efficient is greatly reduced.

In contrast, corotating screws form V-shaped wedge areas that have four to five times more volume. This enables material to be transferred from one screw to the other, resulting in a renewal of material layers and surfaces. The result is a higher degree of mixing.

Self-cleaning screws are often employed to prevent material from adhering to the screw root. Material that adheres to the screw root can degrade, because of a broad residence time distribution, and eventually fall off and be carried out with the product, showing up as contamination.

Self-cleaning action is achieved in both counterrotating and corotating screws, through an opposite roll-off motion or wiping motion. This self-wiping is achieved in different ways in each system, each having different degrees of effectiveness. With counterrotating screws, the roll-off process between the screw crest and screw root and between the screw flanks simulates the action of a calender.

The necessary shear velocity required to wipe the boundary layers is proportionately lower because of the low relative velocity. Also, the material is drawn into the roller gap and is squeezed onto the surface. In corotating screws, one crest edge wipes the flanks of the other screw with a tangentially oriented, constant relative velocity. There is a higher relative velocity in this arrangement, and hence there is a sufficiently high shear velocity available to wipe the boundary layers. The calender effect does not occur; however, a more efficient and uniform self-cleaning action is achieved.

Wearing of the screw surface will increase proportionally to the relative velocity of the shafts. To minimize this effect intermeshing, self-wiping, counterrotating machines should be operated at relatively low speeds. Since corotating screws have no calendering effect between the crest and the flank of the screws, there is considerably less wear. Therefore, corotating machines can be operated at much higher screw speeds with greater throughputs. A typical corotating, self-wiping, twin-screw extruder with 170-mm-diameter screws can process 9000-11,000 lb/hr of polypropylene at a screw speed of 300 rpm.

Source: N. P. Cheremisinoff, *Polymer Mixing and Extrusion Technology*, Marcel Dekker, Inc. New York, 1987.

N

NATIONAL BUREAU OF STANDARDS (NBS)

The overall goal of the NBS is to strengthen and advance the nation's science and technology and to facilitate their effective application of public benefit. The bureau conducts research and provides a basis for the nation's physical measurement system, scientific and technological services for industry and government, a technical basis for increasing productivity and innovation, promoting international competitiveness in American industry, maintaining equity in trade, and technical services, promoting public safety. The Bureau's technical work is performed by the National Measurement Laboratory, the National Engineering Laboratory, and the Institute for Computer Sciences and Technology.

NATIONAL ELECTRICAL MANUFACTURERS ASSOCIATION (NEMA)

NEMA consists of manufacturers of equipment and apparatus for the generation, transmission, distribution, and utilization of electric power. The membership is limited to corporations, firms, and individuals actively engaged in the manufacture of products included within the product scope of NEMA product subdivisions. NEMA develops product standards covering such matters as nomenclature, ratings, performance, testing, and dimensions. NEMA is also actively involved in developing National Electrical Safety Codes and advocating their acceptance by state and local authorities. Along with a monthly news bulletin, NEMA also publishes manuals, guidebooks, and other material on wiring, installation of equipment, lighting, and standards. The majority of NEMA standardization activity is in cooperation with other national organizations. The manufacturers of wires and cables, insulating materials, conduits, ducts, and fittings are required to adhere to NEMA standards by state and local authorities.

NATIONAL FIRE PROTECTION ASSOCIATION (NFPA)

The NFPA has the objective of developing, publishing, and disseminating standards intended to minimize the possibility and effect of fire and explosion. NFPA's membership consists of individuals from business and industry, fire service, health care, insurance, educational, and government institutions. NFPA conducts fire safety education programs for the general public and provides information on fire protection and prevention.

Also provided by the association is the field service by specialists on flammable liquids, electricity, gases, and marine problems. Each year, statistics on causes and occupancies of fires and deaths resulting from fire are compiled and published. NFPA sponsors seminars on the Life Safety Codes, National Electrical Code, industrial fire protection, hazardous materials, transportation emergencies, and other related topics.

NFPA also conducts research programs on delivery systems for public fire protection, arson, residential fire sprinkler systems, and other subjects. NFPA publications include *National Fire Codes Annual, Fire Protection Handbook, Fire Journal,* and *Fire Technology*. Refer to *NFPA Hazard Classifications*.

NATIONAL SANITATION FOUNDATION (NSF)

The NSF is an independent, nonprofit environmental organization of scientists, engineers, technicians, educators, and analysts. NSF frequently serves as a trusted neutral agency for government, industry, and consumers, helping them to resolve differences and unite in achieving solutions to problems of the environment. At NSF, a great deal of work is done on the development and implementation of NSF standards and criteria for health-related equipment. The majority of NSF standards relate to water treatment and purification equipment, products for swimming pool applications, plastic pipe for potable water as well as drain, waste, and vent (DNW) uses, plumbing components for mobile homes and recreational vehicles, laboratory furniture, hospital cabinets, polyethylene refuse bags and containers, aerobic waste treatment plants, and other products related to environmental quality. Manufacturers of equipment, materials, and products that conform to NSF standards are included in official listings and these producers are authorized to place the NSF seal on their products. Representatives from NSF regularly visit the plants of manufacturers to make certain that products bearing the NSF seal do fulfill applicable NSF standards.

NATURAL RUBBER (NBR

Cis - Polyisoprene is derived from rubber trees. A general purpose natural rubber which ranges from soft gum rubbers (erasers) to ebonite. Normally amorphous in the unstrained state and not to be confused with the rigid partially crystalline natural polymer gutta percha (or balata) which is trans-polyisoprene. General properties include excellent mechanical properties with a high resistance to tearing and abrasion; high resilience at 200° C and therefore low heat build-up under the action of mechanical vibrations. Good bonds can be obtained with metals and fabrics. The rubber has good low temperature properties. The working temperature range is approximately -50° C to +100° C. The material shows considerable swelling in mineral oils and degreasing solvents, and it is prone to oxidative and ozone attack. The rubber can be oil-extended.

Typical uses may be described as general, but by and large it has been displaced by synthetic rubbers. Natural rubber is used in engineering applications almost exclusively in its vulcanized (i.e., crosslinked) state, and its characteristics are thus governed by the composite properties of the vulcanizate and not solely by the base polymer properties. Natural rubber can be vulcanized in various ways with sulphur, sulphur donors, activators and accelerators to give sulphur crosslinks of different types or with peroxide systems to give carbon-carbon crosslinks. This versatility is an important attribute, since it is possible in this way to alter the balance of strength, heat resistance and elastic properties to suit the application. Natural rubber has an outstanding combination of strength and resilience qualities. As intermolecular forces are low for polyisoprene, natural rubber exhibits low hysteresial dissipation of energy on repeated deformation, i.e., it has high resilience; this basic property is maintained over a wide temperature range.

Correspondingly the physical component of creep or stress relaxation is low and vulcanizate stiffness is not sharply dependent on deformation frequency. These features hold true for the small or moderate bulk deformations involved in certain engineering applications. However, the behavior is completely different for large deformations - which may exist, for example, at the tips of cracks or at the interface between the rubber and an impacting body. Because of the regularity of the molecular structure, natural rubber crystallizes very readily on stretching and thus becomes a highly hysteresial material under large deformation conditions. It accordingly has high tensile strength, excellent resistance to chipping, cutting and tearing, and high abrasion resistance. The fatigue life is also excellent under some service conditions.

The inherent strength and resilience properties simplify the matching of the rubber to the application. Where optimum elastic properties are required, the rubber can be used in essentially unfilled (gum) vulcanizate form. If fillers are required non-reinforcing types can often be used, instead of the reinforcing fillers that are essential for low-strength rubbers.

Non-reinforcing fillers, for a given increase in vulcanizate stiffness, generally give better permanent set, creep and dynamic properties than reinforcing types. The basic properties of natural rubber also show up to advantage in the fabrication of components.

NFPA HAZARD CLASSIFICATION

The indicated ratings are given in "Fire Protection Guide on Hazardous Materials," National Fire Protection Association. The classifications are defined in Table 1.

Table 1
Explanation of NFPA Hazard Classifications

Class	Definition
Health Hazard (blue)	
4	Materials which on very short exposure could cause death or major residual injury even though prompt medical treatment was given.
3	Materials which on short exposure could cause serious temporary or residual injury even though prompt medical treatment were given.
2	Materials which on intense or continued exposure could cause temporary incapacitation or possible residual injury unless prompt medical treatment is given.
1	Materials which on exposure would cause irritation but only minor residual injury even if no treatment is given.

Table 1 continued

0	Materials which on exposure under fire conditions would offer no hazard beyond that of ordinary combustible material.
Flammability (red)	
4	Materials which will rapidly or completely vaporize at atmospheric pressure and normal ambient temperature, or which are readily dispersed in air and which will burn readily.
3	Liquids and solids that can be ignited under almost all ambient temperature conditions.
2	Materials that must be moderately heated or exposed to relatively high ambient temperatures before ignition can occur.
1	Materials that must be preheated before ignition can occur.
0	Materials that will not burn.
Reactivity (yellow)	
4	Materials which in themselves are readily capable of detonation or of explosive decomposition or reaction at normal temperatures and pressures.
3	Materials which in themselves are capable of detonation or explosive reaction but require a strong initiating source or which must be heated under confinement before initiation or which react explosively with water.
2	Materials which in themselves are normally unstable and readily undergo violent chemical change but do not detonate. Also materials which may react violently with water or which may form potentially explosive mixtures with water.
1	Materials which in themselves are normally stable, but which can become unstable at elevated temperatures and pressures or which may react with water with some release of energy but not violently.
0	Materials which in themselves are normally stable, even under fire exposure conditions, and which are not reactive with water.
Other (white)	
W̶	Materials which react so violently with water that a possible hazard results when they come in contact with water, as in a fire situation. Similar to Reactivity Classification.
Oxy	Oxidizing material; any solid or liquid that readily yields oxygen or other oxidizing gas, or that readily reacts to oxidize combustible materials.

NITRILE RUBBERS (NBR)

Nitrile rubbers are copolymers of butadiene and acrylonitrile. The properties of these special purpose rubbers depend upon the ratio of butadiene to acrylonitrile. General properties include moderate mechanical properties; poor cold resistance; and excellent resistance to swelling in oils and alcohols.

The greatest oil and alcohol resistance occurs in rubbers with a high acrylonitrile content. The best low temperature flexibility occurs with low acrylonitrile content. Approximate working temperature range is from -20° C to >120° C. The monomers are emulsified in soap solutions and a catalyst is added which starts an exothermic polymerisation reaction resulting in the production of latex. This latex can be used directly in coating, impregnation or dipping operations, or can be further processed to give dry rubber in solid form generally known as NBR which is the basic product used by rubber manufacturers in the preparation of a wide range of engineering components.

The two monomers, butadiene and acrylonitrile can be made to combine, in almost any proportions, the properties of the polymers depending on the ratio of monomers used. However, practical considerations limit the range of polymers commercially available to those containing between 18 and 50 percent acrylonitrile and those in wide use to polymers with from 27 to 42 percent. Polymers with 33 to 36 per cent acrylonitrile are normally regarded as standard products and polymers with other monomer ratios are only chosen for their special properties. In the raw state these rubbers are of no use as engineering materials; they must first be modified by the addition of various ingredients such as reinforcing fillers, plasticizers and vulcanizing agents. They are then fabricated to the required shape and vulcanized to impart strength and resilience, the final properties being capable of considerable modification by the compounding technique applied. Compounding is done in order to modify the properties of the polymer in the final article, but also to aid the processing of the polymer in the rubber factory, and to bring about the vulcanization of the polymer.

This rubber can be processed using standard equipment, with the unit operations consisting essentially of mixing, forming, vulcanization and finishing operations. Extrusion is the main process used in the manufacture of hose and in profiles such as automotive weather strip and consists of forcing the material through a die by the action of a rotating screw.

Calendering or calendered sheet is used in fabricating articles, such as fuel tanks and for wrapping large diameter hoses, and providing blanks for molding. Molding is used in the manufacture of a wide variety of mechanical rubber goods for engineering uses. Three processes are available: (i) compression molding; (ii) transfer molding; and (iii) injection molding. Die costs are lowest for compression molding, intermediate for transfer molding and highest for injection molding.

Properties

Oil Resistance: The primary property of nitrile rubber is oil resistance and it is this factor, in conjunction with its excellent physical properties, which account for much of its commercial use. Oils generally attack rubbers by causing them to swell, distorting their shape and reducing their strength. The attack is caused by the chemical affinity of the oil and the rubber, and the former tends to act as a solvent for the latter. The extent of attack in the case of nitrile rubber depends on the acrylonitrile content of the polymer and on the chemical nature of the oil, the higher the aromatic hydrocarbon content of the latter the greater the swell. This swelling tendency of an oil can be roughly measured by determining its aniline point (i.e., the temperature at which the oil is miscible with an equal amount of aniline), low aniline points indicating high aromatic contents and high swelling tendency. Increasing the acrylonitrile content of the polymer reduces swell in oils. Nitrile rubbers are not resistant to all types of oils and solvents, but in general can be expected to withstand mineral lubricating oils, petrols and some synthetic lubricants and refrigerants. They are swollen by aromatic solvents such as benzene, chlorinated solvents and ketones. The apparent resistance of a nitrile rubber compound to swelling in oil is influenced by the presence of the ingredients added during compounding, particularly to plasticizers. These materials which themselves are oily liquids often reduce the swell of nitrile rubber compounds because they are extracted by the immersion fluid and if the latter is one with little swelling tendency of the base rubber a resultant shrinkage may ensue. Thus by proper selection of the base nitrile polymer and the type and amount of plasticizer the rubber compounder can control the volume change within quite wide limits in many media. It should however, be noted that the diffusion of oil into rubber is a slow process and for this reason, the relatively thin samples used in laboratory testing of rubber compounds often give an exaggerated level of swelling compared to that experienced in service.

Low Temperature Properties: Unfortunately the improvement in oil resistance which is obtained by increasing the acylonitrile content of a polymer is accompanied by a decrease in the low temperature flexibility. Like all other rubbers nitrile becomes progressively harder and less flexible as the temperature is reduced until finally the material is quite hard and brittle. This hardening effect is reversible and the material returns to its original flexibility when the temperature is raised and suffers no harm, unless it is flexed while still held at the low temperature.

High Temperature Properties: The effect of elevated temperatures is to cause a reduction in such physical properties of nitrile rubber compounds as tensile modulus, tensile strength, hardness, tear resistance, which are only reversible if the duration of exposure is limited. If this is prolonged, permanent changes take place which result in increased hardness, tensile modulus and loss of elongation at break. Unlike some rubbers which soften and perish with heat, nitrile rubbers harden and a skin of hard material, the thickness of which increases with the length of exposure to high temperature, forms on the surface which ultimately will crack if the rubber is flexed. Since the deterioration is induced by the presence of oxygen the rubber is protected if immersed in oil or other fluid and its effective life is increased several fold. Heat resistance improves as the acrylonitrile content increases. The maximum temperature at which a nitrile rubber article may be used depends on the service life required, the amount of flexing it must withstand during exposure and subsequently, the ease with which a part can be replaced and the consequences of a failure during service. As a result of the wide range of service requirements, temperatures as low as 600° C or as high as 1500° C may be quoted as the maximum allowable.

Weather Resistance: Many rubbers are affected adversely by outdoor exposure, particularly by the traces of ozone, which are always present in the atmosphere and which cause rapid cracking of lightly stressed rubber articles. Nitrile rubber is no exception and thus it is not suitable for use in exposed conditions. However, the incorporation of a proportion of PVC (Polyvinyl chloride) into the nitrile compound results in improved resistance until, with a blend of 70/30 nitrile/PVC, a material with almost complete resistance to ozone attack is produced. The presence of PVC produces stiffer compounds, with lower resiliance and inferior low temperature properties, but the oil and solvent resistance is enhanced.

Fire Resistance: Nitrile rubbers are not flame resistant, but by special compounding a useful degree of resistance can be obtained using PVC blends.

Wear Resistance: The ability to resist the abrasion action of a rubbing surface is a feature of most rubbers and the actual removal of rubber particles from the surface is a complex process. However, compounding variables which affect the resistance of rubbers to abrasion are well known and compound modifications to suit operating conditions is an established technique. Soft resilient compounds are more resistant to abrasion induced by particle impact as in grit blasting while stiffer compounds have greater resistance to scuffing.

Tensile Strength: This property is often used as a criterion of quality of a rubber compound though it leaves much to be desired, since many other factors determine service performance. Nitrile rubbers are capable of producing compounds with tensile strengths as high as 4000 lb./sq. in. measured, as is usual in rubber technology, on the original cross section, the higher acrylonitrile content polymers giving higher results than those containing lower amounts. The highest results are obtained only in the presence of black fillers.

Elongation at Break: The percentage increase in length of a sample at rupture can be as high as 800 percent or as low as 100 percent. High elongations are not obtainable with hard compounds.

Hardness: The hardness of rubber products is measured as an elastic deformation under the influence of a small spherical indentor. The practical range available in nitrile compounds is 30 -1000 International Rubber Hardness Degrees, a scale which approximates closely to the more widely known Shore A hardness. Nitrile rubbers are compatible with phenolic resins in all ratios and these can be used to produce rubber compounds of high hardness and wear resistance, or the rubber may be used to modify the resin properties producing more flexible and impact resistant products. Nitrile rubbers are also capable of being vulcanized by standard methods to the hard rubber or ebonite stage and give products with higher softening points than those based on natural rubber.

Compression Set: Like most other rubbers, nitrile rubber is not perfectly elastic but exhibits some plastic properties. Thus if deformed for a period, particularly at elevated temperatures it does not completely recover but exhibits a permanent set, the extent of which depends on deformation experienced, time and temperature. The resistance of nitrile rubbers to compression set is good and this enables its use in articles which are used under permanent compression such as O-rings and seals.

Resilience: The resilience of nitrile rubber is not as good as that of natural and some other synthetics such as EPDM and generally lies in the region 30-60 percent. This relatively low value means that the energy and shock absorbing properties of the rubber are good. Resilience is affected by acrylonitrile content, those rubbers having low acrylonitrile ratios having higher resilience.

Electrical Properties: Nitrile rubbers do not have good electrical insulating properties, value s of the order of 10^{10} ohm/cm being typical. Power factor and dielectric constant (800 cycles) are 0.03 and approximately 15, respectively. These properties can be used to advantage in applications such as belts and rollers on textile spinning machinery where with suitable compounding the resistivity is reduced to the point where build up of static electricity is prevented in service.

Miscellaneous Physical Properties: The specific gravity of the neat polymer is typically between 0.96 and 1.00, and that of compounds between 1.00 and 1.80. The relative refractive index is 1.521. The specific heat of the polymer is 1. 971 Joules/gm/°C.

Applications

The following is a partial list of common applications for this elastomer.

Hose: as the base polymer used in the inner tube of a wide range of hoses designed for contact with oils, fuels, solvents, and various chemicals. Blends of nitrile and PVC are used for the hose covers on such hoses and with fire hoses.

Seals: The excellent resistance of nitrile rubber to hot oils makes it a good choice for various types of elastomeric seals used in engineering and one of its largest outlets is in rotary lip seals, O-rings and hydraulic packings.

Gaskets: NBR is used either in sheet form, molded or as a binder for cork granules in a wide range of oil resistant gasketing materials.

Rollers: Oil resistant rollers and endless belts made with NBR are used in the printing industry to resist the action of certain inks.

Brake Linings: The compatability of nitrile rubber with phenolic resins is used to advantage in the production of some types of brake linings.

Adhesives: Nitrile rubber both in latex form and as solutions is used in the preparation of adhesives which have to be designed to maintain strength in the presence of oils or solvents.

Footwear: Apart from the obvious use in industrial footwear it is used in the blends with PVC to impart high wear resistance and with phenolic resin to make heel tips for stiletto heels.

Flexible Containers: NBR is used in flexible containers for transport and storage of fluids which can be collapsed when not in use. Nitrile rubber is chosen for its oil and chemical resistance in many cases.

Non-woven Fabrics: In latex form it is used as a binder in the production of non-woven fabrics both for industrial and domestic use.

NOVOLAC RESINS

These are a member of the phenolics. In the presence of acid catalysts, and with the mole ratio of formaldehyde to phenol less than 1, the methylol derivatives condense with phenol to form, first, dihydroxydiphenyl methane and, on further condensation with the methyl bridge formation, fusible and soluble linear low polymers called novolacs are formed, having the following structure:

The *ortho* and *para* links occur in random fashion. Novolac (two-stage) resins are made with an acid catalyst, and only a portion of the necessary formaldehyde is added to the reaction, producing a mole ratio of about 0.8:1. The rest is added later on as hexamethylenetetramine, which decomposes in the final curing stage in the presence of heat and moisture to yield formaldehyde and ammonia, which serve as the catalyst during curing. The novolac resins are typically applied as epoxy coatings for environments exposed to harsh chemicals and solvents. They are used in secondary containment, solvent storage, as pump pads, trenches, and other high exposure areas. Their advantages include excellent resistance to strong acids, alkalis, and most industrial chemicals and solvents. They can be applied in cool damp conditions, and can be recoated in 4 to 6 hours. Quite often they are applied to concrete surfaces. The concrete substrate must be free of dirt, waxes, curing agents, and other foreign materials before application. Steel surfaces must be free of grease and oils. Table 1 provides some general chemical resistance ratings. Refer to *Resole Formation*.

Table 1
Chemical Resistance Ratings for Novolac Epoxy Resins

Reagent	Rating	Reagent	Rating
Acetic Acid 20%	R	Sulfuric Acid 98%	R
Urea	R	FattyAcids	R
Acetone	L	Toluene	R
Bleach	L	Gasoline	R
Methylene Chloride	R	Lactic Acid	R

Table 1 continued

Reagent	Rating	Reagent	Rating
Citric Acid	R	Methyl Ethyl Ketone	L
Diesel Fuel	R	Nitric Acid	R
Sodium Hydroxide 50%	R	Hydrochloric Acid 36%	R
Ethylene Glycol	R	Xylene	L

R - Recommended for Intermittent Immersion. L- Limited recommendation, occasional spills.

NUCLEAR MAGNETIC RESONANCE SPECTROMETER

Nuclear Magnetic Resonance (NMR) is a spectrometric technique for determining chemical structures. When an atomic nucleus with a magnetic moment is placed in a magnetic field, it tends to align with the applied field. The energy required to reverse this alignment depends on the strength of the magnetic field and to a minor extent on the environment of the nucleus, i.e., the nature of the chemical bonds between the atom of interest and its immediate vicinity in the molecule. This reversal is a resonant process and occurs only under select conditions. By determining the energy levels of transition for all of the atoms in a molecule, it is possible to determine many important features of its structure. The energy levels can be expressed in terms of frequency of electromagnetic radiation, and typically fall in the range of 5-600 MHz for high magnetic fields. The minor spectral shifts due to chemical environment are the essential features for interpreting structure and are normally expressed in terms of part-per-million shifts from the reference frequency of a standard such as tetramethyl silane. The most common nuclei examined by NMR are 111 and 11C, as these are the NMR sensitive nuclei of the most abundant elements in organic materials. 1H represents over 99% of all hydrogen atoms, while C is only just over 1 % of all carbon atoms; further, 1H is much more sensitive than 1-C on an equal nuclei basis. Until fairly recently, instruments did not have sufficient sensitivity for routine C NMR, and H was the only practical technique. Most of the time it is solutions that are characterized by NMR, although C NMR is possible for some solids, but at substantially lower resolution than for solutions. In general, the resonant frequencies can be used to determine molecular structures. 1H resonances are fairly specific for the types of carbon they are attached to, and to a lesser extent to the adjacent carbons. These resonances may be split into multiples, as hydrogen nuclei can couple to other nearby hydrogen nuclei. The magnitude of the splittings, and the multiplicity, can be used to better determine the chemical structure in the vicinity of a given hydrogen.

O

OIL CONTENT OF PETROLEUM WAX

A measure of wax refinement, under conditions prescribed by test method ASTM D 721. The sample is dissolved in methyl ethyl ketone, and cooled to -32° C to precipitate the wax, which is then filtered out. The oil content of the remaining filtrate is determined by evaporating the solvent and weighing the residue. Waxes with an oil content generally of 1.0 mass percent or less are known as refined waxes.

Refined waxes are harder and have greater resistance to blocking and staining than waxes with higher oil content. Waxes with an oil content up to 3.0 mass percent are generally referred to as scale waxes, and are used in applications where the slight color, odor, and taste imparted by the higher oil content can be tolerated. Semi-refined slack waxes may have oil contents up to 30 mass percent, and are used in non-critical applications. The distinction between scale and slack waxes at intermediate oil content levels (2-4 mass percent) is not clearly defined, and their suitability for particular applications depends upon properties other than oil content alone.

OSCILLATING DISC RHEOMETER (ODR)

The ODR has proved to be a useful tool in the study of rubber cross-linking kinetics. It is readily used to determine cure time. In addition, it is a useful tool in developing vulcanizates by comparing different cure package systems. The ODR should be set at an oscillation rate of 3 to 100 cpm and an arc of 3°.

A larger arc or a higher oscillation rate will cause heat buildup in the sample. When this happens, the instrument measures the curing profile at a higher temperature than that set, and the results are meaningless. It has been reported that 900 cpm can cause a temperature differential as great as 80° C). For the most accurate results, 3 cpm should be used. ODR torque is directly related to the number of cross-links formed during vulcanization.

The cross-linking half-life can be determined from the ODR plot in two ways. We examine the case of a peroxide cure. A simplified method to determine half-life is based on the concept that the time to 95% cure is 4.35 half-lives. The measurement of t_{95} is not precise, since various compounds heat up at different rates and the macro die takes longer than the micro die. In addition, a variable preheat period may be set on the ODR.

To make this method work, total heat-up time (including preheat time) must be subtracted from t_{95}. A more accurate method to determine cross-linking half-life is to plot log A Torque (Torque infinity - Torque at time t) vs time t. The slope of this straight line (m) can be used to calculate half-life. This method automatically corrects for sample heat-up time. Refer to Figures 1 and 2.

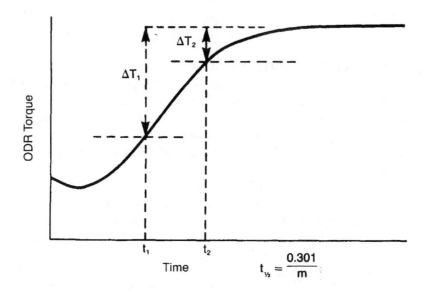

Figure 1. ODR torque rheometer curve.

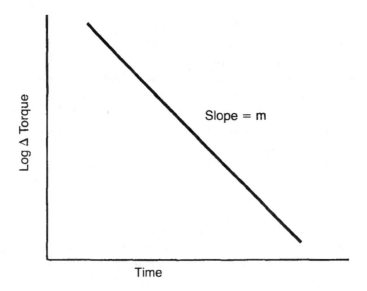

Figure 2. Analysis of the slope.

The ODR is a convenient instrument for determining the quantity of peroxide necessary to cure a compound to the same state of cure as an existing cure system does. This method is applicable only when two formulations of the same stock, differing only in cure system, are compared at the same temperature. The method is based on the fact that two compounds of the same basic recipe have equivalent torques at the same state of cure. The procedure involves the following steps:

Step 1 - Run the cure profile of the reference stock;

Step 2 - From the ODR plot, determine the optimum cure time (time B in Figure 3);

Step 3 - Run the cure profile on the compound. The stock curve must cross the reference stock curve. If the stock curve does not cross, additional curative must be added to the formulation. The exact vulcanizing agent level added to the stock must be known;

Step 4 - Determine the time, D, when the peroxide stock has the same torque, C, as the reference stock, A, cured to time B. Then determine the number of cross-linking half-lives (n) by dividing time D by the half-life;

Step 5 - Determine the percent peroxide reacted at time D from the number of half-lives, using Figure 4; % reacted = 100% - % remaining;

Step 6 - Calculate the quantity of peroxide used to time D by multiplying the percent reacted by the initial level of curative. This level of curative will give an equivalent cure when cured to a cure plateau (7 to 10 half-lives for peroxides).

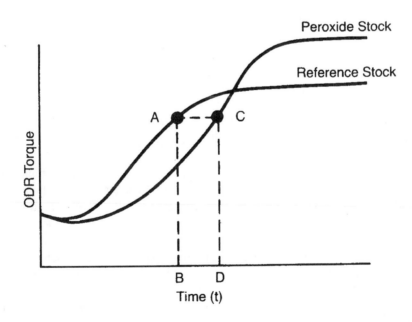

Figure 3. Curative package determination.

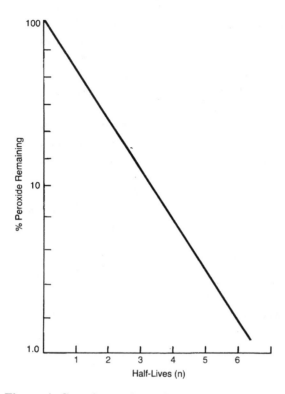

Figure 4. Curative package determination.

OZONE CRACKING

Ozone levels are dependent on the time of day and also the season. Levels reach a maximum in late morning and drop to nearly zero at night. Summer months show considerably higher ozone levels than winter months. Indoors, ozone is formed by fluorescent lighting. Only a few parts per hundred million of ozone in air can cause rubber cracking, which may destroy the usefulness of elastomer products. Ozone degradation mechanisms are illustrated in Figure 1.

Ozone will attack any elastomer with backbone unsaturation. Degradation results from the reaction of ozone with rubber double bonds. Unstretched rubber reacts with ozone but is not cracked. In this case, since only the double bonds at the surface are attacked, the degradation is confined to a thin surface layer (0.5 μm thick). Occasional "frosting" (a white or gray bloomlike appearance) may occur in transparent unstretched rubbers, but elongation is required to induce the characteristic ozone cracking.

As the stressed polymer chains cleave under ozone attack, new high-stress surface is exposed. The localized continuation of this process results in visible cracking, which is always perpendicular to the applied stress. Ozone cracking was first recognized as a serious problem after World War II. Tires on moth-balled military vehicles were found to be so badly cracked that they were unserviceable.

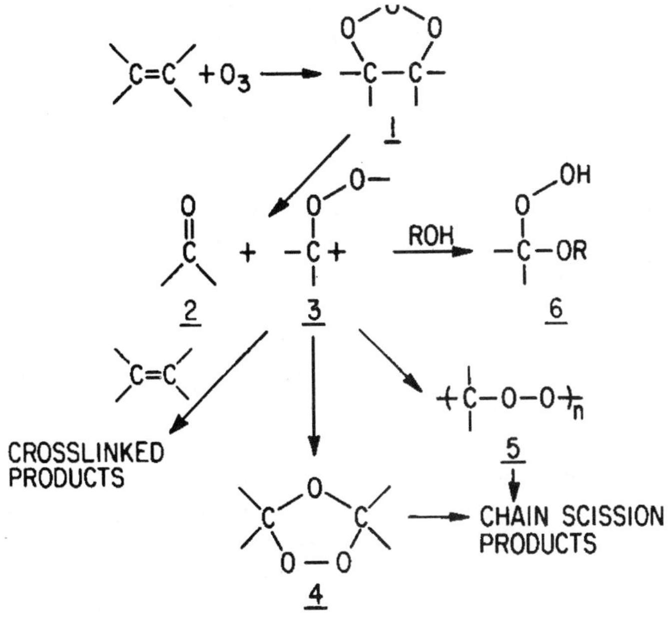

Figure 1. Ozone degradation mechanisms.

A government-sponsored research program led to the discovery of chemicals that prevented ozone cracking when added to rubber compounds. Commercial antiozonants have been available since the early 1950s. Since that time, the ozone degradation problem has worsened as atmospheric ozone concentrations have gradually increased, especially in urban industrial areas.

Ozone cracking is a physicochemical phenomenon, and many factors are involved in explaining the effect of ozone attack on elastomers. From a chemical point of view, ozone attack on olefinic double bonds causes chain scission and the formation of the decomposition products seen in the figure below. These chemical reactions are believed to be similar for both small olefins and unsaturated rubbers.

The first step is the formation of a relatively unstable primary ozonide (or molozonide), which cleaves to an aldehyde or ketone and a carbonyl oxide.

Subsequent recombination produces a secondary ozonide (or just ozonide). In small olefins, ozonide formation is generally a facile process. In stretched rubber, however, ozonide formation is more difficult, since the cleaved intermediates may be forcefully separated to relieve the stress. Interestingly, both crosslinking and chain scission products may form during rubber ozonation. Rubbers containing trisubstituted double bonds (e.g., polyisoprene, IR, and butyl rubber, IIR) are more prone to yield chain scission products. Several pathways can lead to chain scission.

Ozonides are reasonably stable in neutral environments, but they will decompose readily under the influence of heat or various reducing agents to yield such chain scission products as aldehydes, ketones, acids, and alcohols. Polymeric peroxides may be formed initially from the carbonyl oxide, but these are unstable and will eventually decompose to yield chain scission products.

The rate of chain scission is increased in the presence of active hydrogen (e.g., water), probably due to reaction with carbonyl oxides to form reactive hydroperoxides. Crosslinking products may also be formed, especially with rubbers containing disubstituted double bonds (e.g., polybutadiene, BR, and styrene-butadiene rubber, SBR).

It is proposed that this is due to attack of carbonyl oxides, in their biradical form, on the rubber double bonds. Typical diene rubbers (polyisoprene and polybutadiene) have rate constants several orders of magnitude greater than polymers having a saturated backbone (polyolefins). Other unsaturated elastomers having high reaction rates with ozone include styrene-butadiene (SBR) and acrylonitrile-butadiene (NBR) rubbers. As an example, Polychloroprene (CR) is less reactive than other diene rubbers, and it is therefore inherently more resistant to attack by ozone.

The physical aspects of ozone cracking are clearly seen in the case of the predominantly saturated butyl rubber (IIR), a copolymer of isobutylene and isoprene containing - 1-3% isoprene. Although the chemistry is the same as for its diene rubber counterpart, IR, butyl rubber is much more resistant to ozone cracking. Unsaturated elastomers must be stretched for ozone cracking to occur.

Typically, ozone cracking initiates at sites of high stress (flaws) on the rubber surface. Thus in general the rubber article should be designed to minimize potential sites of high elongation such as raised lettering. Similarly, clean molds should be used to reduce the incidence of surface flaws.

Ozone attack leads to chain scission and the formation and propagation of cracks. As crack growth proceeds, fresh surfaces are continuously exposed for further ozone attack. Degradation continues until failure relieves the inherent stress. It has been calculated that only about 1 % of all the ozone that reacts with the rubber is responsible for the formation of cracks.

The study of ozone crack growth has been an active area of research since the early 1960s. Early work showed that ozone crack growth proceeds at a linear rate for typical elastomers and is proportional to the concentration of ozone. The rates in NBR and IIR are roughly an order of magnitude less than those for NR and SBR. This is not due to low unsaturation, but rather because of high internal hysteresis. In other studies it was shown that the rate of crack growth at relatively low temperatures (near Tg) is dependent on segmental mobility, but at higher temperatures (far from Tg) segmental mobility is not an important factor.

Also, at higher temperatures the rate is not strongly dependent on the degree of unsaturation of the elastomer. In the presence of an added plasticizer (which increases chain mobility), the cut growth rates for several elastomers (NR, SBR, NBR, and IIR) are rather similar.

A key result of the early crack growth studies was the "critical stress" effect, i.e., no crack growth occurs unless a specific stress value is exceeded. In practical terms these stress values correspond to threshold tensile strains of 3-5%, depending on stiffness. It has been found that critical stress values are largely unchanged by temperature, plasticization, and ozone concentration. Polychloroprene has a higher critical stress value than other diene rubbers, consistent with its reduced reactivity to ozone.

The following is a list of commercially available antioxidants. All of these additives are in the chemical family of phenolic antioxidants.

Good-Rite
TMQ
Antioxidant 2246
Aminox
Dilaury/Thiodipropionate
Flexzone
Irganox
Agerite DPPD and HP-S, MA, Resin D
Akrochem Antioxidant 33
Vanox

Depending on the form of these antioxidants, they can be added to the polymer during synthesis. For example, in a solution polymerization process, the antioxidant can be metered into the cement (i.e., solution or slurry phase) of the polymer, prior to finishing stages. Normal practice is for the compounder to add additional or separate antioxidants during the compounding and processing stages in making the end-use product. Often relatively small quantities are required to be effective. Information on the above antioxidants can be found in the following reference: N. P. Cheremisinoff, *Hazardous Chemicals in the Polymer Industry*, Marcel Dekker, Inc., New York, 1995.

There are polymers which have high ozone resistance. ENB terpolymers are but one example. Refer to EP Rubbers for a brief discussion.

Refer to *Antiozonants*. (Source: *Handbook of Polymer Science and Technology, Volume 2: Performance Properties of Plastics and Elastomers*, N. P. Cheremisinoff, Macrel Dekker Publishers, New York, 1989).

P

PACKING GROUP

This designation has been given by the U.S. DOT and is assigned to all hazardous materials being shipped. A packing group designation defines the relative hazard of a chemical shipment. The packing group appears as an upper case Roman Numeral **I**, **II** or **III**, depending on the degree of hazard. The meanings of these designations are as follows: **I** refers to Most Hazardous (or Most Regulated); **II** refers to Moderately Hazardous (or Moderately Regulated); **III** refers to Least Hazardous (or Least Regulated). The reader should refer to Section 172.101, part f of Title 49 of the US Code of Federal Regulations (parts 100 to 177) when engaged in the shipment of hazardous materials.

PAEK (POLYARYLETHERKETONE)

The following data represent average properties of this plastic.

Specific Gravity	1.37
Shrinkage, in/in,1/8 in. thick	0.0040
Shrinkage, in/in,1/4 in. thick	0.0060
Water Absorption,% 24hrs	0.120
Mechanical Properties	
Impact, Izod, Notched (Ft-Lb/in)	0.90
Impact, Izod, Unnotched (Ft-Lb/in)	8.00
Tensile Strength (Psi)	16,000
Tensile Elongation (%)	3.000
Tensile Modulus (Psi x 10^6)	0.90
Flexual Strength (Psi)	27,000
Flexural Modulus (Psi x 10^6)	0.80
Compressive Strength (Psi)	NA
Hardness (Rockwell R)	NA
Electrical Properties	
Dielectric Strength (V/Mil)	380
Dielectric Constant (@ 1 MC dry)	NA
Dissipation Factor (@ 1 MC dry)	NA
Arc Resistance (sec)	NA
Thermal Properties	
Heat Deflection Temp 264 psi (°F)	420
Heat Deflection Temp 66 psi (°F)	NA
Thermal Expansion	NA
Thermal Conductivity	NA
Wear	
Wear Factor	NA
U L yellow card	NA

PBT (POLYBUTYLENE TEREPHTALATE)

Polybutylene terephthalate is an engineering thermoplastic manufactured by condensation polymerization of 1,4-butanediol and dimethyl terephthalate in the presence of tetrabutyl titanate. It is highly crystalline, and its rapid rate of crystallization gives short injection molding cycles. PBT has high mechanical strength, a high heat deflection temperature, low moisture absorption, good dimensional stability, and excellent electrical properties. Two types of PBT are produced, one with number average molecular weights of 23,000-30,000 and another with high molecular weights of 36,000-50,000. Both products are crystalline, melt at 224° C, and are generally processed at around 250° C. PBT is particularly amenable to reinforcement with glass. It processes well in injection molding, blow molding, or extrusion operations. Major uses include automotive exterior parts, under-the-hood parts, electrical parts, such as connectors and fuse cases, small appliances, and pump housings. The basic structure of this thermoplastic is as follows:

$$\left[CO-\bigcirc-COO\wedge\!\!\wedge\!\!\wedge O \right]_n$$

Table 1 provides some average properties of this polymer. In reviewing the general properties of this polymer, note the use of the following legend: A = amorphous - Cr = crystalline - C = clear - E = excellent - G = good - P = poor - O = opaque - T = translucent- R = Rockwell - S = Shore.

Table 1

Average Properties of PBT

STRUCTURE: Cr	FABRICATION		
	Bonding	**Ultrasonic**	**Machining**
	G	G	E
Specific Density: 1.31	**Deflection Temperature (° F)**		
	@ 66 psi: 300 @ 264 psi: 140		
Water Absorption Rate (%): 0.15	**Utilization Temperature (° F)**		
	min: -76 max: 330		
Elongation (%): 15	**Melting Point (° F)**: 430		
Tensile Strength (psi): 5600	**Coefficient of Expansion**: 0.000033		
Compression Strength (psi): 10000	**Arc Resistance**: 130		

Table 1 continued

Flexural Strength (psi): 11500	Dielectric Strength (kV/mm): 15		
Flexural Modulus (psi): 11500	Transparency: O		
Impact (Izod ft. lbs/in): 1	UV Resistance: G		
Hardness: R118	**CHEMICAL RESISTANCE**		
	Acids	**Alkalis**	**Solvents**
	G	G	G

PEEK (POLYETHERETHERKETONE)

PEEK is an important thermoplastic used in a variety of applications including automotive parts. The following data represent average properties of this plastic.

Specific Gravity	1.32
Shrinkage, in/in,1/8 in. thick	0.0100
Shrinkage, in/in,1/4 in .thick	0.0140
Water Absorption,% 24hrs	0.150
Mechanical Properties	
Impact, Izod, Notched (ft-lb/in)	1.20
Impact, Izod, Unnotched (ft-lb/in)	16.00
Tensile Strength (Psi)	13,000
Tensile Elongation (%)	150.000
Tensile Modulus (Psi x 10^6)	0.60
Flexual Strength (Psi)	17,500
Flexural Modulus (Psi x 10^6)	0.50
Compressive Strength (Psi)	NA
Hardness (Rockwell R)	NA
Electrical Properties	
Dielectric Strength (V/Mil)	400
Dielectric Constant (@ 1 MC dry)	3.40
Dissipation Factor (@ 1 MC dry)	0.004
Arc Resistance (sec)	40
Thermal Properties	
Heat Deflection Temp. @ 264 psi (°F)	340
Heat Deflection Temp. @ 66 psi (°F)	NA
Thermal Expansion (in/in/°F) x 10^{-5}	3.000
Thermal Conductivity	NA

Wear

Wear Factor	NA
U L yellow card	NA

Note - NA refers to not applicable.

PEI (POLYETHERIMIDE)

The following data represent average properties of this plastic.

Specific Gravity	1.27
Shrinkage, in/in, 1/8 in. thick	0.0050
Shrinkage, in/in, 1/4 in. thick	0.0070
Water Absorption, % 24 hrs	0.250
Mechanical Properties	
Impact, Izod, Notched (ft-lb/in)	1.00
Impact, Izod, Unnotched (ft-lb/in)	25.00
Tensile Strength (Psi)	15,000
Tensile Elongation (%)	7.500
Tensile Modulus (Psi x 10^6)	0.43
Flexural Strength (Psi)	21,000
Flexural Modulus (Psi x 10^6)	0.48
Compressive Strength (Psi)	20,000
Hardness (Rockwell R)	120.0
Electrical Properties	
Dielectric Strength (V/Mil)	500
Dielectric Constant (@ 1 MC dry)	3.20
Dissipation Factor (@ 1 MC dry)	0.004
Arc Resistance (sec)	NA
Thermal Properties	
Heat Deflection Temp. @ 264 psi (°F)	392
Heat Deflection Temp. @ 66 psi (°F)	410
Flammability	V0
Thermal Expansion (in/in/F) x 10^{-5}	3.100
Thermal Conductivity	1.5
Wear	
Wear Factor	NA
U L yellow card	NA

Note - NA refers to not applicable.

PEKEKK (POLYETHERKETONEETHERKETONEKETONE)

The following data represent average properties of this plastic.

Specific Gravity	1.37

Shrinkage, in/in,1/8 in. thick	0.0060
Shrinkage, in/in,1/4 in. thick	0.0130
Water Absorption,% 24 hrs	0.200
Mechanical Properties	
Impact, Izod, Notched (ft-lb/in)	1.90
Impact, Izod, Unnotched (ft-lb/in)	17.00
Tensile Strength (Psi)	19,000
Tensile Elongation (%)	4.500
Tensile Modulus (Psi x 10^6)	1.00
Flexural Strength (Psi)	27,500
Flexural Modulus (Psi x 10^6)	0.70
Compressive Strength (Psi)	NA
Hardness (Rockwell R)	NA
Electrical Properties	
Dielectric Strength (V/Mil)	NA
Dielectric Constant (@ 1 MC dry)	NA
Dissipation Factor (@ 1 MC dry)	NA
Arc Resistance (sec)	NA
Thermal Properties	
Heat Deflection Temp. @ 264 psi (°F)	490
Heat Deflection Temp. @ 66 psi (°F)	NA
Flammability	V0
Thermal Expansion (In/In/F) x 10^{-5}	NA
Thermal Conductivity	NA
Wear	
Wear Factor	NA
U L yellow card	NA

Note - NA refers to not applicable.

PENETROMETER

Apparatus for measuring the consistency of lubricating grease, asphalt and rubbers. A standard cone (for grease) or needle (for wax, rubber or asphalt) is lowered onto a test sample, under prescribed conditions, and the depth of penetration is measured.

PEROXIDES

Peroxides are highly efficient vulcanizing agents for elastomers and plastics. In addition they are used as polymerization catalysts in the coatings industry. These materials are commercially available in several physical forms and container types to suit the needs of users. Sulfur vulcanization of rubber was discovered in the early 1800s. The cross-linking of natural rubber with a peroxide was discovered in 1914 by a Russian, who used benzoyl peroxide. However, benzoyl peroxide vulcanizates have lower strength and poorer heat resistance than do sulfur vulcanizates. In 1950 it was found that di-tert-butyl peroxide gives better properties to vulcanizates than does benzoyl peroxide.

Di-tert-butyl peroxide is volatile, resulting in excessive loss during incorporation into the rubber and, therefore, poor efficiency. Shortly thereafter, Hercules developed Di-Cup (dicurnyl peroxide), the first commercially available peroxide to combine high efficiency and good vulcanizate properties with low cost, thus providing a peroxide cross-linker with broad-spectrum utility. This broad utility capability was extended in the middle 1950s as a result of developments leading to the cross-linking of thermoplastics such as polyethylene, which today constitutes the largest use for Di-Cup.

Widespread acceptability of a chemical cross-linker occurs if the cross-linker satisfies a number of conditions, including: it must be safe to handle, nonirritating, and nontoxic at processing temperatures; its decomposition products must be harmless; it must react to give cross-links as the only modification of the polymer; its decomposition characteristics must be such as to give a rapid cure at approximately 150° C, with no tendency to scorch; it must be nonvolatile to prevent loss during mixing; it must be soluble in rubber and plastics, and preferably solid; neither the peroxide nor its products must accelerate aging of the rubber or plastic; it must be effective in the presence of reinforcing fillers.

Commonly available peroxides that are superficially suitable for rubber and plastic vulcanization can be divided into three classes: diacyl peroxides, dialkyl peroxides, and peresters. With diacyl peroxides, only low cross-linking efficiencies are obtained, up to 10 phr (parts per hundred rubber) being required for adequate vulcanization.

A variety of dialkyl peroxides and tertiary butyl perbenzoate lead to high efficiency of cross-linking reactions in gum rubber, but only di-tertiary butyl and dicurnyl peroxides are capable of curing compounds containing reinforcing black fillers. Of these two, the former has the serious drawback of volatility, and hence very low efficiency.

A class of commercially available peroxides most completely satisfying the rest of the conditions described above, and at the lowest cost, are diarylalkyl peroxides such as Di-Cup dicumyl peroxide and VUL-CUP, α,α'-bis(tert-butylperoxy) diisopropylbenzene. Other peroxide types, at least those commercially available today, suffer from one or more of these shortcomings: high cost; low efficiency; fire or explosion hazard or both; nonuniformity and nonreproducibility of cure; low activation temperature; poor scorch resistance; high activation temperatures; long cure times.

The use of peroxide vulcanization systems has always been somewhat controversial among rubber compounders, largely based on its higher cost compared to sulfur. However, this needs to be examined on a compound by compound formulation basis. From the standpoint of thermal stability, the peroxide crosslink has a bond energy of about 82 kilocalories and is as stable as any of the carbon-carbon bonds in the polymer backbone. In contrast, the sulfur cross-link, composed of both carbon-sulfur and sulfur-sulfur bonds, has a lower bond energy (C-S, 66 kilocalories; S-S, 49 kilocalories), and hence results in a weaker crosslink. Peroxide vulcanizates exhibit little or no added color, whereas sulfur systems form dark metallic sulfides.

Peroxide-induced cross-links are stable to oxidation, whereas sulfur cross-links may oxidize, leading to cross-link rupture. Peroxide-cured vulcanizates exhibit higher compression set resistance than their counterpart sulfur-cured vulcanizates. Peroxide vulcanizates generally have better low-temperature flexibility than do sulfur vulcanizates.

Sulfur vulcanizates tend to show equal or greater tensile strength than do peroxide vulcanizates. Also, sulfur vulcanizates tend to have better abrasion resistance and tear strength than do peroxide vulcanizates.

Note that in general, peroxides are a group of compounds with the generic formula R-O-O-R'. All peroxides are hazardous materials. The organic peroxides in particular are considered extremely hazardous. Most peroxides are water reactive. The user should always refer to the MSDS (Material Safety Data Sheet) when handling these materials. General hazardous properties may be found in the following reference: N. P. Cheremisinoff, Handbook of Industrial Toxicology ad Hazardous Materials, Marcel Dekker Inc, New York (1999). Refer to *Peroxide Cross-linking*.

PEROXIDE CROSS-LINKING

Common polymers that can be peroxide cross-linked include:
NR - natural rubber
SBR - styrene-butadiene rubber
BR - polybutadiene rubber
IR - polystyrene rubber
NBR - nitrile rubber
CR - neoprene
EPM - ethylene-propylene copolymer
EPDM - ethylene-propylene terpolymer
Si, VSi - silicone rubber
Acrylic rubber
Polyester
AU, AE - polyurethane
ABS - acrylonitrile-butadiene-styrene
PE - polyethylene
CSM - chlorosulfonated polyethylene
CM - chlorinated polyethylene
Polymers that cannot be peroxide-crosslinked include:
IIR - butyl rubber
Polyisobutylene
CO, ECO - polyepichlorohydrin rubber

PP - polypropylene
PVC - polyvinyl chloride

Figure 1 lists the basic reactions involved in peroxide vulcanization. The comparative ease with which hydrogens are abstracted by oxy radicals is essentially a function of the hydrogen atom's reactivity. The figure lists hydrogen atom functionality in order of descending reactivity. Figure 2 provides a summary of reactions which are detrimental to the basic polymer cross-linking step. Refer to the subject entries *Peroxides*, *Half-life*, *Vulcanization*, *Vulcanizing Agents*, and *Cure Time*.

The following is a partial list of organic peroxides that are key vulcanizing agents:

Di-Cup R
Di-Cup C
Di-Cup 40kE
Dicumyl Peroxide (Di-Cup)

Specific information and applications of these curatives can be found in the following reference: N. P. Cheremisinoff, *Hazardous Chemicals in the Polymer Industry*, Marcel Dekker, Inc., New York, 1995. Note that organic accelerators are most effective when in combination with zinc oxide. Refer to *Vulcanization*, *Half-Life*, and *Sulfur Vulcanization*.

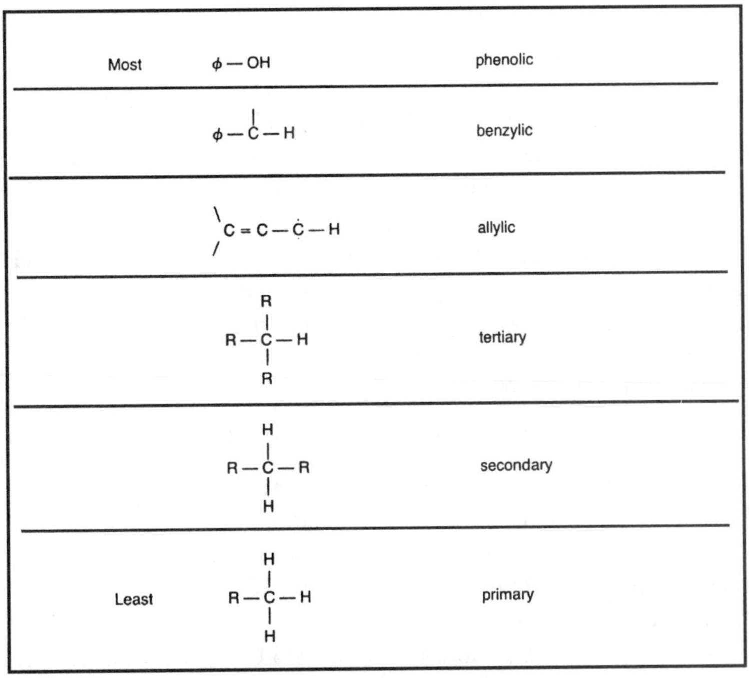

Figure 1. Reactions of peroxide vulcanization.

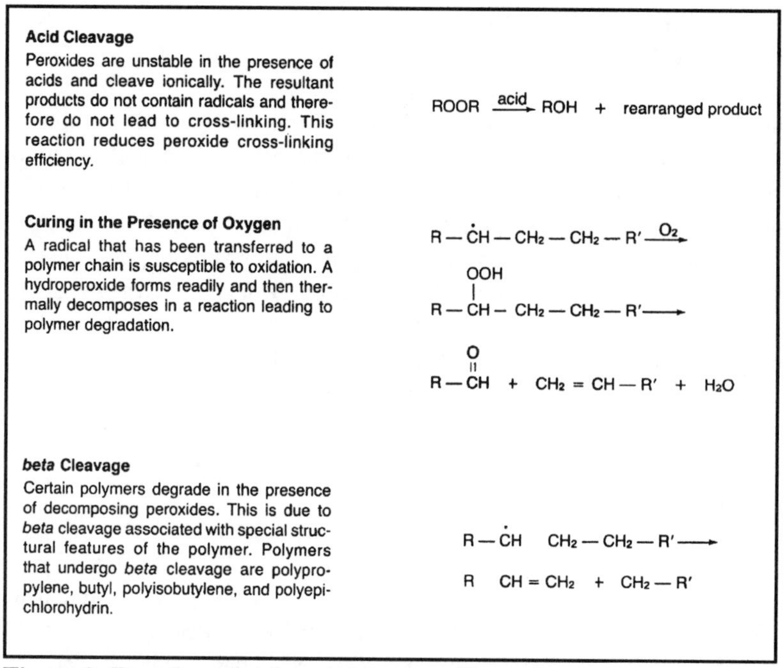

Acid Cleavage

Peroxides are unstable in the presence of acids and cleave ionically. The resultant products do not contain radicals and therefore do not lead to cross-linking. This reaction reduces peroxide cross-linking efficiency.

$$ROOR \xrightarrow{\text{acid}} ROH + \text{rearranged product}$$

Curing in the Presence of Oxygen

A radical that has been transferred to a polymer chain is susceptible to oxidation. A hydroperoxide forms readily and then thermally decomposes in a reaction leading to polymer degradation.

$$R - \overset{\bullet}{C}H - CH_2 - CH_2 - R' \xrightarrow{O_2}$$

$$R - \overset{\overset{OOH}{|}}{C}H - CH_2 - CH_2 - R' \longrightarrow$$

$$R - \overset{\overset{O}{\|}}{C}H + CH_2 = CH - R' + H_2O$$

beta **Cleavage**

Certain polymers degrade in the presence of decomposing peroxides. This is due to *beta* cleavage associated with special structural features of the polymer. Polymers that undergo *beta* cleavage are polypropylene, butyl, polyisobutylene, and polyepichlorohydrin.

$$R - \overset{\bullet}{C}H \quad CH_2 - CH_2 - R' \longrightarrow$$

$$R \quad CH = CH_2 + CH_2 - R'$$

Figure 2. Reactions detrimental to peroxide curing.

PERSONAL PROTECTIVE EQUIPMENT

The items are those recommended by (a) manufacturers, either in technical bulletins or in Material Safety Data Sheets, (b) the Manufacturing Chemists Association, or (c) the National Safety Council, for use by personnel while responding to fire or accidental discharge of the chemical. They are intended to protect the lungs, eyes, and skin. There are many plastics and rubber compounding ingredients, as well as the monomers themselves, which are toxic and pose fire and explosion hazards. The user must consult a Material Safety Data Sheet (MSDS) for the specific properties and safety hazards associated with the chemicals handled. An MSDS will provide specific recommendations on the required personal protective equipment that should be used when handling a certain chemical.

PES (POLYETHERSULFONE)

The following data represent average properties of this plastic.

Specific Gravity	1.37
Shrinkage, in/in, 1/8 in. thick	0.0070
Shrinkage, in/in, 1/4 in. thick	0.0080
Water Absorption, % 24 hrs	0.430
Mechanical Properties	
Impact, Izod, Notched (ft-lb/in)	1.60
Impact, Izod, Unnotched (ft-lb/in)	12.00

Tensile Strength (Psi)	12,000
Tensile Elongation (%)	50.000
Tensile Modulus (Psi x 10^6)	0.35
Flexural Strength (Psi)	18,500
Flexural Modulus (Psi x 10^6)	0.37
Compressive Strength (Psi)	15,000
Hardness (Rockwell R)	120.0
Electrical Properties	
Dielectric Strength (V/Mil)	400
Dielectric Constant (@ 1 MC dry)	3.50
Dissipation Factor (@ 1 MC dry)	0.006
Arc Resistance (sec)	70
Thermal Properties	
Heat Deflection Temp. @ 264 psi (°F)	398
Heat Deflection Temp. @ 66 psi (°F)	405
Flammability	V0
Thermal Expansion (In/In/F) x 10^{-5}	3.100
Thermal Conductivity	1.9
Wear	
Wear Factor	NA
U L yellow card	NO

PET (POLYETHYLENE TEREPHTALATE)

Poly(ethylene terephthalate) (PET) is semicrystalline polyester commonly used in packaging and fiber applications. Some typical properties are: Glass Transition Temperature, Tg = 76°C; Tm = 250°C; Amorphous density at 25°C = 1.33 g/cc; Crystallinity at 25° C = 1.50 g/cc; Molecular weight of repeat unit = 192.2 g/mole. As noted, polyethylene terephthalate is widely used in synthetic fibers and is also used in film and molding applications. Crystallization of PET can be achieved upon heating to 190° C and orientation. In thin-film applications, transparency is achieved by rapid quenching with a water-cooled roll. Highly crystalline PET with 30% glass reinforcement is widely used as a molding resin. This material has a high heat deflection temperature (227° C at 264 psi) because the melting point of PET crystallites is 20° C higher than that of PBT crystallites. The melting point of commercial PET is 265° C; highly crystalline PET melts at 271° C. The structural formula for this polymer is as follows:

The production of polyethylene terephthalate is conducted in two steps. Dimethyl terephthalate is heated with ethylene glycol to give a mixture consisting of dihydroxyethyl terephthalate and higher oligomers. Further heating to 270° C under vacuum in the presence of a catalyst produces the final polymer. Refer to the following reaction scheme:

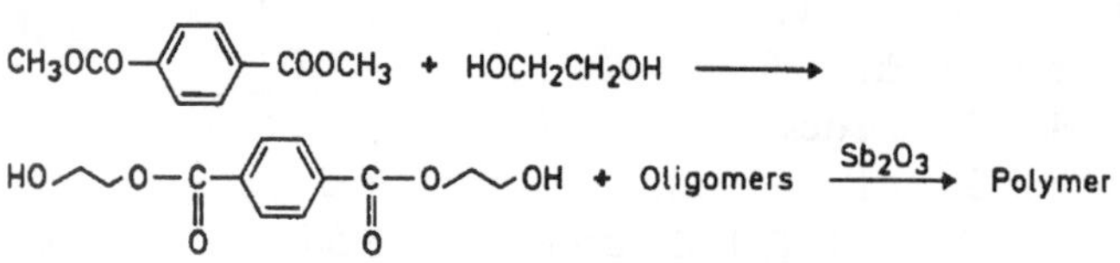

Terephthalic acid is obtained by air oxidation of p-xylene, and ethylene glycol is produced from ethylene oxide and water. The strength of PET in its oriented form is outstanding. Biaxially oriented PET film is used in magnetic tape in X-ray and other photographic film applications, in electrical insulation, and in food packaging, including boil-in-bag food pouches. Production of PET bottles for carbonated beverages by blow molding has gained prominence because PET has low permeability to carbon dioxide. Because of its excellent thermal stability it is also used as coating material for microwave and conventional ovens. Table 1 provides average property values reported by suppliers. In reviewing the general properties of this polymer provided in the table below, note the use of the following legend: A = amorphous - Cr = crystalline - C = clear - E = excellent - G = good - P = poor - O = opaque - T = translucent- R = Rockwell - S = Shore.

Table 1

Average Properties of PET

STRUCTURE: Cr	FABRICATION		
	Bonding	**Ultrasonic**	**Machining**
	G	G	E
Specific Density: 1.37	**Deflection Temperature (° F)** @ 66 psi: 330 @ 264 psi: 212		
Water Absorption Rate (%): 0.15	**Utilization Temperature (° F)** min: -4 max: 212		
Elongation (%): 70	Melting Point (° F): 480		
Tensile Strength (psi): 6600	Coefficient of Expansion: 0.000039		
Compression Strength (psi): 14000	Arc Resistance: 80		
Flexural Strength (psi): 16000	Dielectric Strength (kV/mm): 20		

Table 1 continued

Flexural Modulus (psi): 400000	Transparency: T		
Impact (Izod ft. lbs/in): 0.8	UV Resistance: G		
Hardness: R120	**CHEMICAL RESISTANCE**		
	Acids	Alkalis	Solvents
	G	G	G

PETG (POLYETHYLENE TEREPHTALATE GLYCOL)

Table 1 provides average properties of this polymer. In reviewing the general properties of this polymer, note the use of the following legend: A = amorphous - Cr = crystalline - C = clear - E = excellent - G = good - P = poor - O = opaque - T = translucent- R = Rockwell - S = Shore.

Table 1

Average Properties of PETG

STRUCTURE: A	FABRICATION		
	Bonding	Ultrasonic	Machining
	G	G	E
Specific Density: 1.27	Deflection Temperature (° F) @ 66 psi: 159 @ 264 psi: 245		
Water Absorption Rate (%): 0.13	Utilization Temperature (° F) min: -40 max: 140		
Elongation (%): 110	Melting Point (° F): 190		
Tensile Strength (psi): 4000	Coefficient of Expansion: 0.000051		
Compression Strength (psi):	Arc Resistance: ND		
Flexural Strength (psi): 10000	Dielectric Strength (kV/mm): 18		
Flexural Modulus (psi): 300000	Transparency: C		
IMPACT (IZOD ft. lbs/in): 1.9	UV Resistance: P		
HARDNESS: R105	**CHEMICAL RESISTANCE**		
	Acids	Alkalis	Solvents
	G	G	P

PHENOLICS

The phenolic-based resin compounds are among the most widely used and well-known thermosets. Phenols react with aldehydes to give condensation products if there are free positions in the benzene ring - *ortho* and *para* to the hydroxyl group. Formaldehyde is the most reactive and is used almost exclusively in commercial production. The reaction is always catalyzed, either by acids or bases. The nature of the product greatly depends on the type of catalyst and the mole ratio of the reactants. Refer to *Novolac Resins*.

PIPE OR TUBE EXTRUSION

The die configuration used for the extrusion of pipe or tubing consists of a die body with a tapered mandrel and an outer die ring which control the dimensions of the inner and outer diameters, respectively. The process involves thicker walls than are involved in blown-film extrusion. As such, it is advantageous to cool the extrudate by circulating water through the mandrel as well as by running the extrudate through a water bath. The extrusion of rubber tubing differs from thermoplastic tubing. For thermoplastic tubing, dimensional stability results from cooling below Tg (glass transition temperature) or Tm (melting temperature).

The rubber tubing gains dimensional stability due to a cross-linking reaction at a temperature above that in the extruder. The high melt viscosity of the rubber being extruded ensures a constant shape during the cross-linking. Refer to Figure 1. (Source: Chanda, M. and S. K. Roy, Plastics Technology Handbook, Marcel Dekker, Inc., New York, NY, 1987). See also *Plunger-TypeTransfer Molding, Compression Molding and Screw Transfer Molding, Extruders and Extrusion.*

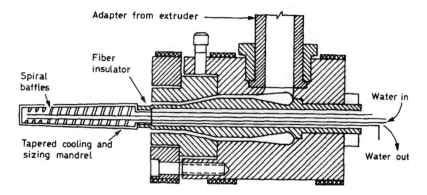

Figure 1. Pipe or tubing extrusion die configuration.

PLASMA POLYMERIZATION

Thin polymer films with unique chemical and physical properties can be produced by this technique. Films are prepared by vapor-phase deposition and can be formed on any substrate with good adhesion between the film and substrate. The thin polymer films usually highly cross-linked and pinhole free, providing good barrier properties. These films are finding great potential in biomaterials applications and in the microelectronics industry.

PLASTICS TECHNICAL EVALUATION CENTER (PLASTEC)

PLASTEC is one of 20 information analysis centers sponsored by the Department of Defense to provide the defense community with a variety of technical information services applicable to plastics, adhesives, and organic matrix composites. For the last 21 years, PLASTEC has served the defense community with authoritative information and advice in such forms as engineering assistance, responses to technical inquiries, special investigations, field troubleshooting, failure analysis, literature searches, state-of-the-art reports, data compilations, and handbooks.

PLASTEC has also been heavily involved in standardization activities. In recent years, PLASTEC has been permitted to serve private industry. The significant difference between a library and technical evaluation center is the quality of the information provided to the user. PLASTEC uses its database library as a means to an end to provide succinct and timely information which has been carefully evaluated and analyzed. Examples of the activity include recommendation of materials, counseling on designs, and performing trade-off studies between various materials, performance requirements, and costs. Applications are examined consistent with current manufacturing capabilities, and the market availability of new and old materials alike is considered. PLASTEC specialists can reduce raw data to the user's specifications and supplement them with unpublished information that updates and refines published data.

PLASTEC works to spin-off the results of government-sponsored R & D to industry and similarly to utilize commercial advancements to the government's goal of highly sought technology transfer. PLASTEC has a highly specialized library to serve the varied needs of their own staff and customers. PLASTEC offers a great deal of information and assistance to the design engineer in the area of specifications and standards on plastics.

PLASTEC has a complete visual search microfilm file and can display and print the latest issues of specifications, test methods, and standards from Great Britain, Germany, Japan, U.S., and International Standards Organization. Military and Federal specifications and standards and industry standards such as ASTM, NEMA, and UL are on file and can be quickly retrieved.

PLASTICITY

The property of an apparently solid material that enables it to be permanently deformed under the application of force, without rupture. (Plastic flow differs from fluid flow in that the *shear stress* must exceed a *yield point* before any flow occurs.)

PLASTICIZERS

Plasticizers are used to change the glass transition temperature (Tg) of a polymer. Polyvinyl chloride (PVC), for instance, is often mixed with nonvolatile liquids for this reason. Vinyl siding used on homes requires an unplasticized, rigid PVC with a Tg of 85° to 90° C (185° to 195° F). A PVC garden hose, on the other hand, should remain flexible even at 0° C. A mixture of 30 parts di(2-ethylhexyl) phthalate (also called dioctyl phthalate, or DOP) with 70 parts PVC will have a Tg of about -10° C, making it suitable for use as a garden hose. Although other polymers can be plasticized, PVC is unique in accepting and retaining plasticizers of widely varying chemical composition and molecular size. The plasticizer may also change the flammability, odor, biodegradability, and cost of the finished product.

PLASTISOLS AND ORGANOSOLS

Coating materials composed of *resins* suspended in a hydrocarbon liquid. An organosol is plastisol with an added solvent, which swells the resin particles, thereby increasing viscosity. Applications include spray coating, dipping, and coatings for aluminum, fabrics, and paper.

PLATE-AND-FRAME HEAT EXCHANGER

The plate-and-frame heat exchanger has emerged as a viable alternative to shell-and-tube exchangers for many applications throughout the chemical process industries. Such units are comprised of a series of plates, mounted in a frame and clamped together. Space between adjacent plates form flow channels, and the system is arranged so that hot and cold fluids enter and exit through flow channels at the four corners, as illustrated in Figure 1. Within the exchanger, an alternating gasket arrangement diverts the hot and cold fluids from each inlet into an alternating sequence of flow channels. In this arrangement, each cell of heat transfer media is separated by a thin metal wall, allowing heat to transfer easily from one media to the other. The plate-and-frame's highly efficient countercurrent flow typically yields heat transfer coefficients three to five times greater than other types of heat exchangers. As a result, a more-compact design is possible for a given heat-exchange capacity, relative to other exchanger configurations. In this design, a corrugated chevron or herringbone pattern is pressed into each plate for several reasons. First, the pattern gives the entire exchanger strength and rigidity. It also extends the effective surface area of plates and increases turbulence in the flow channels. Together, these effects boost heat transfer. Depending on the applications, plate selection is optimized to yield the fewest total number of channel plates. Because the plates can be easily removed, service and maintenance costs are typically lower than that of shell-and-tube exchangers.

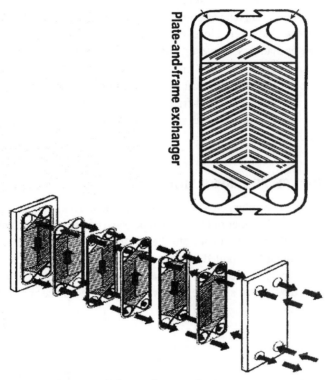

Figure 1. Plate-and-frame heat exchanger.

The following selection criteria should be reconciled:

- maximum design or working pressure is limited to 300 psi;
- temperature limits and fluids must be compatible with gasket materials;
- plate materials must be compatible with process media;
- the narrow passageways in the plate-and-frame can cause high pressure drops, making the exchanger incompatible with low-pressure, high-volume gas applications;
- rapid fluctuations in steam pressures and temperatures can be detrimental to gasket life. For this reason, applications that use steam favor shell-and-tube exchangers;

- in applications where process media contain particulate matter, or when large amounts of scaling can occur, careful consideration should be given to the free channel space between adjacent plates;

- the plate-and-frame design is best suited for applications with a large temperature cross or small temperature approach.

The temperature approach is the difference between the inlet temperature of the cold fluid and the outlet temperature of the hot fluid. Certain exchanger designs operate better at different temperature approaches. Plate-and-frame exchangers, for example, work well at a very close temperature approach, on the order of AF. For shell-and-tube exchangers, however, the lowest possible temperature approach is 10-ME. As for cleanliness of the process fluids, shell-and-tube exchangers have tube diameters that can accommodate a certain amount of particulate matter without clogging or fouling. Plate-and-frame exchangers, however, have narrow passageways, making them more susceptible to damage from precipitation or particulate fouling.

PLUNGER-TYPE TRANSFER MOLDING

Figure 1 illustrates plunger-type transfer molding (also known as *auxiliary raw transfer molding*). The taper of the sprue in pot-type transfer is such that, when the mold is opened, the sprue remains attached to the disc of material left in the pot, known as cull, and is thus pulled away from the molded part, whereas the latter is lifted out of the cavity by the ejector pins (left-most figure). In plunger-type transfer molding, the cull and the sprue remain with the molded piece when the mold is opened.

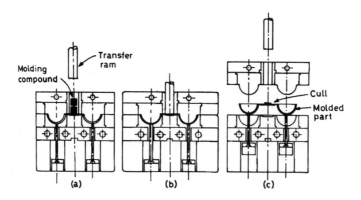

Figure 1. The molding cycle of a plunger-type transfer mold.

Moving from left to right in Figure 1, an auxiliary ram exerts pressure on the heat-softened material in the pot and forces it into the mold. When the mold is opened (middle figure), the cull and sprue remain with the molded piece. (Source: Chanda, M. and S. K. Roy, Plastics Technology Handbook, Marcel Dekker, Inc., New York, NY, 1987). See also *Plunger-Type Transfer Molding, Compression Molding and Screw Transfer Molding.*

PMMA (POLYMETHYLMETHACRYLATE (ACRYLIC))

PMMA is characterized by crystal clear transparency, unexcelled weatherability, and a useful combination of stiffness, density, and moderate toughness. The heat deflection temperatures range from 165 to 212° F, with a service temperature of 200° F. Improvement of the mechanical properties of PMMA can be achieved by orientation of heat-cast sheets. The structural formula for PMMA is as follows:

$$\left[CH_2 - \underset{\underset{COOCH_3}{|}}{\overset{\overset{CH_3}{|}}{C}} \right]_n$$

Polymethyl methacrylate can be modified by copolymerization of methyl methacrylate with other monomers, such as acrylates, acrylonitrile, styrene, and butadiene. Blending with SBR improves impact resistance. The 2-hydroxyethyl ester of methacrylic acid is used as the monomer of choice for the manufacture of soft contact lenses. Copolymerization with ethylene glycol dimethacrylate produces a hydrophilic network polymer, also called hydrogel. Hydrogel polymers are glassy and brittle when dry but become soft and plastic after swelling in water. The most important properties of a hydrogel are its equilibrium water content and oxygen permeability.   The following are glass transition temperatures of several polymethacrylates, which provide a strong argument for its broad application as an engineering material.

Ester Group	Tg (°C)
Methyl	105
Ethyl	65
n-Butyl	20
n-Decyl	-70
n-Hexadecyl	-9

Table 1 provides average properties of PMMA. In reviewing the general properties of this polymer, note the use of the following legend: A = amorphous - Cr = crystalline - C = clear - E = excellent - G = good - P = poor - O = opaque - T = translucent- R = Rockwell - S = Shore.

Table 1. Average Properties of PMMA

STRUCTURE: A	FABRICATION		
	Bonding	**Ultrasonic**	**Machining**
	E	E	E
Specific Density: 1.16	**Deflection Temperature (° F)** @ 66 psi: 187 @ 264 psi: 170		
Water Absorption Rate (%): 0.3	**Utilization Temperature (° F)** min: -94 max: 176		
Elongation (%): 48	Melting Point (° F): 212		
Tensile Strength (psi): 7000	Coefficient of Expansion: 0.000055		
Compression Strength (psi): 11500	Arc Resistance: 80		
Flexural Strength (psi): 10500	Dielectric Strength (kV/mm): 17		
Flexural Modulus (psi): 310000	Transparency: C		
Impact (Izod ft. lbs/in): 1.1	UV Resistance: G		
Hardness: R120	**CHEMICAL RESISTANCE**		
	Acids	**Alkalis**	**Solvents**
	G	E	P

POLAR COMPOUND

A chemical compound whose molecules exhibit electrically positive characteristics at one extremity and negative characteristics at the other. Polar compounds are used as additives in many petroleum products. Polarity gives certain molecules a strong affinity for solid surfaces; as lubricant additives (oiliness agents), such molecules plate out to form a tenacious, friction-reducing film. Some polar molecules are oil-soluble at one end and water-soluble at the other end; in lubricants, they act as emulsifiers, helping to form stable oil-water emulsions. Such lubricants are said to have good metal-wetting properties. Polar compounds with a strong attraction for solid contaminants act as detergents in engine oils by keeping contaminants finely dispersed.

POLYACRYLONITRILE

Polyacrylonitrile is most commonly used in fiber form. Since it softens only slightly below its thermal degradation temperature, it must be processed by wet or dry spinning rather than melt spinning. Some typical properties are: Glass Transition Temperature, Tg = 85° C; Melting Temperature, Tm = 317° C; Amorphous density at 25° C = 1.184 g/cc; Molecular weight of repeat unit = 53.06 g/mole.

POLYAMIDE

The commercial development of polyamides marks one of the most historic developments in polymer engineering. Commercialization of polyamides started with Carothers in the early 1930s when Hill, a coworker of Carothers, stuck a glass rod into the melt of a polyamide obtained by melt condensation of the salt of a dicarboxylic acid and a diamine (AABB- type polycondensation), and pulled out some fibers. In 1938 DuPont started the manufacture of nylon-6,6. In a parallel development, Schlack in Germany developed AB-type polyamides by homopolymerization of cyclic lactams, and nylon-6, derived from caprolactam was introduced in 1939. New melt-spinning processes were developed, and the golden age of synthetic fibers began. Nylon's first target was the silk hosiery market. After overwhelming acceptance of nylon hosiery, the tricot and home furnishings markets were penetrated mainly because of nylon's excellent knittability and durability. Industrial applications, especially in carpeting quickly followed. Aromatic nylons were commercially introduced in 1961. New families of nylon molding compounds have been introduced, and the domination of the nylon market by the classical six-carbon derived products is challenged by the many new products introduced in the 1980s and 1990s. Major end uses of the nylon fibers are in home furnishings, apparel, and tire cord.

The general structure of aliphatic polyamides is as follows:

$$-\left[NH(CH_2)_6 NH - \overset{O}{\underset{\|}{C}} - (CH_2)_4 - \overset{O}{\underset{\|}{C}} \right]_n - \qquad -\left[NH(CH_2)_5 - \overset{O}{\underset{\|}{C}} \right]_n -$$

| Nylon 6,6 | Nylon 6 |

The monomers used in the production of these materials are adipic acid, hexamethylenediamine and caprolactam. Polymerization is traditionally done via bulk polycondensation. The commercial success of nylon-6 and nylon-6,6 is due to outstanding properties and an economically attractive raw material base. Nylon-6,6 is produced by melt condensation of adipic acid with hexamethylenediamine. Both monomers contain six carbon atoms (nylon-6,6) and therefore are obtainable from either benzene or other petrochemical products. Adipic acid is produced by air oxidation of cyclohexane to give a mixture of cyclohexanol and cyclohexanone, followed by further oxidation with nitric acid. Hexamethylenediamine is produced by the reduction of adiponitrile, obtained from butadiene and hydrogen cyanide, and by the oxidative coupling of acrylonitrile. However, not all caprolactam is made from cyclohexane; phenol accounts for nearly half of caprolactam as the alternative raw material.

Two basic types of condensation polymers are possible. Nylon-6,6 is the typical example of the polycondensation of two bifunctional monomers, adipic acid and hexamethylenediamine. This type of polycondensation is generally referred to as the AABB reaction. In contrast, polycondensation of one difunctional monomer is referred to as an AB reaction. The AABB-type polycondensation has the advantage that the stoichiometry is fixed by salt formation. In contrast, AB polyamides are usually prepared by ring-opening polymerization of a cyclic lactam. This reaction requires only a catalytic amount of water, because the water required for ring opening is replenished by the simultaneously occurring polycondensation:

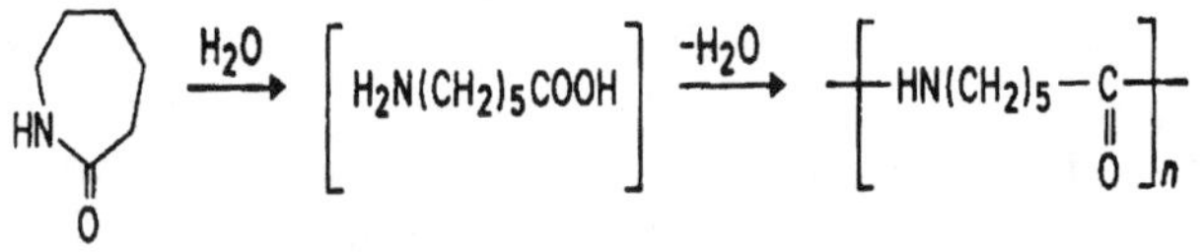

In addition to nylon-6,6, other AABB products are nylon-6,10 and nylon 6,12. Both are based on hexamethylenediamine but use longer chain aliphatic dicarboxylic acids (senacic acid and dodecanoic acid, respectively).

Aromatic polyamides were introduced early on because of their superior heat resistance. The early versions of these polymers were produced from isophthaloyl chloride and m-phenylenediamine. These are known as Nomex fibers. The general structure is as follows:

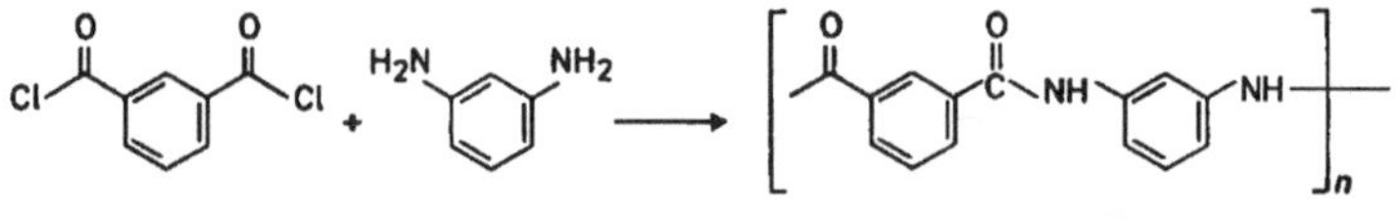

Nomex

The chemistry of these polymers was extended to terphthaloyl chloride and p-phenylenediamine, resulting in the commercialization of Kevlar fiber. Kevlar (structure shown below) is ideally suited for ballistic vests and high impact helmets because of its strength and the fact that it is as strong as steel but at one-fifth the weight. Fibers have a glass transition temperature above 300° C and can be heated without decomposing at temperatures as high as 500° C. Kevlar must be solvent spun from concentrated sulfuric acid since it has poor solubility in many conventional solvents.

The properties of synthetic fibers range from low-modulus, high-elongation fibers like Lycra to high-modulus, high-tenacity fibers such as Kevlar. In between, almost any combination of properties can be built into the macromolecules. A breakthrough in fiber strength and stiffness was achieved with Kevlar and graphite fibers. These fibers have created superior composite materials, generally referred to as fiber-reinforced plastics (FRPs). Their low density, high specific strength (strength/density), and high stiffness (modulus/density) make them useful as metal replacement materials in some engineering applications. Glass fiber reinforced boats and other recreation vehicles are examples. Still other examples are fishing rods which are filament wound with graphite and Kevlar fibers, and golf club shafts, tennis rackets, skis, ship masts, and various other products.

Another important group are the polyamide imides and polyimides. The substitution of tricarboxylic acid anhydride for dicarboxylic acids during synthesis results in thermally more stable polyamide imides. The polycondensation reaction proceeds in two steps, with the solution of the intermediate polycarboxylic acid used in applications where the final product can be generated by a thermal or baking process. Uses include heat-resistant wire and cable coatings. The basic synthesis of polyamide imides is as follows:

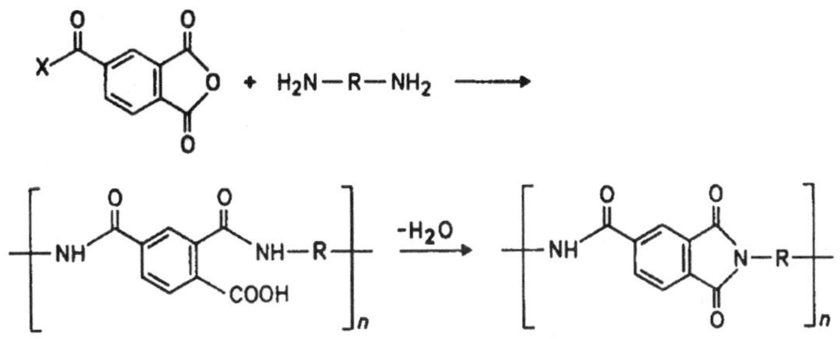

The extension of this reaction to tetrafunctional monomers (e.g., pyromelittic acid anhydride or benzophenone tetracarboxylic acid anhydride), results in the formation of polyimides.

PA 6 (Nylon-6)

Table 1 provides average properties on Nylon-6. In reviewing the general properties of this polymer, note the use of the following legend: A = amorphous - Cr = crystalline - C = clear - E = excellent - G = good - P = poor - O = opaque - T = translucent- R = Rockwell - S = Shore.

Table 1

Average Properties of Nylon-6

STRUCTURE: Cr	FABRICATION		
	Bonding	**Ultrasonic**	**Machining**
	G	P	P
Specific Density: 1.13	**Deflection Temperature (° F)** @ 66 psi: 350 @ 264 psi: 190		
Water Absorption Rate (%): 1.8	**Utilization Temperature (° F)** min: -94 max: 210		
Elongation (%): 200	**Melting Point (° F): 420**		
Tensile Strength (psi): 5800	**Coefficient of Expansion: 0.000044**		
Compression Strength (psi): 7000	**Arc Resistance: 130**		
Flexural Strength (psi): 5000	**Dielectric Strength (kV/mm): 17**		
Flexural Modulus (psi): 170000	**Transparency: T**		
Impact (Izod ft. lbs/in): 3	**UV Resistance: G**		
Hardness: R92	**CHEMICAL RESISTANCE**		
	Acids	**Alkalis**	**Solvents**
	P	P	G

PA 66 (Nylon-66)

Table 2 provides average properties on Nylon-66. In reviewing the general properties of this polymer, note the use of the following legend: A = amorphous - Cr = crystalline - C = clear - E = excellent - G = good - P = poor - O = opaque - T = translucent- R = Rockwell - S = Shore.

Table 2

Average Properties of Nylon-66

STRUCTURE: Cr	FABRICATION		
	Bonding	**Ultrasonic**	**Machining**
	G	P	G
Specific Density: 1.14	**Deflection Temperature (° F)** @ 66 psi: 390 @ 264 psi: 250		

Table 2 continued

Water Absorbtion Rate (%): 1.3	Utilization Temperature (° F) min: -110 max: 250		
Elongation (%): 150	Melting Point (° F): 490		
Tensile Strength (psi): 7600	Coefficient of Expansion: 0.000044		
Compression Strength (psi): 8700	Arc Resistance: 130		
Flexural Strength (psi): 5800	Dielectric Strength (kV/mm): 18		
Flexural Modulus (psi): 200000	Transparency: T		
Impact (Izod ft. lbs/in): 2.1	UV Resistance: G		
Hardness: R100	**CHEMICAL RESISTANCE**		
	Acids	Alkalis	Solvents
	P	P	G

PA 11 - (Nylon-11)

Table 3 provides average properties on Nylon-11. In reviewing the general properties of this polymer, note the use of the following legend: A = amorphous - Cr = crystalline - C = clear - E = excellent - G = good - P = poor - O = opaque - T = translucent- R = Rockwell - S = Shore.

Table 3

Average Properties of Nylon-11

STRUCTURE: Cr	FABRICATION		
	Bonding	Ultrasonic	Machining
	G	P	P
Specific Density: 1.04	Deflection Temperature (° F) @ 66 psi: 266 @ 264 psi: 302		
Water Absorbtion Rate (%): 0.4	UTILIZATION TEMPERATURE (° F) min: -94 max: 176		
Elongation (%): 300	Melting Point (° F): 365		
Tensile Strength (psi): 8000	Coefficient of Expansion: 0.000072		
Compression Strength (psi): 7900	Arc Resistance: 120		

Table 3 continued

Flexural Strength (psi): 6000	Dielectric Strength (kV/mm): 17
Flexural Modulus (psi): 145000	Transparency: O
Impact (Izod ft. lbs/in): 1.8	UV Resistance: G
Hardness: R108	**CHEMICAL RESISTANCE**

	Acids	Alkalis	Solvents
	P	P	G

PA12G (Nylon-12, 30% Glass Filled)

Table 4 provides average properties of glass filled polyamide. In reviewing the general properties of this polymer, note the use of the following legend: A = amorphous - Cr = crystalline - C = clear - E = excellent - G = good - P = poor - O = opaque - T = translucent- R = Rockwell - S = Shore.

Table 4. Average Properties of Nylon-12, 30% Glass Filled

STRUCTURE: Cr	FABRICATION		
	Bonding	**Ultrasonic**	**Machining**
	G	G	P
Specific Density: 1.23	**Deflection Temperature (° F)** @ 66 psi: 230 @ 264 psi: 300		
Water Absorption Rate (%): 0.3	**Utilization Temperature (° F)** min: -110 max: 230		
Elongation (%): 5	Melting Point (° F): 350		
Tensile Strength (psi): 13500	Coefficient of Expansion: 0.000016		
Compression Strength (psi): 16000	Arc Resistance: 120		
Flexural Strength (psi): 4400	Dielectric Strength (kV/mm): 17		
Flexural Modulus (psi): 360000	Transparency: O		
Impact (Izod ft. lbs/in): 4.4	UV Resistance: G		
Hardness: R110	**CHEMICAL RESISTANCE**		
	Acids	**Alkalis**	**Solvents**
	P	P	G

PA 66M (Nylon-66, 40% Mineral Filled)

Table 5 provides average properties of mineral filled polyamide. In reviewing the general properties of this polymer, note the use of the following legend: A = amorphous - Cr = crystalline - C = clear - E = excellent - G = good - P = poor - O = opaque - T = translucent- R = Rockwell - S = Shore.

Table 5

Average Properties of Nylon-66, 40% Mineral Filled

STRUCTURE: Cr	FABRICATION		
	Bonding	Ultrasonic	Machining
	G	G	P
Specific Density: 1.50	Deflection Temperature (° F)		
	@ 66 psi: 480 @ 264 psi: 400		
Water Absorption Rate (%): 0.5	Utilization Temperature (° F)		
	min: -94 max: 284		
Elongation (%): 6	Melting Point (° F): 490		
Tensile Strength (psi): 8700	Coefficient of Expansion: 0.000022		
Compression Strength (psi): 13000	Arc Resistance: 120		
Flexural Strength (psi): 11000	Dielectric Strength (kV/mm): 18		
Flexural Modulus (psi): 550000	Transparency: O		
Impact (Izod ft. lbs/in): 0.5	UV Resistance: G		
Hardness: R118	CHEMICAL RESISTANCE		
	Acids	Alkalis	Solvents
	P	P	G

POLYARYLSULFONE (PAS)

The following data represent average properties of this plastic.

Specific Gravity	1.37
Shrinkage, in/in, 1/8 in. thick	0.0060
Shrinkage, in/in, 1/4 in. thick	0.0070
Water Absorption, % 24 hrs	1.850

Mechanical Properties

Impact, Izod, Notched (Ft-Lb/in)	1.60
Impact, Izod, Unnotched (Ft-Lb/in)	11.00
Tensile Strength (Psi)	12,000
Tensile Elongation (%)	6.500
Tensile Modulus (Psi x 10^6)	0.39
Flexural Strength (Psi)	16,000
Flexural Modulus (Psi x 10^6)	0.40
Compressive Strength (Psi)	NA
Hardness (Rockwell R)	NA
Electrical Properties	
Dielectric Strength (V/Mil)	380
Dielectric Constant (@ 1 MC dry)	3.50
Dissipation Factor (@ 1 MC dry)	0.002
Arc Resistance (sec)	NA
Thermal Properties	
Heat Deflection Temp. @ 264 psi (°F)	400
Heat Deflection Temp. @ 66 psi (°F)	400
Thermal Expansion	NA
Thermal Conductivity	NA
Wear	
Wear Factor	NA

POLYBLENDS

Polymer blends, or polyblends, may be defined as intimate mixtures of different kinds of polymers with no covalent bonds between them. In general, polymer blends can be divided into two main groups:

- *compatible*
- *incompatible*

The term compatibility has been used in different ways. The "hard" definition is correlated with miscibility on a molecular scale; however, other definitions have been proposed based on the properties-composition relationships. Some define the term compatible as those blends that show synergism in some properties or even when they show some valuable end properties.

Compatibility is a concept depending on the scale of a particular experiment. Thus, a polymer blend can be judged compatible when tested for mechanical properties or incompatible as revealed by structural determinations such as glass transition temperature (Tg), scanning electron microscopy (SEM), etc.

For technological purposes, one can define blends as compatible when they show a synergistic behavior in some valuable properties, incompatible when they show a minimum in some property-composition curves, and semicompatible when the properties are intermediate between those of the two polymer parents and are, in particular, almost linearly dependent on the composition (refer to phase diagram in Figure 1).

Incompatible polymer blends are heterogeneous, but only the compatible blends (on a molecular scale) are homogeneous. Both the structure and the presence of different chemical species are important points in the degradation behavior of polymer blends and strongly differentiate the degradation of these polymeric systems from that of pure, homogeneous materials.

The morphology of the incompatible blends depends on two factors: dispersion degree of the two phases, and shape and dimensions of the dispersed particles. In turn, these factors are determined by the rheological characteristics of the two components and by the mixing conditions.

Representation of a heterogeneous, two-component structure is given by a continuous phase of a polymer with a dispersed phase of another polymer separated by a phase boundary. The shape and dimensions of the dispersed phase and degree of dispersion depend on the theological properties of the pure polymers and on the mixing conditions. This can be modified by adding a compatibilizer agent that, by changing the surface tension of the polymer pair, reduces the dimensions of the dispersed particles, giving rise to a finer structure.

These compatibilizers are, in most cases, copolymers of the two components, which are partly soluble in both phases and act as bridges between the two phases. The compatibilized blends show better properties than those of the same incompatible blends. The improvement depends on the degree of homogeneity given by the compatibilizer agent.

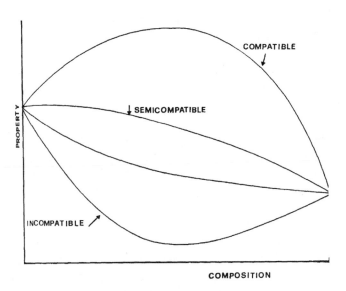

Figure 1. Idealized property-composition (or phase) diagram.

POLYBUTYLENE (PB)

Table 1 provides average properties of polybutylene. In reviewing the general properties of this polymer, note the use of the following legend: A = amorphous - Cr = crystalline - C = clear - E = excellent - G = good - P = poor - O = opaque - T = translucent- R = Rockwell - S = Shore.

Table 1
Average Properties of Polybutylene

STRUCTURE: Cr	FABRICATION		
	Bonding	**Ultrasonic**	**Machining**
	P	P	P
Specific Density: 0.91	**DEFLECTION TEMPERATURE (° F)** @ 66 psi: 220 @ 264 psi: 130		
Water Absorption Rate (%): 0.01	Utilization Temperature (° F) max: 210		
Elongation (%): 350	Melting Point (° F): 255		
Tensile Strength (psi): 5000	Coefficient of Expansion: 0.000075		
Compression Strength (psi): 1300	Transparency: T		
Flexural Strength (psi): 1300	UV Resistance: P		
Flexural Modulus (psi): 26000	**CHEMICAL RESISTANCE**		
Impact (Izod ft. lbs/in): NB	**Acids**	**Alkalis**	**Solvents**
Hardness: R40	P	E	P

POLY(BUTYLENE TEREPHTHALATE) (PBT)

Poly(butylene terephthalate) (PBT) is semicrystalline polyester commonly used in injection molding, composite, and blend applications. Some typical properties are: Glass Transition Temperature, T_g = 66°C; Melting Temperature, T_m = 227° C; Molecular weight of repeat unit = 220.23 g/mole.

POLYCARBONATE (PC)/ABS BLENDS

Polycarbonate (PC) is a clear, colorless polymer used extensively for engineering and optical applications. It is available commercially in both pellet and sheet form. Outstanding properties include impact strength and scratch resistance. The most serious deficiencies are poor weatherability and chemical resistance. Some typical properties are: Glass Transition Temperature, T_g = 145° C; Melting

Temperature, Tm = 225° C; Amorphous density at 25° C = 1.20 g/cc; Molecular weight of repeat unit = 254.3 g/mole. The following data represent average properties of this plastic.

Specific Gravity	1.14
Shrinkage, in/in, 1/8 in. thick	0.0060
Shrinkage, in/in, 1/4 in. thick	0.0070
Water Absorption, % 24 hrs	0.150
Mechanical Properties	
Impact, Izod, Notched (Ft-Lb/in)	8.00
Impact, Izod, Unnotched (Ft-Lb/in)	NO BREAK
Tensile Strength (Psi)	750
Tensile Elongation (%)	100.000
Tensile Modulus (Psi x 10^6)	0.34
Flexural Strength (Psi)	12,000
Flexural Modulus (Psi x 10^6)	0.34
Compressive Strength (Psi)	NA
Hardness (Rockwell R)	116.0
Electrical Properties	
Dielectric Strength (V/Mil)	400
Dielectric Constant (@ 1 MC dry)	3.00
Dissipation Factor (@ 1 MC dry)	0.008
Arc Resistance (sec)	NA
Thermal Properties	
Heat Deflection Temp. @ 264 psi(°F)	230
Heat Deflection Temp. @ 66 psi(°F)	250
Thermal Expansion (in/in/°F) x 10^{-5}	4.000
Thermal Conductivity	1.3
Wear	
Wear Factor	NA
U L yellow card	NO

The basic structure of polycarbonate is shown below:

Polycarbonates are synthesized by interfacial polycondensation of bisphenol A and phosgene. The reaction is conducted in methylene chloride/water, and sodium hydroxide is used as a hydrogen chloride scavenger. Polycarbonates can also be synthesized by means of solution polymerization, using pyridine as the hydrogen chloride scavenger. Still another approach involves reaction of bisphenol A with phenyl carbonate. This is a transesterification reaction in which phenol is produced as a by-product.

Fire-retardant grades of polycarbonates are produced using tetrabromo bisphenol A as comonomer. Polycarbonates possess exceptionally high impact strength, combined with good electrical properties, thermal stability, and creep resistance. Use temperatures range from -60 to 270° F at 264 psi, the material's heat deflection temperature. Additives, such as glass fiber reinforcement, are used frequently, and coatings are used on polycarbonate sheet to improve mar and chemical resistance. Polycarbonates are processed by all the standard thermoplastic methods. Major uses include glazing, lighting, transportation, appliances, signs, electronics, returnable bottles, and solar collector applications. Table 1 provides commercial properties of polycarbonate (PC) reported by several suppliers. In reviewing the general properties of this polymer, note the use of the following legend: A = amorphous - Cr = crystalline - C = clear - E = excellent - G = good - P = poor - O = opaque - T = translucent- R = Rockwell - S = Shore.

Table 1. Average Properties of Polycarbonate

STRUCTURE: Cr	FABRICATION		
	Bonding	**Ultrasonic**	**Machining**
	G	E	E
Specific Density: 1.2	**Deflection Temperature (° F)** @ 66 psi: 287 @ 264 psi: 270		
Water Absorption Rate (%): 0.16	**Utilization Temperature (° F)** min: -145 max: 275		
Elongation (%): 100	**Melting Point (° F)**: 300		
Tensile Strength (psi): 10000	**Coefficient of Expansion**: 0.000039		
Compression Strength (psi): 12000	**Arc Resistance**: 11		
Flexural Strength (psi): 14000	**Dielectric Strength (kV/mm)**: 15		
Flexural Modulus (psi): 330000	**Transparency**: C		
Impact (Izod ft. lbs/in): 2.4	**UV Resistance**: G		
Hardness: R75	**CHEMICAL RESISTANCE**		
	Acids	**Alkalis**	**Solvents**
	G	P	P

POLY(CYCLOHEXANEDIMETHYLENE TEREPHTHALATE) (PCT)

Poly(cyclohexanedimethylene terephthalate) (PCT) is semicrystalline polyester commonly used in injection molding and composite applications. Its high glass transition and melting temperatures relative to other polyesters give it superior thermal performance. Most commercial formulations with PCT include a filler or reinforcement. Some typical properties are: Glass Transition Temperature, T_g = 90 °C; Melting Temperature, T_m = 274° C; Amorphous density at 25° C = 1.20 g/cc; Crystalline density at 25° C = 1.27 g/cc; Molecular weight of repeat unit = 274.32 g/mole.

POLYETHYLENE

Polyethylene is an inexpensive and versatile polymer with numerous applications. Control of the molecular structure leads to low density (LDPE), linear low density (LLDPE) and high density (HDPE) products with corresponding differences in the balance of properties. Some typical properties are: Glass Transition Temperature, T_g = -78° C; Melting Temperature, T_m = 100° C; Amorphous density at 25° C = 0.855 g/cc; Crystallinity at 25° C = 1.00 g/cc; Molecular weight of repeat unit = 28.05 g/mole. The basic structure of PE is as follows:

$$\left[CH_2 CH_2 \right]_n$$

Although it is not the intent of this book to single out any one particular material, PE does deserve extra discussion simply because it is the world's largest tonnage thermoplastic. There are literally many hundreds of grades of PE which differ in their properties in one way or another. The differences arise from variations in the polymer structure due to such variables as the degree of short-chain branching, degree of long-chain branching, average molecular weight, and molecular weight distribution.

Polyethylene is used in a wide variety of applications where its general toughness and resistance to moisture as well as its chemical inertness make it desirable. Since World War II there has been a considerable and continuing expansion in polyethylene production and this, together with increasing competition between manufacturers, has resulted in the material becoming widely available in a range of grades.

The characteristics of polyethylene which lead to its vast use include: low cost, easy processability, excellent electrical insulation properties, high chemical inertness, toughness and flexibility, availability of information regarding its processing and properties, and nontoxicity and odorlessness. There are, however, the following limitations of the polymer: low softening point, susceptibility to oxidation, opacity of the material in bulk, poor scratch resistance, lack of rigidity, low tensile strength, and high gas permeability. Many of these limitations can be overcome by the correct choice of polymer, additive incorporation, processing conditions, and after-treatment.

There are many different methods for the processing of polyethylenes. However, polyethylene films are the most widely processed form. Figures 1 through 3 illustrate some of the methods employed in producing polyethylene films. Polyethylene film is the most important application of LDPE. It is used for making heavy-duty sacks, refuse bags, carrier bags, general packaging, and in the building and construction industry. LDPE and blends of LDPE and polypropylene have also been used as gellants in mineral oils for producing lubricating greases of improved thermal stability.

In addition to their use as sole gellants, polyethylenes are often incorporated as supplements to other types of thickening agents in order to improve the adhesion and cohesion characteristics of lubricating greases. LLDPE has caused some threat to the LDPE market since polymerization plants of LLDPE are cheaper to build and easier to operate and to maintain than are high-pressure plants. There are also some technical advantages to the user since films of LLDPE possess higher impact strength, tensile strength, and extensibility. Such properties will yield film of lower gage but of the same mechanical performance. However, LLDPEs tend to have some undesirable properties such as lower gloss, greater haze, and a narrower heat sealing range. PE is also an important material used in injection molding applications.

There is a much higher percentage of HDPE and polypropylene than LDPE used in injection molding, but on a tonnage basis the LDPE is significant. There is also considerable use of blends of LDPE and HDPE used in the manufacture of various consumer items such as toys, chemical plant components, electrical fittings, seals, bushings, and a variety of articles that were traditionally made from rubber. Blow molding is widely used for HDPE for the production of bottles (most notably detergent bottles).

The excellent electrical insulation properties of PE and its good moisture resistance make it well suited for wire and cable coatings. Uses include power, communications, and control applications. Other major applications of PE include pipes such as domestic water and gas piping, agricultural piping, and ink tubes for ballpoint pens. PE is also used as a filament on rope, fishing nets, and fabrics.

The basic monomer for PE synthesis is ethylene. For LDPE polymerization it is via free radical chain polymerization. For HDPE, synthesis is via Ziegler-Natta or metal oxide-catalyzed chain polymerization.

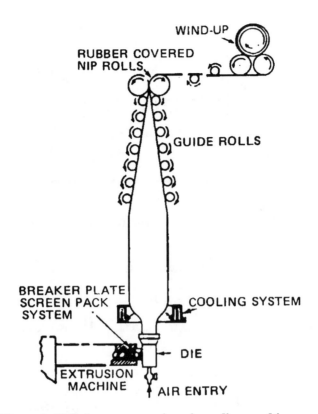

Figure 1. Tubular process using air cooling used in making PE film.

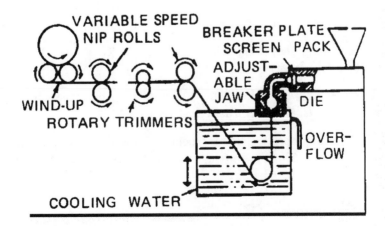

Figure 2. Tubular process for making PE film.

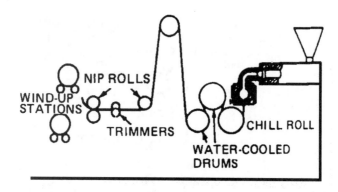

Figure 3. Flat film using chill roll for making PE film.

The following are some general properties of polyethylene.

Crystallinity: Low Linear: 60%; High Density: up to 95%

Pyrolysis: melts and becomes transparent; in the absence of air is stable up to 300° C. Cross-linked types do not melt.

Specific gravity: Low density 0.915 to 0.94; High density 0.94 to 0.97

Refractive index: 1.51 to 1.52

Moisture relations: polyethylenes are resistant to water; increase in weight after 1 year less than 0.2%.

Thermal properties: Specific heat 0.55 at 20° C, rising to 0.7 at 120° C

Coefficient of linear expansion: 0.00017 to 0.00022 for low density; 0.00013 to 0.00020 for high density.

Melting point (loss of crystallinity): 109 to 125° C for low density 130 to 135° C for high density.

Electrical properties: Resistivity 10^{17} to 10^{19} (ohm cm); Dielectric strength 20 to 160 (kV/mm); Dielectric constant 2.28.

Mechanical properties: Elastic modulus 15.4 to 125.0 (daN/mm^2); Strain rate (mm/min) 50-500; Yield strength (daN/mm^2) 2.28-2.36; Ultimate tensile strength (daN/mm^2) 3.43-1.33.

Hardness: low density, 50-60 high density, 65-70.

Impact: low density, 0.92 high density, 0.96.

Refer to *High-Pressure Polymerization, LDPE, LLDPE, HDPE,* and *HMWHDPE.* (Source: *Handbook of Polymer Science and Technology: Volume 2 - Performance Properties of Plastics and Elastomers*, N. P. Cheremisinoff - editor, Marcel Dekker Inc., New York, 1989).

POLY(ETHYLENE 2,6-NAPHTHALENEDICARBOXYLATE) (PEN)

Poly(ethylene 2,6-naphthalenedicarboxylate) (PEN) is a semicrystalline polyester capable of being used at higher temperatures than most other polyesters due to its higher Tg and Tm. Some typical properties are: Glass Transition Temperature, Tg = 120°C; Melting Temperature, Tm = 270°C; Molecular weight of repeat unit = 242.23 g/mole.

POLY(ETHERETHERKETONE) (PEEK)

Poly(etheretherketone) (PEEK) is a high performance thermoplastic generally used with fiber reinforcements such as glass, carbon, or Kevlar. Glass Transition Temperature, Tg = 143°C; Melting Temperature, Tm = 334°C; Molecular weight of repeat unit = 288.31 g/mole.

POLYIMIDE (TPI)

The following data represent average properties of this plastic.

Specific Gravity	1.42
Shrinkage, in/in,1/8 in. thick	0.0130
Shrinkage, in/in,1/4 in. thick	0.0260
Water Absorption,% 24 hrs	0.150
Mechanical Properties	
Impact, Izod, Notched (ft-lb/in)	1.40
Impact, Izod, Unnotched (ft-lb/In)	10.00
Tensile Strength (Psi)	18,000
Tensile Elongation (%)	3.300
Tensile Modulus (Psi x 10^6)	1.00
Flexural Strength (Psi)	32,000
Flexural Modulus (Psi x 10^6)	0.90
Compressive Strength (Psi)	NA
Hardness (Rockwell R)	126.0
Electrical Properties	
Dielectric Strength (V/Mil)	400
Dielectric Constant (@ 1 MC dry)	3.10
Dissipation Factor (@ 1 MC dry)	0.003
Arc Resistance (sec)	NA

Thermal Properties

Heat Deflection Temp 264 psi (°F)	525
Heat Deflection Temp 66 psi (°F)	NA
Flammability	V0
Thermal Expansion (in/in/F) x 10^{-5}	NA
Thermal Conductivity	NA
Wear	
Wear Factor	NA
U L yellow card	NO

POLY(METHYL METHACRYLATE) (PMMA)

Poly(methyl methacrylate) (PMMA) is a clear, colorless polymer used extensively for optical applications. It is available commercially in both pellet and sheet form. Outstanding properties include weatherability and scratch resistance. The most serious deficiencies are low impact strength and poor chemical resistance. Some typical properties are: Glass Transition Temperature, $T_g = 114°C$; Melting Temperature, noncrystalline; Amorphous density at $25°C = 1.17$ g/cc; Molecular weight of repeat unit $= 100.12$ g/mole. The basic structure of PMMA is as follows:

$$\left[CH_2 - \underset{\underset{COOCH_3}{|}}{\overset{\overset{CH_3}{|}}{C}} \right]_n$$

PMMA is characterized by crystal clear transparency, unexcelled weatherability, and a useful combination of stiffness, density, and moderate toughness. The heat deflection temperatures range from 165 to 212° F, with a service temperature of 200° F. Further improvement of the mechanical properties of PMMA can be achieved by orientation of heat-cast sheets. Polymethyl methacrylate can be modified by copolymerization of methyl methacrylate with other monomers, such as acrylates, acrylonitrile, styrene, and butadiene. Blending with SBR improves impact resistance. The 2-hydroxyethyl ester of methacrylic acid is used as the monomer of choice for the manufacture of soft contact lenses. Copolymerization with ethylene glycol dimethacrylate produces a hydrophilic network polymer, also called hydrogel. Hydrogel polymers are glassy and brittle when dry but become soft and plastic after swelling in water. The most important properties of a hydrogel are its equilibrium water content and oxygen permeability.

POLYPROPYLENE (PP)

Polypropylene is an inexpensive polymer with a wide variety of uses. Glass Transition Temperature, $T_g = -10°$ C; Melting Temperature, $T_m = 173°$ C; Amorphous density at $25°$ C $= 0.85$ g/cc; Crystallinity at $25°$ C $= 0.95$ g/cc;

Molecular weight of repeat unit = 42.08 g/mole. Polypropylene is a linear hydrocarbon containing little or no unsaturation, resembling polyethylene; however the -CH_3 groups cause stiffening and less stability with regard to oxidation. The following are some general properties.

Plasticized by elastomers (e.g., nitrile rubbers) and best by copolymerization.

Solvents: at elevated temperatures aromatic hydrocarbons (e.g. xylene, tetralin, decalin).

Combustion: burns with blue-based non-smokey flame.

Specific gravity: films 0.89, general purpose 0.9 to 0.91, fibers 0.9-0.92

Refractive index: 1.49

Water absorbtion: 0.01 to 0.03% in 24 hours

Water retention: 5%

Specific heat: 0.46 at 40° C and 0.5 at 100° C

Coefficient of linear expansion: 0.00011 at 20 -60° C, 0.00015 to 0.00017 at 60 to 100° C, 0.00021 at 100-140° C

Electrical properties: Resistivity: 10^{16} (ohm cm); Dielectric strength: >32 (kV/mm) Dielectric constant: 2.25

Mechanical properties: Young modulus: 120 (daN/mm^2); Extension at break: 20-300%; Flexural yield strength: 4.2 to 4.9 (daN/mm^2); Compressive yield strength: general purpose, 3.8-4.6 (daN/mm^2) high impact, 3.1 (daN/mm^2); Flexural modulus: 110 (daN/mm^2); Impact strength: (Izod ft lb/in) general purpose, 0.4-2.2 high impact, 1.5-12.

Hardness (Rockwell): general purpose, R80-R90 high impact, R28-R90

Friction: (polymer/polymer) 0.1-0.3

The structural formula for PP is as follows:

$$\left[-CH_2CH \underset{\overset{|}{CH_3}}{} - \right]_n$$

Table 1 provides average properties for the *polypropylene homopolymer*. In reviewing the general properties of this polymer, note the use of the following legend: A = amorphous - Cr = crystalline - C = clear - E = excellent - G = good - P = poor - O = opaque - T = translucent- R = Rockwell - S = Shore.

Table 1. Average Properties of Polypropylene Homopolymer

STRUCTURE: Cr	FABRICATION		
	Bonding	Ultrasonic	Machining
	P	P	G
Specific Density: 0.90	DEFLECTION TEMPERATURE (° F) @ 66 psi: 210 @ 264 psi: 140		
Water Absorption Rate (%): 0.01	Utilization Temperature (° F) min: -15 max: 265		
Elongation (%): 200	Melting Point (° F): 338		
Tensile Strength (psi): 4500	Coefficient of Expansion: 0.00006		
Compression Strength (psi): 6500	Arc Resistance: 180		
Flexural Strength (psi): 7000	Dielectric Strength (kV/mm): 23		
Flexural Modulus (psi): 190000	Transparency: T		
Impact (Izod ft. lbs/in): 1	UV Resistance: P		
Hardness: R95	CHEMICAL RESISTANCE		
	Acids	Alkalis	Solvents
	G	E	G

Table 2 provides average properties for the *polypropylene copolymer*.

Table 2. Average Properties of Polypropylene Copolymer

STRUCTURE: Cr	FABRICATION		
	Bonding	Ultrasonic	Machining
	P	P	G
Specific Density: 0.90	Deflection Temperature (° F) @ 66 psi: 174 @ 264 psi: 140		
Water Absorption Rate (%): 0.01	Utilization Temperature (° F) min: -22 max: 248		

Table 2 continued

Elongation (%): 200	Melting Point (° F): 320
Tensile Strength (psi): 4000	Coefficient of Expansion: 0.000065
Compression Strength (psi): 5800	Arc Resistance: 180
Flexural Strength (psi): 6500	Dielectric Strength (kV/mm): 22
Flexural Modulus (psi): 135000	Transparency: T
Impact (Izod ft. lbs/in): 6	UV Resistance: P
Hardness: R80	**CHEMICAL RESISTANCE**		

Acids	Alkalis	Solvents
G	E	G

Table 3 provides average properties for the *polypropylene with 40 % talc filled*.

Table 3
Average Properties of Talc-Filled Polypropylene

STRUCTURE: Cr	**FABRICATION**		
	Bonding	Ultrasonic	Machining
	P	P	E
Specific Density: 1.22	**Deflection Temperature (° F)** @ 66 psi: 230 @ 264 psi: 175		
Water Absorption Rate (%): 0.02	**Utilization Temperature (° F)** min: -14 max: 275		
Elongation (%): 18	Melting Point (° F): 338		
Tensile Strength (psi): 4000	Coefficient of Expansion: 0.000037		
Compression Strength (psi): 5800	Arc Resistance: ND		
Flexural Strength (psi): 6800	Dielectric Strength (kV/mm):		
Flexural Modulus (psi): 464000	Transparency: O		
Impact (Izod ft. lbs/in): 2.8	UV Resistance: P		
Hardness: R95	**CHEMICAL RESISTANCE**		
	Acids	Alkalis	Solvents
	G	E	G

Table 4 provides average properties for the *glass filled (30 %) polypropylene.*

Table 4
Average Properties of Glass-Filled Polypropylene

STRUCTURE: Cr	FABRICATION		
	Bonding	**Ultrasonic**	**Machining**
	P	P	P
Specific Density: 1.13	**Deflection Temperature** (° F) @ 264 psi: 260		
Water Absorption Rate (%): 0.03	**Utilization Temperature** (° F) min: -14 max: 284		
Elongation (%): 3	**Melting Point** (° F): 338		
Tensile Strength (psi): 8800	**Coefficient of Expansion**: 0.000023		
Compression Strength (psi): 8500	**Arc Resistance**: 130		
Flexural Strength (psi): 12100	**Dielectric Strength (kV/mm)**: 24		
Flexural Modulus (psi): 700000	**Transparency**: O		
IMPACT (IZOD ft. lbs/in): 2.2	**UV Resistance**: P		
Hardness: R110	**CHEMICAL RESISTANCE**		
	Acids	**Alkalis**	**Solvents**
	P	E	P

POLYOXYMETHYLENE (POM)

Polyoxymethylene (POM) is commonly used as a direct replacement for metals due to its stiffness, dimensional stability and corrosion resistance. Copolymers including ethylene oxide are quite common, primarily because they reduce the propensity for depolymerization at normal processing temperatures. Some typical properties are: Glass Transition Temperature, Tg = -30° C; Melting Temperature, Tm = 183° C; Amorphous density at 25° C = 1.25 g/cc; Crystalline density at 25° C = 1.54 g/cc; Molecular weight of repeat unit = 30.03 g/mole. There are a number of grade variations of this material commercially available.

POLYSTYRENE CRYSTAL (XPS)

Table 1 provides average properties of polystyrene. PS has a variety of consumer uses, including hot and cold drink disposable cups, plates, and many decorative type items. In reviewing the general properties of this polymer, note the use of the following legend: A = amorphous - Cr = crystalline - C = clear - E = excellent - G = good - P = poor - O = opaque - T = translucent- R = Rockwell - S = Shore. Refer also to *HIPS* (*High Impact Polystyrene*).

Table 1
Average Properties of Polystyrene

STRUCTURE: A	FABRICATION		
	Bonding	**Ultrasonic**	**Machining**
	E	E	G
Specific Density: 1.05	**Deflection Temperature** (° F) @ 66 psi: 211 @ 264 psi: 200		
Water Absorption Rate (%): 0.06	**Utilization Temperature** (° F) min: -92 max: -150		
Elongation (%): 47	**Melting Point** (° F): 212		
Tensile Strength (psi): 5900	**Coefficient of Expansion**: 0.00004		
Compression Strength (psi): 14500	**Arc Resistance**: 70		
Flexural Strength (psi): 6100	**Dielectric Strength (kV/mm)**: 24		
Flexural Modulus (psi): 475000	**Transparency**: C		
Impact (Izod ft. lbs/in): 0.8	**UV Resistance**: P		
Hardness: R80	**CHEMICAL RESISTANCE**		
	Acids	**Alkalis**	**Solvents**
	G	E	P

Polystyrene is made by bulk or suspension polymerization of styrene. It is commonly available in crystal, high impact, and expandable grades. Its major characteristics include transparency, ease of coloring and processing, and low cost. Styrene monomer is produced from benzene and ethylene.

Impact polystyrene is produced commercially by dispersing small particles of butadiene rubber in styrene monomer. This is followed by mass prepolymerization of styrene and completion of the polymerization either in mass or in aqueous suspension. During prepolymerization, styrene starts to polymerize by itself, forming droplets of polystyrene with phase separation. When nearly equal phase volumes are obtained, phase inversion occurs, and the droplets of polystyrene become the continuous phase in which the rubber particles are dispersed. The impact strength increases with rubber particle size and concentration, while gloss and rigidity are decreasing. The stereochemistry of the polybutadiene has a significant influence on properties and a 36% cis- 1,4-polybutadiene provides optimal properties. End uses for all types of polystyrene are packaging, housewares, toys and recreational products, electronics, appliances, furniture, and building and construction (insulation).

POLYSULFONE (PSO)

Polysulfones are a family of engineering thermoplastics that exhibit excellent high-temperature properties. Many variations of this material have a continuous use temperature of 150° C and a maximum temperature of around 170° C. Polysulfones are produced by the Friedel-Crafts reaction of sulfonyl chloride groups with aromatic nuclei, or by reacting 4,4'-dichlorodiphenylsulfone with alkali salt of bisphenol A. The latter polycondensation is conducted in highly polar solvents, such as dimethylsulfoxide or sulfolane. These materials can be injection molded into complex shapes and can compete with many metals. The following is the chemical structure:

The following data represent average properties of this plastic.

Specific Gravity	1.24
Shrinkage, in/in,1/8 in. thick	0.0070
Shrinkage, in/in,1/4 in. thick	0.0080
Water Absorption,% 24hrs	0.220
Mechanical Properties	
Impact, Izod, Notched (ft-lb/in)	13.00
Impact, Izod, Unnotched (ft-lb/in)	60.00
Tensile Strength (Psi)	10,000
Tensile Elongation (%)	75.000
Tensile Modulus (Psi x 10^6)	0.36
Flexural Strength (Psi)	15,000
Flexural Modulus (Psi x 10^6)	0.39
Compressive Strength (Psi)	14,000
Hardness (Rockwell R)	120.0
Electrical Properties	
Dielectric Strength (V/Mil)	25
Dielectric Constant (@ 1 MC dry)	3.00
Dissipation Factor(@ 1 MC dry)	0.003
Arc Resistance (sec)	122
Thermal Properties	
Heat Deflection Temp 264 psi (°F)	345

Heat Deflection Temp 66 psi (°F)	358
Flammability	V1
Thermal Expansion (in/in/°F) x 10^{-5}	3.100
Thermal Conductivity	1.8
Wear	
Wear Factor	NA
U L yellow card	NO

POLYSULPHIDES

Also called Thioplasts or Thiokol. These are special purpose rubbers. General properties include poor mechanical properties; excellent resistance to oils and degreasing solvents. The swelling resistance increases as the sulphur content increases. Good oxidative, ozone, and weathering resistance. Approximate working temperature range -50° C to + 90° C.

POLYURETHANES

These are special purpose elastomers with properties which are in general somewhere between rubbers and thermoplastics. There are three main types which may be considered to be thermoplastics, casting liquids, or millable types. The harder grades can be machined. The very soft polyurethanes have poor mechanical properties. The harder polyurethanes have moderate mechanical properties with very high tear and abrasion resistance. They have very good resistance to mineral oils and degreasing solvents. Very good resistance to ozone and oxidation. High heat buildup. Approximate working temperature range -70° C to +90° C.

The following data represent average properties of polyurethane rigid plastics.

Specific Gravity	1.27
Shrinkage, in/in, 1/8 in. thick	0.0025
Shrinkage, in/in, 1/4 in. thick	0.0050
Water Absorption, % 24hrs	0.150
Mechanical Properties	
Impact, Izod, Notched (Ft-Lb/in)	2.00
Impact, Izod, Unnotched(Ft-Lb/in)	7.00
Tensile Strength (Psi)	12,000
Tensile Elongation (%)	6.600
Tensile Modulus (Psi x 10^6)	0.50
Flexural Strength (Psi)	16,000
Flexural Modulus (Psi x 10^6)	0.50
Compressive Strength (Psi)	NA
Hardness (Rockwell R)	119.0
Electrical Properties	
Dielectric Strength (V/Mil)	510
Dielectric Constant(@ 1 MC dry)	3.40

Dissipation Factor(@ 1 MC dry)	0.012
Arc Resistance (sec)	NA
Thermal Properties	
Heat Deflection Temp 264 psi (°F)	175
Heat Deflection Temp 66 psi (°F)	NA
Flammability	HB
Thermal Expansion (In/In/F)x 10^{-5}	NA
Thermal Conductivity	NA
Wear	
Wear Factor	NA
U L yellow card	NO

Polyurethanes are part of the polyolefins family, which are by far the most important addition polymers. In the process of generating the polymer backbone, carbon-carbon bonds are formed and the process is for all practical purposes irreversible. In contrast, polyurethanes and polyacetal are addition polymers in which reversible adducts are formed. In the case of the polyurethanes, reversal of the generated carbon-oxygen bonds occurs above 120° C. The reverse process is usually catalyzed by the same catalysts that are used to facilitate polymer formation. Polyacetal, another important engineering thermoplastic, is also thermally labile, and unzipping is observed upon heating. The latter polymer has been stabilized by end-group capping, which allows the use of this nonpetrochemical-based polymer.

Polyurethanes are the most versatile group of polymers in this group, because products ranging from soft linear thermoplastic elastomers to hard thermoset rigid foams are readily produced from liquid monomers. Polyurethanes are divided into three broad groups: flexible foam, rigid foam, and elastomers. The basic building blocks for polyurethanes are diisocyanates and macroglycols, also called polyols. The commonly used isocyanates are tolylene diisocyanate (TDI), methylenediphenyl isocyanate (MDI) and polymeric isocyanate (PMDI) mixtures manufactured by phosgenating polyamines derived from the acid-catalyzed condensation of aniline with formaldehyde. MDI and PMDI are coproducts, and separation is achieved by distilling part of the MDI from the reaction mixture.

Specialty aliphatic isocyanates, such as derivatives of hexamethylene diisocyanate, are used in light-stable polyurethane coatings. The macroglycols used as coreagents are either polyether or polyester based. Polyether diols are low molecular weight polymers based on ring-opening polymerization of alkylene oxides, and commonly used polyester polyols are polyadipates. A polyol produced by ring-opening polymerization of caprolactone, initiated with low molecular weight glycols, is also used. Flexible foams are manufactured from TDI and higher molecular weight polyether triols. Low-density flexible foams are used in furniture and bedding applications, higher-density flexible foams are used in automotive seating applications, and semiflexible foams are used for automotive interior padding. Low-density rigid polyurethane foams are manufactured from PMDI and polyether polyols. These foams are the most efficient commercially available insulation materials and are widely used in the construction and transportation industries. High-density rigid polyurethane foams are used in structural parts. A good example is molded solar panels. Urethane elastomers have the rigidity of plastics and the resiliency of rubber and are often referred to as elastoplastics. Elastomer-type polyurethane formulations also find wide use as coatings, adhesives, and sealants.

POLY(VINYL ALCOHOL)

Poly(vinyl alcohol) is completely water soluble and thus is used as a thickener in some suspensions and emulsions. Although usually amorphous, it can be drawn into a semicrystalline fiber. However, the melting point of the crystallites is above the thermal degradation temperature. Some typical properties are: Glass Transition Temperature, Tg $= 85°$ C; Amorphous density at $25°$ C $= 1.26$ g/cc; Crystalline density at $25°$ C $= 1.35$ g/cc; Molecular weight of repeat unit $= 44$ g/mole.

POLY(VINYL CHLORIDE) (PVC)

Poly(vinyl chloride) (PVC) is one of the most widely used polymers in the world. This polymer is usually plasticized with low or medium molecular weight materials such as dioctyl phthalate, trioctyl phosphate, and poly(propylene glycol) esters. The properties can be finely tuned from rigid to soft and flexible by varying the plasticizer content from a few percent to more than 60%. The largest use is for rigid pipe; other uses include window frames, flooring, cable insulation, and other household and construction applications.

Some typical properties are: Glass Transition Temperature, Tg $= 85°$ C; Melting Temperature, Tm $= 240°$ C; Amorphous density at $25°$ C $= 1.385$ g/cc; Crystalline density at $25°$ C $= 1.52$ g/cc; Molecular weight of repeat unit $= 62.5$ g/mole. Poly(vinyl chloride) (PVC) is one of the most investigated polymers, as far as thermal degradation is concerned. Its degradation occurs in two stages: dehydrochlorination and, at higher temperatures, scission of the resulting polyene.

The evolving molecules of HCl can react with macromolecules or macroradicals of the other components of blends, which can lead to a destabilization as well as to a stabilization of the blends. The presence of PVC in a blend induces destabilization, and then a more rapid degradation, in other polymers such as poly(vinyl acetate) (PVA). In its turn, its degradation rate increases in the presence of PVA, polyacrylamide (PAM), polyacrylonitrile (PAN), chlorinated rubber, etc. On the contrary, in a few cases, stabilization to some extent is achieved by PVC blended with PAN and some acrylic polymers.

The mechanism invoked in order to explain the increase of degradation rate for blends with PVC mainly involves the diffusion of HCl into the other polymeric phase and its reactions with macroradicals, macromolecules, or degradation products. The blend PVC/PVA shows many peculiar interesting features. The two polymers are immiscible and the blend results in a biphasic structure.

PVC is also destabilized when blended with PAN. In these blends PAN is, on the contrary, stabilized. Also in this case, the explanation has been given in terms of interactions between products of reactions and macromolecules or macroradicals. Polymethyl methacrylate (PMMA), poly-n-butyl methacrylate (PBMA), polyethyl acrylate (PEA), polystyrene (PS) and poly-ot-methylstyrene (PAMS) show the same qualitative effect on the degradation of PVC. In particular, stabilization of the PVC is observed, while the degradation behavior of the other component of the blend may be different. The stabilizing effect of these polymers on PVC has been correlated with reactions of chlorine radicals with the chains of the other polymeric phase. Thus, chlorine radicals cannot play their role in the dehydrochlorination, which is shifted toward higher temperatures. As for the stabilization of the other component, this effect has been associated with reactions of HCl with residues of dehydrochlorinated PVC, which give rise to "blocking" of chain cleavage. Finally, PVC shows a stabilizing effect on rubbery polymers, such as polybutadiene (PBD), at least at low temperature.

All the discussed blends are incompatible, and their heterogeneous nature could be responsible for their peculiar degradation behavior. Since most polymers are immiscible, the influence of compatibility on the degradation behavior of polymer blends has not been extensively studied. Moreover, some contrasting results have been reported. Refer to *Polyblends*.

Table 1 provides average properties of rigid polyvinyl chloride (RPVC). In reviewing the general properties of this polymer, note the use of the following legend: A = amorphous - Cr = crystalline - C = clear - E = excellent - G = good - P = poor - O = opaque - T = translucent- R = Rockwell - S = Shore.

Table 1
Average Properties of Rigid Polyvinyl Chloride (RPVC)

STRUCTURE: A	FABRICATION		
	Bonding	**Ultrasonic**	**Machining**
	G	G	G
Specific Density: 1.35	**Deflection Temperature (° F)** @ 66 psi: 170 @ 264 psi: 162		
Water Absorption Rate (%): 0.2	**Utilization Temperature (° F)** min: 14 max: 140		
Elongation (%): 20	**Melting Point (° F):** 176		
Tensile Strength (psi): 6500	**Coefficient of Expansion:** 0.000045		
Compression Strength (psi): 11000	**Arc Resistance:** 70		
Flexural Strength (psi): 12100	**Dielectric Strength (kV/mm):** 18		
Flexural Modulus (psi): 400000	**Transparency:** C		

Table 1 continued

Impact (Izod ft. lbs/in): 5	UV Resistance: G		
Hardness: R105	**CHEMICAL RESISTANCE**		
	Acids	Alkalis	Solvents
	P	E	G

POLYVINYLIDENE CHLORIDE (PVDC)

Table 1 provides average properties of PVDC. In reviewing the general properties of this polymer, note the use of the following legend: A = amorphous - Cr = crystalline - C = clear - E = excellent - G = good - P = poor - O = opaque - T = translucent- R = Rockwell - S = Shore.

Table 1
Average Properties of PVDC

STRUCTURE: Cr	**FABRICATION**		
	Bonding	Ultrasonic	Machining
	G	G	P
Specific Density: 1.7	**Deflection Temperature (° F)** @ 66 psi: 158 @ 264 psi: 140		
Water Absorption Rate (%): 0.1	Utilization Temperature (° F) max: 175		
Elongation (%): 250	Melting Point (° F): 169		
Tensile Strength (psi): 4300	Coefficient of Expansion: 0.00011		
Compression Strength (psi): 8000	Transparency: C		
Flexural Strength (psi): 5800	UV Resistance: P		
Flexural Modulus (psi): 72500	**CHEMICAL RESISTANCE**		
Impact (Izod ft. lbs/in): 12.5	Acids	Alkalis	Solvents
Hardness: R55	P	E	P

POLYVINYLIDENE FLUORIDE (PVDF)

Table 1 provides average properties of PVDF. In reviewing the general properties of this polymer, note the use of the following legend: A = amorphous - Cr = crystalline - C = clear - E = excellent - G = good - P = poor - O = opaque - T = translucent- R = Rockwell - S = Shore.

Table 1

Average Properties of PVDF

STRUCTURE: Cr	FABRICATION		
	Bonding	Ultrasonic	Machining
	P	P	E-G
Specific Density: 1.77	Deflection Temperature (° F) @ 66 psi: 300 @ 264 psi: 190		
Water Absorption Rate (%): 0.03	Utilization Temperature (° F) min: -40 max: 310		
Elongation (%): 50	Melting Point (° F): 340		
Tensile Strength (psi): 5800	Coefficient of Expansion: 0.000022		
Compression Strength (psi): 10000	Arc Resistance: 60		
Flexural Strength (psi): 7800	Dielectric Strength (kV/mm): 22		
Flexural Modulus (psi): 260000	Transparency: T		
Impact (Izod ft. lbs/in): 6.6	UV Resistance: G		
Hardness: R95	CHEMICAL RESISTANCE		
	Acids	Alkalis	Solvents
	P	G	

POM (ACETAL)

The following data represent average properties of this plastic.

Specific Gravity	1.41
Shrinkage, in/in, 1/8 in. thick	0.0200
Shrinkage, in/in, 1/4 in .thick	0.0300
Water Absorption, % 24hrs	0.220

Mechanical Properties

Impact, Izod, Notched (Ft-Lb/in)	1.00
Impact, Izod, Unnotched(Ft-Lb/in)	20.00
Tensile Strength (Psi)	8,800
Tensile Elongation (%)	60.000
Tensile Modulus (Psi x 10^6)	0.41
Flexural Strength (Psi)	13,000
Flexural Modulus (Psi x 10^6)	0.38
Compressive Strength (Psi)	4,500

Hardness (Rockwell R)	107.0
Electrical Properties	
Dielectric Strength (V/Mil)	500
Dielectric Constant (@ 1 MC dry)	3.70
Dissipation Factor (@ 1 MC dry)	0.006
Arc Resistance (sec)	200
Thermal Properties	
Heat Deflection Temp. @ 264 psi (°F)	230
Heat Deflection Temp. @ 66 psi (°F)	316
Flammability	HB
Thermal Expansion (in/in/°F) x 10^{-5}	4.700
Thermal Conductivity	1.7
Wear	
Wear Factor	NA
U L yellow card	NO

POSTCURING

During postcuring operations, the curing agent decomposition can lead to either cross-linking or chain scission. Figure 1 illustrates the effect of postcure on stocks that have been press-cured for an increasing length of time. Stocks cured for short times contain an unused curing agent.

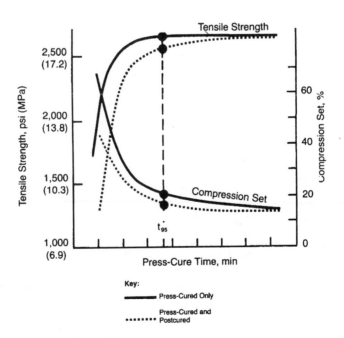

Figure 1. EPDM cure performance.

Postcuring in the presence of oxygen leads to chain scission and reduced compound performance in the case of peroxides. When the stocks are cured to a cure plateau, all peroxide is gone, and there is no change in performance. The response of physical properties to the postcuring of partially press-cured vulcanizates varies from polymer to polymer. The example used in Figure 1 is an EPDM stock. This elastomer exhibits theoretical behavior and is predictable. The more unsaturated polymers, such as NBR, deviate from this ideal response. Postcuring leads to both thermal and oxidative changes in the polymer. Figure 2 shows the result of postcuring. NBR vulcanizates after they have been partially press-cured.

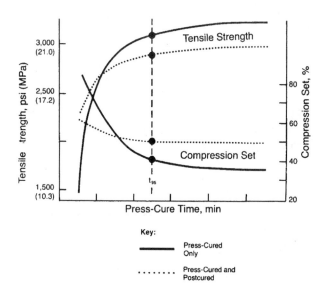

Figure 2. NBR peroxide cure performance.

POTENTIAL ENERGY

Potential Energy is defined as the stored energy possessed by a system as a result of the relative positions of the components of that system. For example, if a ball is held above the ground, the system comprising the ball and the earth has a certain amount of potential energy; lifting the ball higher increases the amount of potential energy the system possesses. Other examples of systems having potential energy include a stretched rubber band, and a pair of magnets held together so that the like poles are touching.

Work is needed to give a system potential energy. It takes effort to lift a ball off the ground, stretch a rubber band, or force two magnets together. In fact, the amount of potential energy a system possesses is equal to the work done on the system. Potential energy also can be transformed into other forms of energy. For example, when a ball is held above the ground and released, the potential energy is transformed into kinetic energy.

Potential energy manifests itself in different ways. For example, electrically charged objects have potential energy as a result of their position in an electric field. An explosive substance has chemical potential energy that is transformed into heat, light, and kinetic energy when detonated. Nuclei in atoms have potential energy that is transformed into more useful forms of energy in nuclear power plants. Refer to *Kinetic Energy*.

PPA (POLYPHTHALAMIDE)

The following data represent average properties of this plastic.

Specific Gravity	1.20
Shrinkage, in/in, 1/8 in. thick	0.0120
Shrinkage, in/in, 1/4 in. thick	0.0160
Water Absorption, % 24 hrs	0.360
Mechanical Properties	
Impact, Izod, Notched (ft-lb/in)	0.30
Impact, Izod, Unnotched(ft-lb/in)	3.20
Tensile Strength (Psi)	11,000
Tensile Elongation (%)	3.000
Tensile Modulus (Psi x 10^6)	0.47
Flexural Strength (Psi)	18,100
Flexural Modulus (Psi x 10^6)	0.47
Compressive Strength (Psi)	NA
Hardness (Rockwell R)	125.0
Electrical Properties	
Dielectric Strength (V/Mil)	530
Dielectric Constant (@ 1 MC dry)	4.30
Dissipation Factor (@ 1 MC dry)	0.027
Arc Resistance (sec)	NA
Thermal Properties	
Heat Deflection Temp. @ 264 psi (°F)	280
Heat Deflection Temp. @ 66 psi (°F)	280
Flammability	HB
Thermal Expansion (in/in/°F) x 10^{-5}	3.000
Thermal Conductivity	NA
Wear	
Wear Factor	NA
U L yellow card	NO

PPE (PHENYLENE ETHER CO-POLYMER)

The following data represent average properties of this plastic.

Specific Gravity	1.43
Shrinkage, in/in, 1/8 in. thick	0.0018
Shrinkage, in/in, 1/4 in. thick	0.0036
Water Absorption,% 24 hrs	0.140

Mechanical Properties

Impact, Izod, Notched ft-lb/in)	2.70
Impact, Izod, Unnotched (ft-lb/in)	15.00
Tensile Strength (Psi)	25,500
Tensile Elongation (%)	3.400
Tensile Modulus (Psi x 10^6)	1.50
Flexual Strength (Psi)	35,500
Flexural Modulus (Psi x 10^6)	1.10
Compressive Strength (Psi)	NA
Hardness (Rockwell R)	125.0

Electrical Properties

Dielectric Strength (V/Mil)	500
Dielectric Constant (@ 1 MC dry)	3.60
Dissipation Factor (@ 1 MC dry)	0.003
Arc Resistance (sec)	NA

Thermal Properties

Heat Deflection Temp. @ 264 psi (°F)	270
Heat Deflection Temp. @ 66 psi (°F)	NA
Flammability	HB
Thermal Expansion (In/In/F) x 10^{-5}	5.000
Thermal Conductivity	NA

Wear

Wear Factor	NA
U L yellow card	NO

PPS (POLYPHENYLENE SULFIDE)

The following data represent average properties of this plastic.

Specific Gravity	1.24
Shrinkage, in/in,1/8 in. thick	0.0070
Shrinkage, in/in,1/4 in. thick	0.0080
Water Absorption,% 24 hrs	0.220

Mechanical Properties

Impact, Izod, Notched (ft-lb/in)	13.00
Impact, Izod, Unnotched (ft-lb/in)	60.00
Tensile Strength (Psi)	10,000
Tensile Elongation (%)	75.000
Tensile Modulus (Psi x 10^6)	0.36

Flexual Strength (Psi)	15,000
Flexural Modulus (Psi x 10^6)	0.39
Compressive Strength (Psi)	14,000
Hardness (Rockwell R)	120.0
Electrical Properties	
Dielectric Strength (V/Mil)	425
Dielectric Constant (@ 1 MC dry)	3.00
Dissipation Factor (@ 1 MC dry)	0.003
Arc Resistance (sec) 122	
Thermal Properties	
Heat Deflection Temp 264 psi (°F)	345
Heat Deflection Temp 66 psi (°F)	358
Flammability	V1
Thermal Expansion (in/in/°F) x 10^{-5}	3.100
Thermal Conductivity	1.8
Wear	
Wear Factor	NA
U L yellow card	NO

PREFORMING

Preforming refers to the process of compressing the molding powder into the shape of the mold before placing it in the mold or to *pelleting,* which consists of compacting the molding powder into pellets of uniform size and approximately known weight. Preforming has many advantages, which include avoiding waste, reduction in bulk factor, rapid loading of charge, and less pressure than uncompacted material. Preformers are basically compacting presses. These presses may be mechanical, hydraulic, pneumatic, or rotary cam machines.

PREPLASTICATING EQUIPMENT

An auxiliary piece of equipment used for providing a measured amount of heated compound. There are two basic types of this equipment used for the preparation of molded thermosets. One type is a nonreciprocating the machine, which is an extruder with a nozzle at the exit end that contains a tapered opening through which the heated compound is forced and compressed by a stationary, rotating screw into a preset size. The extrudate is forced out of the nozzle until it comes in contact with a limit switch, which, when depressed, activates the cut-off knife, producing a heated slug of compound of a desired weight and temperature. Both the weight and temperature are very consistent. The second variation is a reciprocating machine, which has no nozzle and the inside diameter of the barrel determines the diameter of the extrudate. With the cut-off knife in the downward position, the screw rotates and moves backward, against a preset back-pressure, forcing the heated compound against the face of the knife, densifying the compound. As the screw reaches the preset "shot adjustment," the limit switch causes the screw to stop and the cut-off knife to return to its upper position. On reaching its up position, another limit switch is activated, causing the screw to move forward under pressure, acting as a plunger, and forcing the extrudate from the barrel. The screw, moving toward the exit end of the barrel, will contact the forward position of the shot adjustment, activating the cut-off knife, which will then cut off the desired weight slug, ready for delivery into the mold. Either preplasticator, reciprocating or nonreciprocating, is capable of being programmed into the molding cycle and will produce "slugs" of extremely uniform weight, density, and temperature. Simplified diagrams of the two systems are shown in Figures 1 and 2. (Source: Wright, R.E., *Molded Thermosets: A Handbook for Plastics Engineers, Molders, and Designers*, Hanser Publishers, New York, 1991).

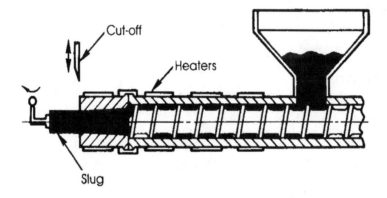

Figure 1. Example of a nonreciprocating screw plasticator.

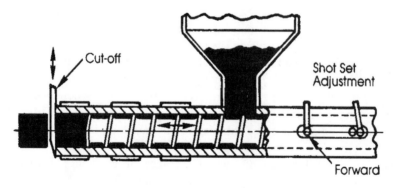

Figure 2. Example of a reciprocating screw plasticator.

PRESSURE FORMING

Pressure forming refers to a type of thermoforming process in which pressures greater than atmospheric are used to force material against the contours of a mold. Deep draw PS deli containers and PP barrier food containers are product examples. See *Thermoforming* and *Vacuum Forming*.

PROCESS OIL

Oil that serves as a temporary or permanent component of a manufactured product. Aromatic process oils have good solvency characteristics; their applications include proprietary chemical formulations, ink oils, and extenders in synthetic rubbers. Naphthenic process oils are characterized by low pour points and good solvency properties; their applications include rubber compounding, printing inks, textile conditioning, leather tanning.

PROFILING

Many rubber products require uncured components that are not rectangular in cross-section. In such cases, a calendering operation comprised of at least one roll may have a peripheral design cut into its surface to produce the desired cross-section. This method is useful when long production runs are required, however it becomes expensive in terms of roll change and roll inventory when many different sections are needed. In this instance, the calender roll may consist of a heavy basic mandrel onto which may be clamped solid cylindrical or split cylindrical steel shells into which the appropriate profile design has been cut. This operation is referred to as profiling. Refer to *Calendering* and *Embossing*.

R

REACTION INJECTION MOLDING

Reaction injection molding (RIM) is a processing technique for the formation of polymer parts by direct polymerization in the mold through a mixing activated reaction. A simplified process schematic is shown in Figure 1. Two reactive monomeric liquids, designated in the figure as A and B, are mixed together by impingement and injected into the mold. In the mold, polymerization and usually phase separation occur, the part solidifies, and is then ejected. Primary uses for RIM products include automotive parts, business machine housings, and furniture. The best general reference on reaction injection molding is probably C.W. Macosko, *RIM, Fundamentals of Reaction Injection Molding*, Hanser Publishers, New York (1989) ISBN 3-446-15196-6.

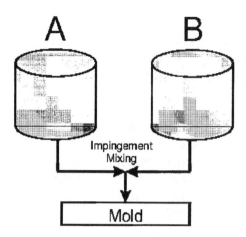

Figure 1. RIM process schematic.

RECYCLING

In many municipalities, a method of disposing of solid waste is in sanitary landfills, in which layers of refuse alternate with layers of soil. However, concerns over the wisdom of such land use has encouraged efforts to dispose of various materials by recycling them for re-use or to derive some positive benefits. Paper as well as glass and aluminum containers have been recycled to some degree for many years, and in more recent years plastic recycling has become common. There are several technical and economic problems in the recycling of plastics; they fall into four general categories: identification, segregation (or sorting), gathering into central stations, and the economics of recovering value. Since plastics used in packaging form a highly visible part (approximately 20 percent by volume but less than 10 percent by weight) of the waste stream, most recycling efforts have focused on containers. Almost all bottles, food trays, cups, and dishes made of the major commodity plastics now bear an identifying number enclosed in a triangle together with an abbreviation. In addition to such labeling, in many localities consumers are encouraged to return empty beverage containers to the place of purchase by being required to pay a deposit on each unit at the time of purchase. This system helps to solve two of the major problems associated with economical recycling, since the consumer seeking return of the deposit does the sorting and the stores gather the plastics into central locations. An added attraction of deposit laws is a notable decrease in roadside litter. However, while such measures have helped to raise dramatically the recycling rate of plastic bottles--especially those made of polyethylene terephthalate (PET) and high-density polyethylene (HDPE)--less than 5 percent of all plastic products are recycled after first use. (On the other hand, most plastics are used in long-term applications such as construction, appliances, and home furnishings, for which efficient recycling is difficult.) Recycling at the industrial scale is practiced within certain companies successfully as part of waste minimization programs. Figures 1 through 3 show photographs of a bottle carton recycling operation being practiced by a Coca Cola and beer bottling facility in Macedonia. In this example, damaged PVC bottle cartons are stockpiled and then sent to a grinder. The grinding machine sends the PVC particles through a cyclone classifier, which collects and bags a narrow cut size of the recycled plastic.

Figure 1. PVC bottle cartons stockpiled at the site.

Figure 2. Stockpiled regrind for recycling.

When a sufficient stockpile of these bagged PVC particles are collected, they are sold back to or traded to the carton manufacturer for new cartons. In polymer synthesis, recycling and various waste minimization techniques have been practiced for many years. For example, in the production of many types of rubber (e.g.,

EPDMs), off-spec production lots are very often either blended in with normal production so that the entire production falls within the product spec-range, or it may be sold into secondary or so-called scrap markets.

Figure 3. Plastics grinding operation.

REFINED WAX

Low-oil-content wax, generally with an oil content of 1.0 mass percent or less, white in color, and meeting Food and Drug Administration standards for purity and safety. Refined waxes are suitable for the manufacture of drugs and cosmetics, for coating paper used in food packaging, and for other critical applications. Also called fully refined wax.

REFRACTIVE INDEX

Ratio of the velocity of light at a specified wave length in air to its velocity in a substance under examination. The refractive index can be determined by test method ASTM D642, using a refractometer and a monochromatic light source. Refractive index is an excellent test for uniform composition of solvents, rubber process oils, and other petroleum products. It may also be used in combination with other simple tests to estimate the distribution of naphthenic, paraffinic, and aromatic carbon atoms in a process oil. Table 1 provides average literature reported values of the refractive index for several common polymers.

Table 1
Refractive Index of Polymers

Polymer	Ref. Index	Polymer	Ref. Index
Fluorcarbon (FEP)	1.34	Inonomers	1.51
Polytetrafluoro-Ethylene (TFE)	1.35	Polyethylene (Low Density)	1.51
Chlorotrifiuoro-Ethylene (CTFE)	1.42	Nylons (PA) Type II	1.52
Cellulose Propionate	1.46	Acrylics Multipolymer	1.52
Cellulose Acetate Butyrate	1.46-1.49	Polyethylene (Medium Density)	1.52
Cellulose Acetate	1.46-1.50	Styrene Butadiene Thermoplastic	1.52-1.55
Methylepentene Polymer	1.485	PVC (Rigid)	1.52-1.55
Ethyl Cellulose	1.47	Nylons (Polyamide) Type 6/6	1.53
Acetal Homopolymer	1.48	Urea Formaldehyde	1.54-1.58
Acrylics	1.49	Polyethylene (High Density)	1.54
Cellulose Nitrate	1.49-1.51	Styrene Acrylonitrile Copolymer	1.56-1.57
Polypropylene (Unmodified)	1.49	Polystyrene (Heat and Chemical)	1.57-1.60
Polyallomer	1.492	Polycarbonate (Unfilled)	1.586
Polybutylene	1.50	Polystyrene (General Purpose)	1.59

REINFORCEMENTS

Reinforcements are used to enhance the mechanical properties of a plastic or elastomer. Finely divided silica, carbon black, talc, mica, and calcium carbonate, as well as short fibers of a variety of materials, can be incorporated as particulate fillers. Incorporating large amounts of particulate filler during the making of plastics such as polypropylene and polyethylene can increase their stiffness. The effect is less dramatic when temperature is below the polymer's T_g.

RELATIVE THERMAL INDEX

The Relative Thermal Index is based on a method developed by Underwriters Laboratories. It is defined as the temperature at which a physical property retains 50 % of its initial value for a period of 60,000 hours. Relative thermal index temperatures (in °C) are given for some common plastics in Table 1.

Table 1
Relative Thermal Index of Polymers

Polymer	Thermal Index, °C	Polymer	Thermal Index, °C
ABS	60	UF	100
PA, PP	65	UP, SR	105
PC, PETP, PBTP	75	PCTFE, FEP, PF, MF, PSU	150
EP	90	PTFE, PES	180

RESIN

Refers to any of a class of amorphous solids or semisolids. Natural resins occur as plant exudations (e.g., of pines and firs), and are also obtained from certain scale insects. They are typically yellow to brown in color, tasteless, and translucent or transparent. Oleoresins contain essential oils and are often sticky or plastic; other resins are exceedingly hard, brittle, and resistant to most solvents. Resins are used in varnish, shellac, and lacquer and in medicine. Synthetic resins, e.g., bakelite, are widely used in making plastics. Refer to *Varnish* and *Shellac*.

RESIN TRANSFER MOLDING

Resin Transfer Molding (RTM) is accomplished with the use of a reinforcement system that has been preshaped to fit the mold cavity and a low-pressure resin in liquid form that is injected into a closed mold. The injection process can be assisted with the use of a vacuum system to provide for evacuation of any volatiles and to gain molded density.

RESOLE FORMATION

A member of the Novolac resin family. In the presence of alkaline catalysts and with formaldehyde, the methylol phenols can condense either through methylene linkages or through ether linkages. In the latter case, subsequent loss of formaldehyde may occur with methylene bridge formation. If the reactions leading to their formation are carried further, large numbers of phenol nuclei can condense to give a network formation. In the production of a single stage phenolic resin, all the necessary reactants for the final polymer (phenol, formaldehyde, and catalyst) are charged into a resin kettle and reacted together. The ratio of formaldehyde to phenol is about 1.25:1, and an alkaline catalyst is used. The resole (single stage) resin based molding compounds will not outgas ammonia during the molding process as do the novolac (two stage) compounds. This leads the resole compounds to be used for applications where their lack of outgassing is most beneficial. Their chief drawback is that they are not easy to control during their production because they are susceptible to the effects of temperatures in excess of 7.2 °C. Production of these compounds is usually scheduled for the cooler months of the year, and they must be stored in sealed containers in areas where the temperature is maintained at the 7.2° C maximum limit. Proper mold venting practices are employed to overcome the ammonia outgassing problem. Resoles have outstanding property profiles and are readily molded in all thermosetting methods. A major drawback is that they tend to be costly. In addition, they have a limited color range (basically black, brown and dark green). (Source: Wright, R.E., *Molded Thermosets: A Handbook for Plastics Engineers, Molders, and Designers*, Hanser Publishers, New York, 1991).

RHEOLOGY

Rheology is the study of the deformation and flow of matter in terms of stress, strain, temperature, and time. The rheological properties of a grease are commonly measured *by penetration* and *apparent viscosity*.

ROTATIONAL MOLDING

In order to make a hollow article, a split mold can be partially filled with a plastisol or a finely divided polymer powder. Rotation of the mold while heating converts the liquid or fuses the powder into a continuous film on the interior surface of the mold. When the mold is cooled and opened, the hollow part can be removed. Among the articles produced in this manner are many toys such as balls and dolls.

RUBBER COMPOUNDING

The objective of the mixing process is to produce a compound with its ingredients thoroughly incorporated and dispersed so that it will process easily in the subsequent forming operations, cure efficiently, and develop the necessary properties for end use, all with the minimum expenditure of machine time and energy. The four properties that affect subsequent operations are viscosity, dispersion, scorch stability, and cure rate. The common features among all internal mixers are (a) the ability to exert a high localized shear stress to the material being mixed (a nip action) and (b) a lower shear rate stirring (homogenizing action). The effectiveness of dispersive mixing results from the combination of high-shear stress and large shear deformation.

There are two basic designs of rotors in internal mixers, nonintermeshing (for example, Banbury, Bolling and Werner & Pfleiderer type) and intermeshing (for example, Intermix and Werner & Pfleiderer type). Representative rotor designs are shown in Figures 1 through 3. Intermeshing rotors provide good heat transfer and are better for heat-sensitive compounds with lengthy mixing cycles.

In general, rotor design is relatively unimportant as a determinant of internal mixer efficiency. This is a result of the importance of elongational flow in the mixing process. Elongational flow is the result of converging flow lines regardless of rotor design. In the rubber industry, the best known internal mixer is the Banbury mixer (Figure 1). This mixer consists of a completely enclosed mixing chamber in which two spiral-shaped rotors operate, a hopper at the top to receive compounding ingredients for mixing and a door at the bottom for discharging the mixed batch of compound. The rotors are driven by an electric motor while pressure is applied from the top by a plunger or ram. The two rotors subject the compound to a certain amount of shear by revolving in opposite directions and at slightly different speeds. The build-up of the shearing action, however, occurs between the rotors and the chamber wall. Water or steam is usually circulated through the hollow rotors and the chamber wall to provide cooling or heating. At the specified mixing time or temperature, the compound is discharged onto a two-roll mill where the material is sheeted off to auxiliary equipment, such as a slab cooling system.

Another design is the Bolling mixer. In this configuration, as the ram is pushed down toward the mixing chamber, the ingredients are forced between helically fluted rotors. As in the Banbury mixer, the bulk of shearing action occurs between the rotors and the chamber wall. A so-called spiral flow arrangement inside the shell of the chamber wall is designed for circulation of steam to provide heat around the shell through baffles cast into the shell liner. Separate channels running through the shell liner provide water for cooling.

A third design is the Shaw Intermix. Here the great bulk of shearing action occurs between the rotors rather than between the rotors and the chamber wall. Werner & Pfleiderer GK-E mixers have a similar mixing action. An older and still widely used system is mixing mills. The open two-roll mill consists of parallel and horizontal rolls rotating in opposite directions. The rotation of the rolls pulls the ingredients through the nip (or bite), which is the clearance between the rolls. The remaining surface of the roll is used as a means of transportation for returning the stock to the nip for further mixing. The back roll is usually rotating faster than the front roll by a ratio called the friction ratio. Most of the work is accomplished on the slower front roll during incorporation of ingredients. Cold or hot water, steam or hot oil may be circulated through the hollow rolls to modify the temperature of the material coming into direct contact with the roll surfaces during mixing operations.

In general, the rubber industry is moving toward continuous internal mixers and mixing extruders. A masterbatch line for tire compounds that consists of an internal mixer dumping the compound into the hopper of a continuous mixer, or mixing extruder, can be considered continuous. The hopper holds two to three batches at a time and the output from the pelletizer or roller die is essentially continuous. The continuous mixer or mixing extruder refines or homogenizes the product from the batch mixer and therefore allows shorter mixing cycles.

The Farrel Continuous Mixer (FCM) is a true internal mixer, with rotors and mixing action similar to the Banbury mixer. The machine does not work on the extruder principle. In operation, raw materials are fed automatically from feed hoppers into the FCM, where the first section of the rotor acts as a screw conveyor, propelling the ingredients to the mixing section. The action within this mixing section is similar to that within a Banbury mixer, incorporating intensive shear of material between rotor and chamber wall and a rolling action of the material itself. Interchange of material between the two bores of the mixing section is an inherent feature of the design of the rotors.

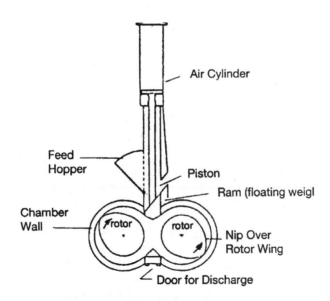

Figure 1. The rubber internal mixer.

The amount and quality of mixing is flexible and can be controlled by the adjustment of speed, feed, rates, and orifice opening. As the feed screw is constantly starved and the mixing action is rotary, there is little thrust or extruding action involved. Production rates and temperature are controlled by the rotor speed and the discharge orifice. Refer to *Internal Batch Mixing* and *Internal Mixing Procedures*. (Source: N. P. Cheremisinoff, *Guidebook to Extrusion Technology*, PTR Prentice Hall, New Jersey, 1993).

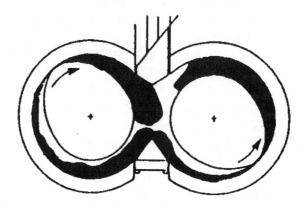

Figure 2. Flow lines and filling configuration of an internal mixer.

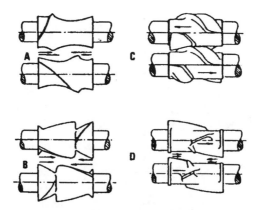

Figure 3. Rotor designs: (A) Banbury two-wing design; (B) four-wing design; (C) Shaw Intermix three-wing design; (D) Werner & Pfleiderer four-wing design.

RUBBER OIL

Any petroleum process oil used in the manufacture of rubber and rubber products. Rubber oils may be used either as rubber extender oils or as rubber process oils. Rubber extender oils are used by the synthetic rubber manufacturer to soften stiff elastomers and reduce their unit volume cost while improving performance characteristics of the rubber. Rubber process oils are used by the manufacturer of finished rubber products (tires, footwear, tubing, etc.) to speed mixing and compounding, modify the physical properties of the elastomer, and facilitate processing of the final product.

RUBBER OXIDATION

Essential to the understanding of rubber oxidation is the understanding of the nature of atmospheric oxygen. The atmospheric oxygen molecule, O_2, demonstrates paramagnetism, which is weak attraction to a magnetic field. Simple Lewis structures could not be internally consistent and explain the paramagnetism of oxygen. However, the relatively simple rules of molecular orbital theory show that the O_2 molecule has two unpaired electrons after bonding. These two unpaired electrons can, for the purposes of understanding oxygen attack on rubber, be considered two free radical species existing in the oxygen molecule.

The ability of oxygen to attack unsaturation in hydrocarbon molecules is much easier to understand. Details about the molecular orbital theory as applied to the O_2 molecule are found in most beginning chemistry textbooks. The kinetic mechanism is illustrated below. In this figure RH represents the rubber hydrocarbon polymer. Steps 2-6 in the mechanism rely on the free radical nature of oxygen to propagate and terminate the free radicals generated in the first stage. Examination of the figure shows that generation of free radicals on the polymer is key to the entire process. Such free radicals can be generated on the polymer through application of heat or by mechanical action causing high shear and rupture of individual polymer chains.

At elevated temperatures, generation of free radicals on the polymer due to carbon-hydrogen bond cleavage or due to carbon-carbon bond cleavage is likely. However, many elastomers are observed to oxidize at relatively low temperatures, say below 60° C, where carbon-hydrogen and carbon-carbon bond cleavage are highly unlikely. Thus it is believed that trace structural impurities in the polymer could account for the oxidation of rubber at relatively mild temperatures. The ease of oxidation of rubber at low temperatures is likely due to the ease of homolytic cleavage of the oxygen-oxygen bonds in the peroxides present in the polymer creating the initiating free radical species necessary for the oxidation mechanism.

Due to the high reactivity of free radicals formed from peroxides, only small traces of such peroxides need be present in the polymer to initiate the low-temperature oxidation of the polymer. Mechanical shear of the polymer during processing and bailing and localized heat during drying and packing of the rubber polymer can cause carbon-hydrogen and carbon-carbon bond cleavage. The resultant free radicals formed during processing of the raw rubber will react to form the trace levels of peroxides necessary to account for the ease of rubber oxidation at low temperatures. Therefore, the oxidation of rubber hydrocarbon polymers resembles the oxidation mechanism for low molecular weight hydrocarbons with the polymer having its own internal source of peroxide initiators present for low-temperature oxidation.

In Figure 1 propagation step 4 can be rewritten as initiation step 1 for rubber polymer containing peroxides formed during processing. It is reasonable to assume that such initiating peroxides are present in even the most carefully prepared raw rubber polymer. Thus it is extremely important to compound rubber for prolonged life in the presence of oxygen through the use of antioxidants and by understanding the effects of pro-oxidants present in the raw polymer or in the rubber compound.

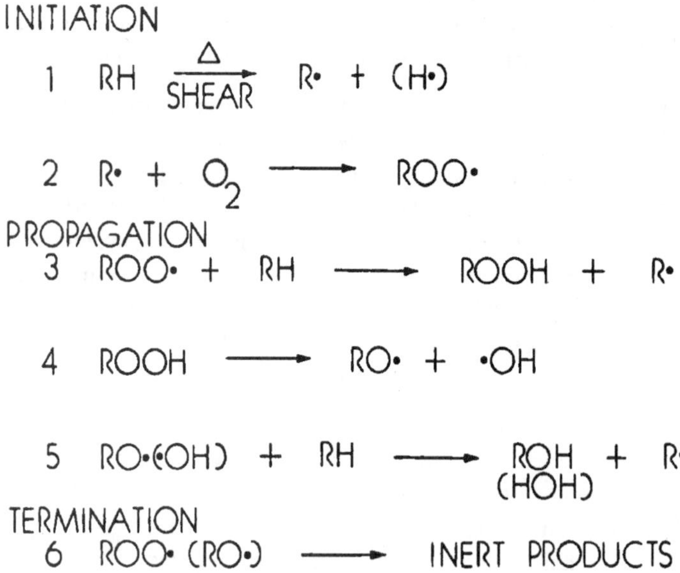

INITIATION

1   $RH \xrightarrow[SHEAR]{\Delta} R\cdot + (H\cdot)$

2   $R\cdot + O_2 \longrightarrow ROO\cdot$

PROPAGATION

3   $ROO\cdot + RH \longrightarrow ROOH + R\cdot$

4   $ROOH \longrightarrow RO\cdot + \cdot OH$

5   $RO\cdot(\cdot OH) + RH \longrightarrow ROH + R\cdot$
    (HOH)

TERMINATION

6   $ROO\cdot (RO\cdot) \longrightarrow$ INERT PRODUCTS

Figure 1. Oxidation of rubber hydrocarbons.

Table 1 provides a general list of common elastomers that are resistant to oxidative degradation. (Source: *Handbook of Polymer Science and Technology: Volume 2 - Performance Properties of Plastics and Elastomers*, N. P. Cheremisinoff - editor, Marcel Dekker Inc., New York, 1989).

Table 1

General Resistance of Elastomers to Oxidative Degradation

Elastomer	ASTM Designation
Elastomers resistant to oxidation	
Acrylic	ACM
Chloro-sulfonyl-polyethylene	CSM
Ethylene propylene diene	EPDM
Fluoroelastomers	FKM
Ethylene oxide epichlorohydrin	ECO
Silicones	MQ, VMQ, FVMQ
Polyester urethanes	AU
Polyether urethanes	EU
Butyl rubber	IIR
Halobutyl rubber	BIIR, CIIR
Elastomers not resistant to oxidation	
Natural rubber	NR
Isoprene rubber	IR
Styrene-butadiene rubber	SBR
Poly(butadiene)	BR
Nitrile rubber	NBR
Neoprene	CR

Refer to additional subject entries on *Antiozonants*, *Antioxidants*, *Peroxide Cure*, *Half-Life*, and *Sulfur Vulcanization*.

S

SAN (STYRENE ACRYLONITRILE COPOLYMER)

Table 1 provides average properties data. In reviewing the general properties of this polymer, note the use of the following legend: A = amorphous - Cr = crystalline - C = clear - E = excellent - G = good - P = poor - O = opaque - T = translucent- R = Rockwell - S = Shore.

Table 1
Average Properties of SAN

STRUCTURE: A	FABRICATION		
	Bonding	Ultrasonic	Machining
	E	E	E
Specific Density: 1.08	Deflection Temperature (° F) @ 66 psi: 209 @ 264 psi: 195		
Water Absorption Rate (%): 0.2	Utilization Temperature (° F) min: -94 max: 185		
Elongation (%): 4	Melting Point (° F): 239		
Tensile Strength (psi): 500000	Coefficient of Expansion: 0.000006		
Compression Strength (psi): 10000	Arc Resistance: 100		
Flexural Strength (psi): 16000	Dielectric Strength (kV/mm): 18		
Flexural Modulus (psi): 16000	Transparency: C		
Impact (Izod ft. lbs/in): 1.2	UV Resistance: P		
Hardness: R120	CHEMICAL RESISTANCE		
	Acids	Alkalis	Solvents
	G	E	P

SCANNING ELECTRON MICROSCOPY

When the domain size is in the range of < 1 μm to 10 nm, scanning electron microscopy (SEM) and/or transmission electron microscopy (TEM) is necessary. A schematic of a scanning electron microscope is shown in Figure 1. Samples in the SEM can be examined "as is" for general morphology, as freeze fractured surfaces or as microtome blocks of solid bulk samples. Contrast is achieved by any one or combination of the following methods: *Solvent etching* - When there exists a large solubility difference in a particular solvent of the polymers being studied, e.g., PP/EP blends.

O_5O_4 Staining - When there exists at least 5% unsaturation in the polymers being investigated, e.g., NR/EPDM, BIIR/Neoprene.

RuO_4 Staining - When there is no solubility differences or unsaturation this possibility is explored, e.g., knit explored line between two DVA's (dynamic vulcanized alloys). In addition, the SEM can be used to study liquids or temperature sensitive polymers on a cryostage.

The SEM is also used to do X-ray/elemental analysis. This technique is qualitative. X-ray analysis and mapping of the particular elements present is useful for the identification of inorganic fillers and their dispersion in compounds as well as inorganic impurities in gels or on surfaces and curatives, e.g., aluminum, silicon, or sulfur in rubber compounds and Cl and Br in halobutyl blends. Refer to *Transmission Electron Microscopy.* (Source: Cheremisinoff, N.P. *Polymer Characterization: Laboratory Techniques and Analysis*, Noyes Publishers, New Jersey, 1996).

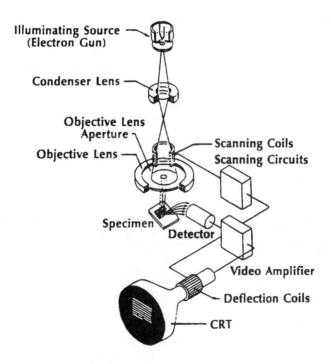

Figure 1. Schematic of SEM.

SCORCH

Premature vulcanization of a rubber compound (called scorch) can occur if the reaction temperature of the vulcanizing ingredients is reached before the desired time. If this temperature is reached in the mixing process before the proper viscosity and level of dispersion is obtained, then the addition of accelerators and vulcanizing agents will cause scorchiness and poor processability.

SEALING COMPONENTS

Automotive sealing components represent a large porting of the market for specialty elastomers like EPDM. These parts require high performance attributes in such areas as surface appearance, ozone and UV resistance, glass adhesion, compressive strength, and others. Figures 1 through 4 provide examples of the typical applications. Parts manufacturers typically produce these sealing components on high speed extrusion lines. The operations are normally fully integrated, with up-front compounding lines, followed extrusion and curing lines. Table 1 provides an overview of the automotive sealing components that are in use today. (Source: N. P. Cheremisinoff, *Guidebook to Extrusion Technology*, Prentice Hall Publishers, Inc., New Jersey, 1993).

Table 1
Overview of Sealing Component

Component Type	Systems/Materials
Static Window Seal	
Front window seal	Small profiles. Typically RIM PU or PVC and/or cross-linked rubbers
Backlight	
Q-light gasket	
Dynamic Window Seal	
Glass channel	Traditionally, flocked thermoset, 70 Shore A. Low friction coating (PA film/HDPE coating)
Belt line seal indoor	Bihardness technology - stiff 55 Shore D (snap-on) and soft 70 Shore A.
Belt line seal outdoor	Aluminum carrier (light-rigid-nonrusty). Coextruded with 85 Shore A cross-linked rubber.

Table 1 continued

Component Type	Systems/Materials
Dynamic Door Sealing	
Sponge door	Rubbers but with low load deflection in range -30° C to +100° C; plus low friction and smooth surface.
Floor seal	Bihardness technology - 55 Shore D and 70 Shore A hardness.
Secondary door seal combination with flexible gutter	Bihardness technology or sponge/solid foot plus metal carrier.
Hood Seal	55 Shore D and 70 Shore A.
Boot Seal	Sponge/solid foot, metal carrier.

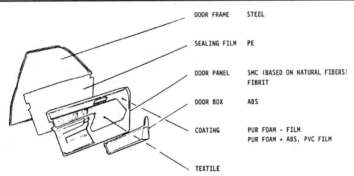

DOOR FRAME   STEEL

SEALING FILM   PE

DOOR PANEL   SMC (BASED ON NATURAL FIBERS)
               FIBRIT

DOOR BOX   ABS

COATING   PUR FOAM - FILM
               PUR FOAM + ABS, PVC FILM

TEXTILE

Figure 1. Shows design components of internal door panel.

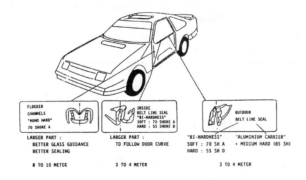

FLOCKED CHANNELS "HOMO HARD" 70 SHORE A

INSIDE BELT LINE SEAL "BI-HARDNESS" SOFT : 70 SHORE A HARD : 55 SHORE D

OUTDOOR BELT LINE SEAL

LARGER PART : BETTER GLASS GUIDANCE BETTER SEALING

LARGER PART : TO FOLLOW DOOR CURVE

"BI-HARDNESS" SOFT : 70 SH A HARD : 55 SH D   "ALUMINIUM CARRIER" + MEDIUM HARD (85 SH)

8 TO 10 METER   3 TO 4 METER   3 TO 4 METER

Figure 2. Shows dynamic window sealing.

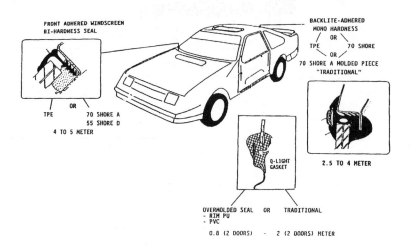

Figure 3. Shows static window sealing on automotive parts.

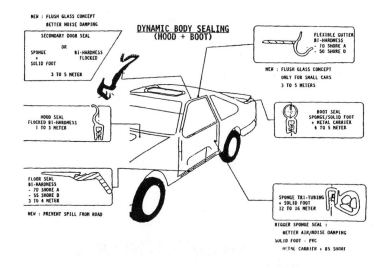

Figure 4. Shows details on dynamic body sealing applications.

SEAL SWELL (RUBBER SWELL)

Swelling of rubber (or other elastomer) gaskets, or seals, when exposed to petroleum, synthetic lubricants, or hydraulic fluids. Seal materials vary widely in their resistance to the effect of such fluids. Some seals are designed so that a moderate amount of swelling improves sealing action. Refer to *Swell and Shrinkage Tests*.

SEMI-CRYSTALLINE THERMOPLASTICS

The molecules in semi-crystalline thermoplastic polymers exist in a more ordered fashion when compared to amorphous thermoplastics. The molecules align in an ordered crystalline form as shown for polyethylene in Figure 1. The crystalline structure is part of an amellar *crystal* which in turn forms the *spherulites*. The sperulitic structure is the largest domain with a specific order and has a characteristic size of 50 to 500 μm. This size is much larger than the wavelength of visible light, making semi-crystalline materials translucent and not transparent. The crystalline regions are very small with molecular chains comprised of both crystalline and amorphous regions. The degree of crystallinity in a typical thermoplastic will vary from grade to grade. For example, in polyethylene the degree of crystallinity depends on the branching and the cooling rate. A low density polyethylene (LDPE) with its long branches can only crystallize to about 40-50%, whereas a high density polyethylene (HDPE) crystallizes to up to 80%. The density and strength of semicrystalline thermoplastics increase with the degree of crystallinity. (Source: Osswald, T.A. and G. Menges, *Material Science of Polymers for Engineers*, Hanser Publishers, New York, 1996).

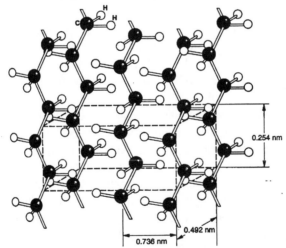

Figure 1. Crystalline structure of polyethylene.

SCREW CONFIGURATIONS

There are a large number of different types of screw configurations employed with extrusion machines. Only the more common configurations are considered. The simplest configuration is a simple continuous-flight screw with constant pitch (Refer to Figure 1). The more sophisticated screw designs include flow disrupters or mixing sections.

These mixer screws have mixing sections which are designed as mechanical means to break up and rearrange the laminar flow of the melt within the flight channel, which results in more thorough melt mixing and more uniform heat distribution in the metering section of the screw.

Mixer screws have also been used to mix dissimilar materials (e.g., resin and additives or simply dissimilar resins) and to improve extrudate uniformity at higher screw speeds (> 100 rpm).

The fluted - mixing-section - barrier type design has proved to be especially applicable for extrusion of polyolefins. For some mixing problems, such as pigment mixing during extrusion, it is convenient to use rings or mixing pins and sometimes parallel interrupted mixing flights having wide pitch angles. Many variations are possible in screw design to accommodate a wide range of polymers and applications. So many parameters are involved, including such variables as screw geometry, materials characteristics, operating conditions, etc., that the industry now uses computerized screw design, which permits analysis of the variables by using mathematical models to derive optimum design of a screw for a given application.

Various screw designs have been recommended by the industry for extrusion of different plastics. For polyethylene, for example, the screw should be long with an L/D of at least 16:1 or 30:1 to provide a large area for heat transfer and plastication. A constant pitch, decreasing-channel depth, metering-type polyethylene screw or constant-pitch, constant -channel- depth, metering-type nylon screw with a compression ratio between 3 to 1 and 4 to 1 is recommended for polyethylene extrusion, the former being preferable for film extension and extrusion coating. Nylon-6,6 melts at approximately 500° F. Therefore, an extruder with an L/D of at least 16:1 is necessary. A screw with a compression ratio of 4:1 is recommended.

Screws are characterized by their length-diameter ratio (commonly written as L/D ratios). L/D ratios most commonly used for single-screw extruders range from 15:1 to 30:1. Ratios of 20:1 and 24:1 are common for thermoplastics, whereas lower values are used for rubbers. A long barrel gives a more homogeneous extrudate, which is particularly desirable when pigmented materials are handled. Screws are also characterized by their compression ratios - the ratio of the flight depth of the screw at the hopper end to the flight depth at the die end. Compression ratios of single-screw extruders are usually 2:1 to 5:1.

Screw design, configurations and extruders also play a major role in polymer finishing operations, which are such operations as drying, desolventizing, and pelletizing. Machines such as expellers are essentially extruders of special design aimed at removing water that is bound to rubbers that are quenched after polymerization (e.g., quenching is done to cool the reaction down and stop further polymerization and chain growth) and in some cases for solvent recovery in solution processes. Figures 2 through 4 show various screw designs. See also *Multiple-Screw Extruders*.

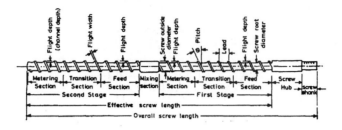

Figure 1. Single-flight, two-stage extrusion screw with mixing section.

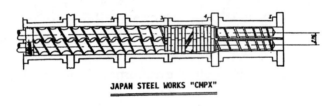

JAPAN STEEL WORKS "CMPX"

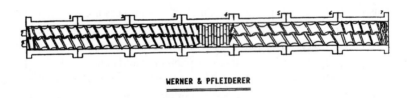

WERNER & PFLEIDERER

Figure 2. Examples of screw designs.

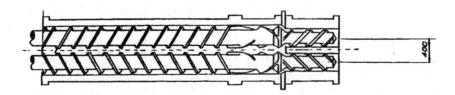

Figure 3. Japan Steel Works counter-meshing twin screw design.

FARREL CORP.

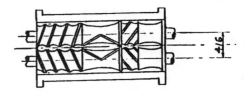

KOBE STEEL CO.

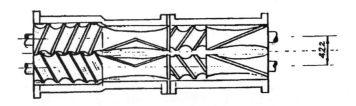

Figure 4. Illustrates Farrel Corp. and Kobe Steel Co. screw designs.

SCREW TRANSFER MOLDING

This is a variation of transfer molding. In this process the molding material is preheated and plasticized in a screw chamber and dropped into the pot of an inverted plunger mold. The screw-transfer- molding technique is a fully automatic operation. The optimum temperature of a phenolic mold charge is $115 \pm 11°C$, the same as that for pot-transfer and plunger molding techniques. For transfer molding, generally pressures of three times the magnitude of those required for compression molding are needed. For example, usually a pressure of 9000 psi and upward is required for phenolic molding material (the pressure referred to here is that applied to the powder material in the transfer chamber). The principle of transferring the liquefied thermosetting material from the transfer chamber into the molding cavity is similar to that of the injection molding of thermoplastics. Therefore, the same principle must be employed for working out the maximum area which can be molded - that is, the projected area of the molding multiplied by the pressure generated by the material inside the cavity must be less than the force holding the two halves together. Otherwise, the molding cavity plates will open as the closing force is overcome. Transfer molding has an advantage over compression molding in that the molding powder is fluid when it enters into the mold cavity. The process therefore, enables the production of intricate parts and molding around thin pins and metal inserts. By transfer molding, metal inserts can be molded into the component in predetermined positions held by thin pins, which would, however, bend or break under compression-molding conditions. Typical articles made by the transfer molding process are terminal-block insulators with many metal inserts and intricate shapes, such as cups and caps for cosmetic bottles. Refer to Figure 1. (Source: Chanda, M. and S. K. Roy, *Plastics Technology Handbook*, Marcel Dekker, Inc., New York, NY, 1987).

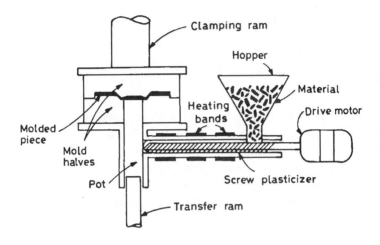

Figure 1. Screw-transfer molding machine.

SEALANTS

Sealants are materials which are used at a juncture of two or more substrates which adhere to the substrates and seal (prevent or control passage) against moisture, gases dust, etc. Sealants typically have a polymeric base and may additionally contain fillers and pigments, resin and adhesion promoters, plasticizers, protective chemicals, and curing agents. The most common types of sealants are solvent release sealing compounds (these have a putty-like consistency); pre-formed sealing tapes (100 % solids systems); hot melt (also called "hot-flow") sealants, which are thermoplastic systems; and curing sealants (one component, two component). Butyl rubber, polyisobutylene, ethylene propylene rubbers and parapol polubutene are typically used in the first three classifications. Acrylics are typically used in water based solvents.

SHEAR STRESS

Frictional force overcome in sliding one "layer" of fluid along another, as in any fluid flow. The shear stress of a petroleum oil or other Newtonian fluid at a given temperature varies directly with shear rate (velocity). The ratio between shear stress and shear rate is constant; this ratio is termed viscosity. The higher the viscosity of a Newtonian fluid, the greater the shear stress as a function of rate of shear. In a non-Newtonian fluid - such as a grease or a polymer containing oil (e.g., multi-grade oil)--shear stress is not proportional to the rate of shear. A non-Newtonian fluid may be said to have an apparent viscosity, a viscosity that holds only for the shear rate (and temperature) at which the viscosity is determined. Refer to *Brookfield viscosity*.

SHEET EXTRUSION

Sheet extrusion is also commonly referred to as the flat-film process. In this process the polymer melt is extruded through a slot die (T-shaped or "coat hanger" die). The die has relatively thick wall sections on the final lands (as compared to the extrusion coating die) to minimize deflection of the lips from internal melt pressure. The die opening (for polyethylene) may be 0.015 to 0.030 in. even for films that are less than 0.003 in. thick. The reason is that the speed of various driven rolls used for taking up the film is high enough to draw down the film with a concurrent thinning. By convention, the term film is used for material less than 0.010 in. thick, and sheet for that which is thicker. Following extrusion, the film may be chilled below Tm (melting point) or Tg (glass transition temperature) by passing it through a water bath or over two or more chrome-plated chill rollers which have been cored for water cooling. A schematic drawing of a chill-roll (also called cast-film) operation is shown in Figure 1. The polymer melt extruded as a web from the die is made dimensionally stable by contacting several chill rolls before being pulled by the powered carrier rolls and wound up. The chrome-plated surface of the first roll is highly polished so that the product obtained is of extremely high gloss and clarity.

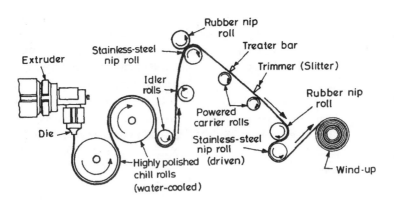

Figure 1. Chill-roll film extrusion.

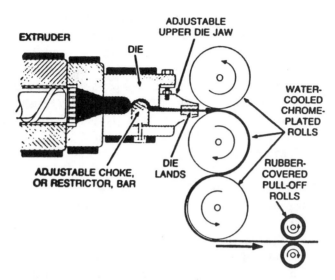

Figure 2. Cross-section of sheet extrusion die and three-roll stack.

In flat-film extrusion (particularly at high takeoff rates), there is a relatively high orientation of the film in the machine direction (i.e., the direction of the extrudate flow) and a very low one in the transverse direction. Biaxially oriented film can be produced by a flat-film extrusion by using a tenter. Polystyrene, for example, is first extruded through a slit die at about 190° C and cooled to about 120° C by passing between rolls. Inside a temperature-controlled box the moving sheet, rewarmed to 130° C, is grasped on either side by tenter-hooks which exert a drawing tension (longitudinal stretching) as well as a *widening tension* (lateral stretching). Stretch ratios of 3:1 to 4:1 in both directions are commonly employed for biaxially oriented polystyrene film.

Biaxial stretching leads to polymers of improved tensile strength. Commercially available oriented polystyrene film has a tensile strength of 10,000 to 12,000 psi, compared to 6,000 to 8,000 psi for unstretched material. Biaxial orientation effects are important in the manufacture of films and sheet. Biaxially stretched polypropylene, poly(ethyleneterephthalate) (e.g., Melinex) and poly(vinylidene chloride) (Saran) produced by flat-film extrusion and tentering are strong films of high clarity. In biaxial orientation, molecules are randomly oriented in two dimensions just as fibers would be in a random mat; the orientation-induced crystallization produces structures which do not interfere with the light waves. With polyethylene, biaxial orientation often can be achieved in blown-film extrusion.

In sheet extrusion, the key process factors are:

1. extruder output and melt uniformity versus extruder size and screw design
2. die design
3. chill roll design and control; thick sheet cooling
4. temperature and speed control

Process problems can vary, with the most significant ones being machine surging; dimples, pits or gels on the sheet surface; lines perpendicular to the extrusion direction; parabolic lines in the extrusion direction; rough, grainy or orange peel surface; low gloss; curling and warping; streak, discolorations, plate out; poor melt thermoforming. Figure 2 shows a cross section of a sheet extrusion die and three roll stack. (Source: Chanda, M. and S. K. Roy, Plastics Technology Handbook, Marcel Dekker, Inc., New York, NY, 1987). See also *Plunger-TypeTransfer Molding, Compression Molding and Screw Transfer Molding, Extruders and Extrusion* and *Thermoforming*.

SHELL AND TUBE TYPE HEAT EXCHANGER

Heat exchangers are a standard piece of equipment in any polymerization solution process. The standard to which heat exchangers are designed and constructed to is TEMA. The Tubular Exchanger Manufacturers Association, or TEMA, is a group of leading manufacturers, who have pioneered the research and development of heat exchangers for over fifty-five years. Founded in 1939, TEMA has grown to include a select group of member companies. Although it may be easy to choose a TEMA member as a supplier, it is not easy for manufacturers to become a TEMA member.

Member companies must meet stringent criteria to even qualify for TEMA membership, and are periodically examined by TEMA to ensure that the manufacturer meets membership criteria, and designs and manufacturers according to TEMA standards. Members adhere to strict specifications. TEMA Standards and Software have achieved worldwide acceptance as the authority on shell and tube heat exchanger mechanical design. These tools give engineers a valuable edge when designing and manufacturing all types of heat exchangers. Seven editions of TEMA Standards have been published, each updating the industry on the latest developments in technology.

TEMA has also developed engineering software that complements the TEMA Standards in the areas of flexible shell elements (expansion joints) analysis, flow induced vibration analysis and fixed tubesheet design and analysis. This state-of-the-art software works on an IBM PC or compatible, and features a materials data-bank of 38 materials, as well as user-friendly, interactive input and output screens. The programs handle many complex calculations, so users can focus on the final results. Many companies can manufacture heat transfer equipment, but not all can assure its safe, effective design and quality construction. That's why the TEMA Heat Exchanger Registration System was instituted in 1994. For quality assurance, one need only look for the TEMA Registration Plate attached to the heat exchanger. Each plate includes a unique TEMA registration number.

Before a company can even become a member of TEMA and participate in the registration system, it must have a minimum of 5 years of continuous service in the manufacture, design and marketing of shell and tube heat exchangers. All TEMA companies must have in-house thermal and mechanical design capabilities, and thoroughly understand current code requirements and initiate strict quality control procedures. Additionally, all welding must be done by the company's own personnel, and the company must have its own quality control inspectors. These criteria ensure the highest level of technical expertise, which gives TEMA members a meaningful advantage when designing or fabricating heat exchangers. The Tubular Exchanger Manufacturers Assn. has established heat exchanger standards and nomenclature. Every shell-and-tube device has a three-letter designation; the letters refer to the specific type of stationary head at the front end, the shell type, and the rear-end head type, respectively (a fully illustrated description can be found in the TEMA standards).

The shell and tube heat exchanger (the most common design) consists of a shell, usually a circular cylinder, with a large number of tubes, attached to an end plate and arranged in a fashion where two fluids can exchange heat without the fluids coming in contact with one another. The most common types of heat exchangers configurations are illustrated in the Figure 1.

There are many textbooks that describe the fundamental heat transfer relationships, but few discuss the complicated shell side characteristics. On the shell side of a shell and tube heat exchanger, the fluid flows across the outside of the tubes in complex patterns. Baffles are utilized to direct the fluid through the tube bundle and are designed and strategically placed to optimize heat transfer and minimize pressure drop.

A measure of the complexity of predicting shell side heat transfer can be obtained by considering the path of shell side fluid flow. The flow is partially perpendicular and partially paralleled to the tubes. It reverses direction as it travels around the baffle tips and the flow regime is governed by tube spacing, baffle spacing and leakage flow paths. Throughout the fluid path, there are a number of obstacles and configurations, which cause high localized velocities. These high velocities occur at the bundle entrance and exit areas, in the baffle windows, through pass lanes and in the vicinity of tie rods, which secure the baffles in their proper position.

In conjunction with this, the shell side fluid generally will take the path of least resistance and will travel at a greater velocity in the free areas or by-pass lanes, than it will through the bundle proper, where the tubes are on a closely spaced pitch. All factors considered, it appears a formidable task to accurately predict heat transfer characteristics of a shell and tube exchanger. The problem is further complicated by the manufacturing tolerances or clearances that are specified to allow assembly and disassembly of the heat exchanger.

It is improbable that these clearances will all accumulate to either the positive or negative side, so it is customary to compute heat transfer relationships on the basis of average clearances. Refer to Plate and Frame Heat Exchanger, Spiral Plate Heat Exchanger, and U-Tube Heat Exchanger. (Source: N. P. Cheremisinoff, *Handbook of Chemical Processing Equipment*, Butterworth-Heinemann Publishers, U.K., 2000).

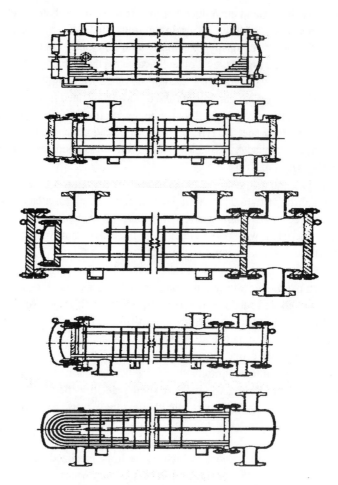

Figure 1. Common shell-and-tube exchanger configurations.

SHELLAC

Solution or a resin exuded by a scale insect, in alcohol or acetone. The color ranges from light yellow to orange; the darker shellacs are the less pure. When bleached it is known as white shellac. Applied to surfaces, e.g., wood, shellac forms a hard coating when the solvent evaporates. Shellac is used as a spirit varnish, as a protective covering for drawings and plaster casts, for stiffening felt hats, and in electrical insulation.

SHEWHART CONTROL CHART

The Shewhart control chart is a common control charting method applied in

SPC (Statistical Process Control). It is a model based on a zero order polynomial. The basis for this model is the assumption that the variations lying inside the control limits are the results of random causes and the variations lying outside the control limits are the results of assignable causes. These two types of variations are referred to as *fluctuations* and *interruptions* in a process operation.

The model assumes that the random causes are inherent in the current process and it is usually uneconomical to investigate and eliminate these variations within the confines of current technology. However, the assignable causes are external to the current process, and it is usually economical to intensively search for and eliminate the root causes and, thus, effectively optimize the application of the current technology. Hence, the approach to quality improvement is the separation of the variations internal to the process from those that are external to the process.

Deming's contribution to quality improvement is the recognition that, while operating personnel can deal with the assignable causes, only management can change the process fundamentally and reduce the effects of the random causes. While the Shewhart Control Chart is a good model for a process with two types of variations (fluctuation and drift) it is not a very effective model for processes which have significant drifts.

One approach that has been employed to provide a degree of separation between interruptions and drift is the use of moving averages as statistics for the Shewhart Control Chart. The benefit is the reduction of the influence of a single data point caused by an interruption. The drawback, however, is the lengthening of response time. The Shewhart control chart is illustrated in Figure 1.

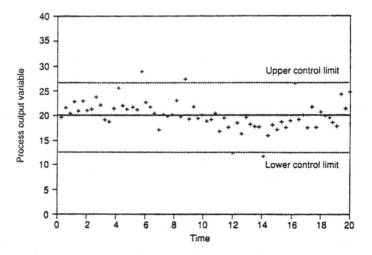

Figure 1. Illustrates the Shewhart control chart.

SHORT-TERM INHALATION LIMITS

The parts of vapor (gas per million parts of contaminated air by volume) at 25°C (77° F) and atmospheric pressure are given. The limits are normally given in milligrams per cubic meter for chemicals that can form a fine mist or dust. The values given are the maximum permissible average exposures for the time periods specified. The term Short Term Exposure Limit (STEL) is also used and is considered interchangeable with Short Term Inhalation Limit. The STEL designation is derived from OSH standards. In some instances the values disagree, or the short-term limits overlap the TLV. These are not errors; the values were supplied by several laboratories, each of which used its own experimental techniques and methods of calculation.

SILICONE

Special purpose rubbers of three main types of substituted siloxanes which, considered as a group, offer exceptional resistance to extremes of temperature. General properties include poor to moderate mechanical properties and swelling resistance to liquids; excellent resistance to extremes of temperature; good resistance to ozone and sunlight; good electrical properties. Approximate working temperature range -100° C to +250° C. Typical uses include high temperature gaskets, fabric coatings, as a sealant.

SOCIETY OF PLASTICS ENGINEERS (SPE)

The SPE promotes scientific and engineering knowledge relating to plastics. SPE is a professional society of plastics scientists, engineers, educators, students, and others interested in the design, development, production, and utilization of plastics materials, products, and equipment. SPE currently has over 22,000 members scattered among its 80 sections. The individual sections as well as the SPE main body arrange and conduct monthly meetings, conferences, educational seminars, and plant tours throughout the year. SPE also publishes *Plastics Engineering, Polymer Engineering and Science, Plastics Composites,* and the *Journal of Vinyl Technology*. The society presents a number of awards each year encompassing all levels of the organization: section, division, committee, and international. SPE divisions of interest are color and appearance, injection molding, extrusion, electrical and electronics, thermoforming, engineering properties and structure, vinyl plastics, blow molding, medical plastics, plastics in building, decorating, mold making, and mold design.

SOCIETY OF PLASTICS INDUSTRY (SPI)

The SPI is a major society, whose membership consists of manufacturers and processors of plastic materials and equipment, The society has four major operating units consisting of the Eastern Section, the Midwest Section, the New England Section, and the Western Section. SPI's Public Affairs Committee concentrates on coordinating and managing the response of the plastics industry to issues like toxicology, combustibility, solid waste, and energy.

The Plastic Pipe Institute is one of the most active divisions, promoting the proper use of plastic pipes by establishing standards, test procedures, and specifications. Epoxy Resin Formulators Division has published over 30 test procedures and technical specifications. Risk management, safety standards, productivity, and quality are a few of the major programs undertaken by the machinery division. SPI's other divisions include Expanded Polystyrene Division, Fluoropolymers Division, Furniture Division, International Division, Plastic Bottle Institute, Machinery Division, Molders Division, Mold Makers Division, Plastic Beverage Container Division, Plastic Packaging Strategy Group, Polymeric Materials Producers Division, Polyurethane Division, Reinforced Plastic/Composites Institute, Structural Foam Division, Vinyl Siding Institute, and Vinyl Formulators Division. The National Plastics Exposition and Conference, held every 3 years by the Society of Plastic Industry, is one of the largest plastic shows in the world.

SOLUBILITY

The value represents the pounds of a chemical that will dissolve in 100 pounds of pure water. Solubility usually increases when the temperature increases. The following terms are used when numerical data are either unavailable or not applicable: The term "Miscible" means that the chemical mixes with water in all proportions. The term "Reacts" means that the substance reacts chemically with water; thus, its solubility has no real meaning. "Insoluble" usually means that one pound of the chemical does not dissolve entirely in 100 pounds of water. (Weak solutions of "Insoluble" materials may still be hazardous to humans, fish, and waterfowl, however.) Table 1 provides solubility parameters of polymers and various solvents.

Table 1. Solubility Parameters of Polymers and Solvents

Polymers	δ (cal$^{1/2}$ cm$^{1/2}$)	Solvents	δ (cal$^{1/2}$ cm$^{1/2}$)
PTFE	6.2	n-Pentane	6.3
PE	8.0	n-Hexane	7.3
PP	7.9	n-Octane	7.6
PIB	8.0	Diisopropylketone	8.0
SBR	8.1-8.5	Cyclohexane	8.2

Table 1 continued

Polymers	δ (cal$^{1/2}$ cm$^{1/2}$)	Solvents	δ (cal$^{1/2}$ cm$^{1/2}$)
NR	8.1	Carbon tetrachloride	8.6
BR	8.5	Toluene	8.9
PS	8.5-9.6	Ethyl acetate	9.1
CR	9.2	Dioxane	9.9
PVAC	9.4	Acetone	10.0
PVC	9.6	Pyridine	10.9
PET	10.7	Ethanol	12.7
CA	11.4	Methanol	14.5
EP	11.0	Glycerol	16.5
POM	11.1	Water	23.4
PA	13.5		
PAN	15.5		

SOLUTION POLYMERIZATION

The conducting of polymerization reactions in a solvent is an effective way to disperse heat; in addition, solutions are much easier to stir than bulk polymerizations. Solvents must be carefully chosen, however, so that they do not undergo chain-transfer reactions with the polymer. Because it can be difficult to remove solvent from the finished viscous polymer, solution polymerization lends itself best to polymers that are used commercially in solution form, such as certain types of adhesives and surface coatings. Polymerization of gaseous monomers is also conducted with the use of solvents, as in the production of polyethylene. Refer to *Bulk Polymerization* and *Suspension Polymerization*.

SOLVENT

Compound with a strong capability to dissolve a given substance. The most common petroleum solvents are mineral spirits, xylene, toluene, hexane heptane, and naphthas. Aromatic-type solvents have the highest solvency for organic chemical materials, followed by naphthenes and paraffins. In most applications the solvent disappears, usually by evaporation, after it has served its purpose. The evaporation rate of a solvent is very important in manufacture: rubber cements often require a fast-drying solvent, whereas rubber goods that must remain tacky during processing require a slower-drying solvent. Solvents have a wide variety of industrial applications, including the manufacture of paints, inks, cleaning products, adhesives, and petrochemicals.

SPECIFIC GRAVITY

The ratio of the mass of a given volume of product and the mass of an equal volume of water, at the same temperature. The standard reference temperature is 15.6° C (60° F). Specific gravity is determined by test method ASTM D792: the higher the specific gravity, the heavier the product. Specific gravity of a liquid can be determined by means of a hydrometer, a graduated float weighted at one end, which provides a direct reading of specific gravity, depending on the depth to which it sinks in the liquid. A related measurement is density, an absolute unit defined as mass per unit volume--usually expressed as kilograms per cubic meter (kg/m^3). Petroleum products may also be defined in terms of API gravity (also determinable by ASTM D 1298). Table 1 provides average literature reported values of the specific gravities of several common polymers.

Table 1. Specific Gravities of Common Polymers

Polymer	Specific Gravity
Polybutylene	0.60
Polymethylpentane	0.83
Ethylene-Propylene	0.86
Polypropylene	0.90-0.92
Polyethylene	
LDPE	0.91-0.93
LLDPE	0.91-0.94
HDPE	0.96-0.97
Polybutene	0.91-0.92
Natural Rubber	0.91
Butyl Rubber	0.92
Styrene-butadiene	0.93
Polyaminde	
PA-12	1.02
PA-11	1.04

Table 1 continued

Polymer	Specific Gravity
PA-6	1.12-1.13
PA-66	1.13-1.15
ABS	1.04-1.07
Ploystyrene	1.05
Polyacronitrile	1.17
Polyvinyl Acetate	1.19
Polycarbonate	1.2
Polychloroprene Rubber	1.23
Polysulphone	1.24
Polyethylene Terephtalate	1.34-1.39
PVC	1.37-1.39
POM	1.41-1.43
Polytetrafluorethylene	2.27

SPECIFIC HEAT

Ratio of the quantity of heat required to raise the temperature of a substance one degree Celsius (or Fahrenheit) and the heat required to raise an equal mass of water one degree.

SPIRAL-PLATE HEAT EXCHANGER

A spiral-plate exchanger is fabricated from two relatively long strips of plate, which are spaced apart and wound around an open, split center to form a pair of concentric spiral passages. Spacing is maintained uniformly along the length of the spiral by spacer studs welded to the plate.

For most services, both fluid-flow channels are closed by welding alternate channels at both sides of the spiral plate (see Figure 1). In some applications, one of the channels is left completely open on both ends and the other closed at both sides of the plate (see Figure 2). These two types of construction prevent the fluids from mixing.

Spiral-plate exchangers are fabricated from any material that can be cold worked and welded. Materials commonly used include: carbo steel, stainless steel, nickel and nickel alloys, titanium, Hastelloys, and copper alloys. Baked phenolic-resin coatings are sometimes applied.

Electrodes can also be wound into the assembly to anodically protect surfaces against corrosion. Spiral-plate exchangers are normally designed for the full pressure of each passage. The maximum design pressure is 150 psi because the turns of the spiral are of relatively large diameter, each turn must contain its design pressure, and plate thicknesses are somewhat limited. For smaller diameters, however, the design pressure may sometimes be higher. Limitations of the material of construction govern design temperatures.

The spiral assembly can be fitted with covers to provide three flow patterns: (1) both fluids in spiral flow, (2) one fluid in spiral flow and the other in axial flow across the spiral, and (3) one fluid in spiral flow and the other in a combination of axial and spiral flow. For spiral flow in both channels, the spiral assembly includes flat covers at both sides (see Figure 3). In this arrangement, the fluids usually flow counter-currently, with the cold fluid entering at the periphery and flowing toward the core, and the hot fluid entering at the core and flowing toward the periphery.

For this arrangement, the exchanger can be mounted with the axis of the spiral either vertical or horizontal. This arrangement finds wide application in liquid-to-liquid service, and for gases or condensing vapors if the volumes are not too large for the maximum flow area of 72 square inches.

For spiral flow in one channel and axial flow in the other, the spiral assembly includes conical covers, dished heads, or extensions with flat covers. In this arrangement, the passage for axial flow is open on both sides and the spiral flow channel is sealed by welding on both sides of the plate. Exchangers with this arrangement are suitable for services in which there is a large difference in the volumes of the two fluids. This includes liquid-liquid service, heating or cooling gases, condensing vapors, or boiling liquids.

Fabrication can provide for single pass or multipass on the axial-flow side. This arrangement can be mounted with the axis of the spiral either vertical or horizontal. It is usually vertical for condensing or boiling.

For combination flow, a conical cover distributes the axial fluid to its passage (see Figure 4). Part of the open spiral is closed at the top, and the entering fluid flows only through the center part of the assembly. A flat cover at the bottom forces the fluid to flow spirally before leaving the exchanger.

This type is most often used for condensing vapors and is mounted vertically. Vapors flow first axially until the flow volume is reduced sufficiently for final condensing and subcooling in spiral flow.

A modification of combination flow is the column-mounted condenser. A bottom extension is flanged to mate the column flange. Vapor flows upward through a large central tube and then axially across the spiral, where it is condensed. Subcooling may be achieved by falling-film cooling or by controlling a level of condensate in the channel. In the latter case, the vent stream leaves in spiral flow and is further cooled. The column mounted condenser can also be designed for updraft operation and allows condensate to drop into an accumulator with a minimum amount of subcooling. (Source: N. P. Cheremisinoff, *Handbook of Chemical Processing Equipment*, Butterworth-Heinemann Publishers, U.K., 2000).

Figure 1. Spiral flow in both channels.

Figure 2. Flow is both spiral and axial.

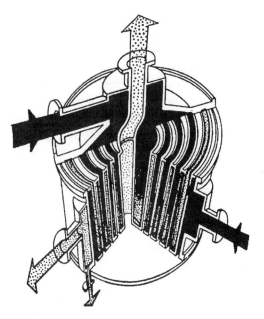

Figure 3. Combination flow used to condense vapors.

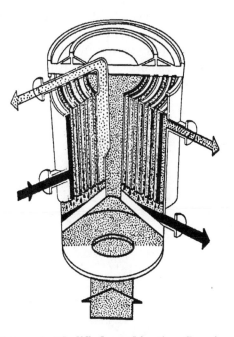

Figure 4. Modified combination flow in a column mounted design.

STABILIZERS

In order for a plastic or elastomer to have a long and useful life in any application, the properties of that material should change as little as possible with time. Stabilizers are added, usually in small quantities, to counter the effects of aging. Because all carbon-based polymers are subject to oxidation, the most common stabilizers are antioxidants. Hindered phenols and tertiary amines are used in plastics in concentrations as low as a few parts per million. For example, butylated hydroxytoluene (BHT) is used in polyolefin packaging films for foods and pharmaceuticals. PVC requires the addition of heat stabilizers in order to reduce dehydrohalogenation (loss of hydrogen chloride [HCl]) at processing temperatures. Zinc and calcium soaps, organotin mercaptides, and organic phosphites are among the many additives found to be effective. Other stabilizers are designed specifically to reduce degradation by sunlight, ozone, and biological agents.

STEPWISE SPC CHART

The Stepwise SPC Chart is a method that combines the functions of the CUSUM Chart and the Shewhart Control Chart. A sequential analysis is used to estimate the current mean of the process. Shewhart Control Charts are then constructed about the step functions representing the current mean. Figure 1 illustrates the Stepwise Control Chart. The step functions, based on sequential analysis, respond to process drift the same way as the CUSUM Chart. Each segment of the control chart then helps to identify the interruptions, in the same way as the Shewhart Control Chart does. With the proper choice of parameters, the Stepwise SPC Chart can effectively separate the three types of variations: drift, fluctuations and interruptions, and thus help us to monitor and analyze the quality of the product. Refer to *EWMA Chart*.

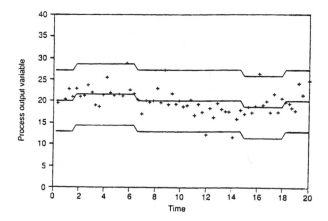

Figure 1. Stepwise SPC chart.

STODDARD SOLVENT

Mineral spirits with a minimum flash point of 37.8° C (100° F), relatively low odor level, and other properties conforming to Stoddard solvent specifications, as described in test method ASTM D 484. Though formulated to meet dry cleaning requirements, Stoddard solvents are widely used wherever this type of mineral spirits is suitable.

STRESS STRAIN BEHAVIOR

From a stress-strain standpoint, one can distinguish four types of polymers:

1. Flexible thermoplastics; i.e., materials that are capable of large plastic deformations, an example is PE;

2. Rigid thermoplastics (e.g., PVC, PS);

3. Rigid thermosets (e.g., EP, PF);

4. Elastomers or rubbers.

Table 1 provides property data which represent average values and can serve as a rough guide.

Table 1

Stress Strain Data for Polymers

Polymer	Spec. Gravity	Tens. Strength (psi)	Tens. Mod. of Elasti-city (psi)	Flex Strength (psi)	Flex. Mod. of Elasti-city (psi)	Compressive Strength (psi)	Rock-well Hard-ness
ABS	1.04	2500-8500	100000-350000	3500-14000	120000-150000	2400-10500	R86-101
Acrylic	1.19	8500	300000-450000	14500	350000-500000	14000-17500	M90
Acetal	1.41	8900-12000	400000-500000	13000-15000	370000-550000	-	R120
Nylon	1.15	9000-12000	250000-380000	12000-14000	170000-400000	5000	R110
LDPE	0.91	1400	30000	1200	5000	4000	10
HDPE	0.94	3500	150000	1400	150000	4600	30-50
PC	1.2	9500	300000	12000	300000-500000	10000-15000	R110

Table 1 continued

Polymer	Spec. Gravity	Tens. Strength (psi)	Tens. Mod. of Elasti-city (psi)	Flex Strength (psi)	Flex. Mod. of Elasti-city (psi)	Compre ssive Strength (psi)	Rock-well Hard-ness
PP	0.9	4500	180000	6500	220000-240000	6700	95
PVC	1.45	7500	300000-550000	12500	400000	10000	R115
TFE	1.52	7800	400000-430000	14600	475000	11400	R118

STYRENE-BUTADIENE RUBBER (SBR)

A copolymer of about 25% styrene and 75% butadiene. A general purpose synthetic rubber which is competitive in properties with NR. Remains amorphous when stretched and therefore the tensile strength of gum (unfilled) SBR is only 10-15% of the tensile strength of gum NBR. Available hardness range 35-100 IRHD. This rubber has very good mechanical properties when reinforced with carbon black; very good abrasion resistance; slightly less resilient than NR. Swelling and adhesion properties are similar to NR. Aging resistance in general slightly superior to NR. Approximate working temperature range - 60° C to +70° C. The rubber can be oil-extended.

STYRENE MALEIC ANHYDRIDE

The copolymerization of styrene with maleic anhydride creates with a copolymer (SMA) which has a higher glass transition temperature than polystyrene and is chemically reactive with certain functional groups. Thus, SMA polymers are often used in blends or composites where interaction or reaction of the maleic anhydride provides for desirable interfacial effects. The anhydride reaction with primary amines is particularly potent.

SULFUR VULCANIZATION

Sulfur vulcanization can be divided into two main categories: unaccelerated and accelerated sulfur vulcanization. Unaccelerated formulations typically consist of sulfur, zinc oxide, and a fatty acid such as stearic acid, while accelerated formulations include an accelerator in the system. A subcategory of accelerated sulfur vulcanization is sulfur-free systems, also referred to as sulfur-donor systems. In these systems the sulfur needed for network formation is supplied by the accelerator, which functions as both an accelerator and sulfur-donor. It should be noted that while unaccelerated sulfur systems are no longer of commercial significance, they are of interest as a starting point to understanding accelerated sulfur vulcanization systems.

It is generally believed that sulfur vulcanization is promoted by a free radical mechanism, although some literature suggests a polar mechanism of vulcanization. There is also evidence that both free radical and polar mechanisms are operative depending on the formulation and curing conditions. A major source of the uncertainty has arisen from the intractable nature of the cured vulcanizate. Cured elastomers are insoluble; this eliminates most analytical techniques for examining polymeric structures.

To avoid these problems, several methods have been used, including model compound work, electron spin resonance (ESR), the use of chemical probes, radical scavengers, solid-state C13 -NMR, and analysis of the extranetwork materials. The chemical probe work has allowed characterization and quantification of the number of mono-, di-, and polysulfidic crosslinks. The model compound work has been useful in providing information for mechanistic studies by allowing comparison of products predicted by a mechanism to the products obtained from model compound work. Electron spin resonance and radical scavenger work has been used to examine whether radical processes are active during vulcanization with various formulations and elastomers. Solid-state NMR work has been used to help elucidate the products of vulcanization. Analysis of the extra-network material has provided insights into the nature of the vulcanization chemistry by analysis of content of various intermediates as cure progresses. A major contributing factor to the disagreement over mechanisms is the possible reactions that sulfur and accelerators can undergo. Sulfur occurs naturally as an eight-membered ring; this ring is capable of both homolytic cleavage to form radicals or heterolytic cleavage to form ions. Also, the precise interaction of the accelerator and sulfur in the vulcanization process has not been definitively elucidated. It is known that accelerator complexes are formed, but the actual sulfurating species has not been determined. This question and the possible reactions of sulfur and accelerators has made it impractical to eliminate either the radical or polar mechanisms. The reaction mechanisms of unaccelerated sulfur vulcanization has been a subject of research interest for several decades. Despite numerous works designed to elucidate the reaction mechanisms, the actual mechanism has not been conclusively proved. Model compound work has been useful in providing information on the nature of the cross-link structures. Model compound work for unaccelerated sulfur vulcanization has included the use of mono- and diolefins.

The polyisoprene model compound work has indicated that the structures in elemental sulfur vulcanization are alkylalkenyl and alkyl-alkyl in nature. These studies observed migration of the olefinic double bonds and formation of conjugated trienes and internally cyclized structures. The polybutadiene model compound work found significant saturation of the double bond and formation of vicinal sulfur structures. Upon addition of zinc oxide to both systems, the amount of alkenyl substitution increased. Addition of zinc oxide dramatically changes the stereochemistry of the vicinal structures. Free-radical sulfurization is believed to be initiated by homolytic scission of the sulfur ring leading to formation of radicals. Proton abstraction by this radical and sulfur addition to the resulting rubber radical gives rise to the propagating sulfurization species. This mechanistic scheme is able to account for alkenyl-alkyl products. Alkyl-alkyl products and vicinal crosslinks are formed if the crosslinked rubber radical added sulfur instead of abstracting a proton.

Isomerization and doublebond migration is also accounted for by proposed schemes. The rubber radical exists in two allylic positions. Sulfur can add to the radical at either position; addition of sulfur at the quaternary carbon results in double-bond migration. Isomerization occurs when the radical returns to the methylene carbon; the double bond is then capable of reforming in either configuration. The proposed polar mechanism for unaccelerated sulfur vulcanization allows for either proton or hydride transfer. The key step is the formation of the three-membered, sulfur-carbon charged ring. This reaction mechanism is also able to predict the formation of alkenyl-alkyl structures. Alkyl-alkyl structures result from the saturated persulfenyl rubber ion reacting with a rubber molecule, followed by proton transfer. Isomerization occurs through the nonsulfurated saturated ion. If this ion undergoes proton transfer then the double bond reforms; this can occur in either configuration. Formation of noncross-linked saturated structures and conjugated trienes are formed from rubber moieties that undergo proton abstraction and then either proton transfer or hydride abstraction. This leads to the formation of cyclic structures.

The sulfur cross-link can cleave at an S-S bond, which is a relatively labile bond. The sulfur atom then becomes bonded to a carbon to form the cyclic structure. Related to the formation of cyclic structures is the reduction of rank of crosslinks with increasing cure time. There are several proposed reaction schemes; the mechanism that is commonly favored is an exchange reaction between the crosslink and the sulfurating intermediates. This reaction terminates upon formation of monosulfidic structures. Based on available evidence, unaccelerated vulcanization is believed to occur by a polar mechanism. A radical mechanism should give much higher levels of migration of the double bond; this arises from the stability of the radical at the quaternary carbon. Electron spin resonance studies and the radical scavenger work also point toward polar reactions. Due to the ability of sulfur to undergo radical reactions, it is not possible to completely eliminate the possibility of radical reactions during unaccelerated sulfur vulcanization. The study of accelerated sulfur vulcanization suffers from the same problems as unaccelerated sulfur vulcanization, including the ability of sulfur to undergo both radical and ionic reactions and the intractable nature of cured vulcanizates. Further complicating the situation is the necessity of understanding how the accelerators and activators interact, and how these interactions affect the vulcanization mechanism. Proposed mechanisms have ranged from radical to polar.

Several researchers have concluded that both radical and polar mechanisms are operative and that the precise nature is dependent on the formulation. A typical accelerated sulfur system contains sulfur, accelerator, zinc oxide, and a saturated fatty acid such as stearic or lauric acid. Other additives can include stabilizers, antioxidants, and fillers such as silica or carbon black. There are also sulfur-free systems; the most widely used sulfurless system is zinc oxide and tetramethyl thiuram disulfide. Additionally, other accelerators can be added to thiocarbamate systems to increase the induction time. The most widely used compound for this application is MBT. Accelerated sulfur vulcanization has been found to consume the accelerator in the system at a rate far greater than the rate of cross-linking. This has led to the proposal that accelerated vulcanization proceeds through an intermediate.

SURFACE GRAFTING

Polymers can be modified by graft polymerization to change their physical and chemical properties for such applications as biomaterials, membranes, and adhesives. The key advantage of these techniques is that the surface of the same polymer can be modified to have very distinctive properties through the choice of different monomers. Grafting is most commonly performed by irradiating the polymer in the presence of a solvent containing a monomer. This direct (mutual irradiation) method, which involves only one step, is very efficient. Grafting can occur at the surface and in bulk. The depth of grafting is mostly controlled by the interaction between the substrate polymer and the solvent. If penetration of the solvent is limited to a very small depth, surface grafting predominates. At the other extreme, if the solution can diffuse uniformly into the polymer film, then homogeneous grafting occurs. This method cannot totally eliminate homopolymerization initiated by free radicals formed during irradiation of the monomer.

For surface modification applications, thick grafting layers are unnecessary and even undesirable because they may change bulk physical properties of the polymer, such as crystallinity and tensile modulus. A two-step method can be used to minimize the formation of the homopolymer. The polymer is preirradiated in air to produce peroxide groups on the surface. Grafting is subsequently initiated thermally in contact with a monomer. Other methods such as corona discharge, ozone treatment, and plasma treatment have also been used to generate peroxide groups on polymer surfaces.

The energy sources most commonly used in radiation grafting are high-energy electrons, T radiation, X-rays, ultraviolet (UV), and visible light. Electron-beam radiation, γ radiation, and X-rays are classified as ionization radiation. The study of the chemical effects caused by ionization radiation is referred to as radiation chemistry. Ionization radiation is not selective and produces a large number of ionized and excited molecules along its track.

The study of the chemical effects produced by UV or visible light is known as photochemistry. In photochemistry, the absorption of a photon causes excitation of the reacting molecule to a definite, usually known excited electronic state. The principal difference between radiation chemistry and photochemistry is that the energy of the ionization radiation used in radiation chemistry is much greater than that of the light source used in photochemistry.

SUSPENSION POLYMERIZATION

In this type of polymerization the monomer is dispersed in a liquid (usually water) by vigorous stirring and by the addition of stabilizers such as methyl cellulose. A monomer-soluble initiator is added in order to initiate chain-growth polymerization. Reaction heat is efficiently dispersed by the aqueous medium. The polymer is obtained in the form of granules or beads, which may be dried and packed/bagges directly for shipment. Refer to *Bulk Polymerization*, *Emulsion Polymerization*, and *Solution Polymerization*.

SWELL AND SHRINKAGE TESTS

Rheological test more often used for product quality control is a measurement of the elastic recovery in polymer melts and their compounds. This is usually accomplished by measuring the swell (or shrinkage) of materials undergoing extrusion. The tendency of polymers, whether thermoplastic or rubber, to enlarge when emerging from an extruder die is called *die swell*.

Die swell normally refers to the ratio of extruded size to die size. This behavior is a measure of the relative elasticity in the flowing polymer stream. Die swell is caused by the release of the residual stresses when the sample emerges from the die. Measurement of this behavior has become widely recognized in the rubber and plastics industries as an important indication of polymer processability.

Traditional measurements reported by various investigators have been made on extrudates either with a micrometer or by weight per unit length. The Monsanto Industrial Chemicals Co. (Akron, OH) has introduced an automated capillary rheometer (the Monsanto Processability Tester) which employs an extrudate swell detector based on interaction with a scanning laser beam positioned immediately below the capillary die exit. This enables measurement of the running die swell.

Figure 1 shows running die swell data for a compound correlated against wall shear for different L/D dies. It is important to note that with some materials a distinction between running die swells and relaxed die swells should be made. Highly elastic materials will tend to swell more over time depending on their elastic recovery energy.

C. W. Brabender Instruments, Inc. (Hackensack, NJ) markets an optical, electronic die swell measuring system that can be used for continuous control and monitoring of extruders and plasticorders. The basic design features are illustrated in Figure 2. An infrared, long-life diode (A) emits homogeneous infrared light through a filter (B), diffuser (C), and lens (D); then through a window (E). The light impinging on the object (F) is scattered.

Only that light reaching the ambient light filter (G) is measured by the photodetector Hl. Light reaching photodetector H2 is diverted through the prism (1) and is uninterrupted. This acts as a reference beam. The electronic instrumentation measures the differences between photodetector HI and H2 and calculates the object thickness to an accuracy of 0.2% of full scale. The object to be measured must be located vertically within the measuring area. The maximum permissible deviation from vertical is 2.5°. Operation at angles >2.5° from vertical will yield greater errors. If the measuring angle is > 2.5° from vertical, the measurement will be increased by the factor 1/cos a, where a is deviation from vertical.

While the upper collimated light serves as the measuring beam, the lower portion impacts on the second photodetector and acts as the reference beam. Thus, the temperature and age affected by drift which occurs on all semiconductor elements is negated. In some applications such as hose and tubing manufacturing, it is important to know both the extrudate diameter and cross-section. The device described can provide two diameter measurements, which will give an out-of-round indication (eccentricity).

Both swell and eccentricity are important to a variety of parts profiles that are made from extrusion operations. Sealing applications for automotive components, wire and cable insulation, flexible tuning, gaskets, and many other parts can be affected by swell and eccentricity characteristics.

Swell is essentially a measure of the viscoelastic properties of the polymer, which in turn is related to the neat polymer properties (molecular weight, molecular weight distribution, degree of crystallinity, degree of long-chain branching, cure performance of the polymer, and compounding ingredients, as well as process conditions.

Polymer designers need to focus on processing performance as quite often polymer grades designed for particular applications or parts often fail due to poor processability. Swell and shrinkage tests can be considered fundamental tests that should be run as a qualitative assessment of processability.

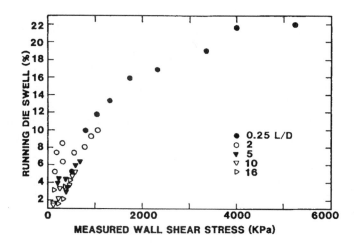

Figure 1. Running die swell data from extrusion tests.

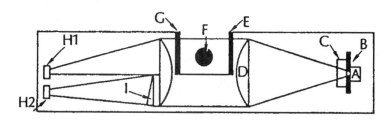

Figure 2. Optical die swell tester.

SYNTHETIC RUBBER

Any petrochemical-based elastomer. Like natural rubber, synthetic rubbers are polymers, consisting of a series of simple molecules, called monomers, linked together to form large chain-like molecules. The chain forms a loose coil that returns to its coiled form after it is extended. See under individual listings: butyl rubber, ethylene-propylene rubber, natural rubber, neoprene rubber, nitrile rubber, polybutadiene rubber, polyisoprene rubber, styrenebutadiene rubber. The terms rubber and elastomer are used interchangeably, generally referring to a material that can be stretched to at least twice its original length after release of the force applied in the stretching.

The properties of synthetic rubbers can be greatly enhanced by the incorporation of additives like carbon black. Refer to *Compounding*, *Mill Mixing*, *Carbon Black*, *Vulcanization*, *Sulfur Vulcanization*, *Peroxide Cure*, and *Half Life*.

T

THERMAL ANALYSIS

Thermal analysis refers to a variety of techniques in which a property of a sample is continuously measured as the sample is programmed through a predetermined temperature profile. Among the most common techniques are thermal gravimetric analysis (TA) and differential scanning calorimetry (DSC). In TA the mass loss versus increasing temperature of the sample is recorded. The basic instrumental requirements are simple: a precision balance, a programmable furnace, and a recorder. Modern instruments, however, tend to be automated and include software for data reduction. In addition, provisions are made for surrounding the sample with an air, nitrogen, or an oxygen atmosphere.

In a DSC experiment the difference in energy input to a sample and a reference material is measured while the sample and reference are subjected to a controlled temperature program. DSC requires two cells equipped with thermocouples in addition to a programmable furnace, recorder, and gas controller. Automation is even more extensive than in TA due to the more complicated nature of the instrumentation and calculations. A thermal analysis curve is interpreted by relating the measured property versus temperature data to chemical and physical events occurring in the sample. It is frequently a qualitative or comparative technique. In TA the mass loss can be due to such events as the volatilization of liquids and the decomposition and evolution of gases from solids. The onset of volatilization is proportional to the boiling point of the liquid. The residue remaining at high temperature represents the percent ash content of the sample. In DSC the measured energy differential corresponds to the heat content (enthalpy) or the specific heat of the sample.

DSC is often used in conjunction with TA to determine if a reaction is endothermic, such as melting, vaporization and sublimation, or exothermic, such as oxidative degradation. It is also used to determine the glass transition temperature of polymers. Liquids and solids can be analyzed by both methods of thermal analysis. The sample size is usually limited to 10-20 mg. Thermal analysis can be used to characterize the physical and chemical properties of a system under conditions that simulate real world applications. It is not simply a sample composition technique. Much of the data interpretation is empirical in nature and more than one thermal method may be required to fully understand the chemical and physical reactions occurring in a sample. Condensation of volatile reaction products on the sample support system of a TA can give rise to anomalous weight changes.

A simple example of the relationship between "structure" and "properties" is the effect of increasing molecular weight of a polymer on its physical (mechanical) state; a progression from an oily liquid, to a soft viscoelastic solid, to a hard, glassy elastic solid. Even seemingly minor rearrangements of atomic structure can have dramatic effects as, for example, the atactic and syndiotactic stereoisomers of polypropylene - the first being a viscoelastic amorphous polymer at room temperature while the second is a strong, fairly rigid plastic with a melting point above 160°C. At high thermal energies conformational changes via bond rotations are frequent on the time scale of typical processing operations and the polymer behaves as a liquid (melt). At lower temperatures the chains solidifies by either of two mechanisms: by ordered molecular packing in a crystal lattice, crystallization, or by a gradual freezing out of long range molecular motions, vitrification. These transformations, which define the principal rheological regimes of mechanical behavior: the melt, the rubbery state, and the semicrystalline and glassy amorphous solids, are accompanied by transitions in thermodynamic properties at the glass transition temperature, the crystalline melting, and the crystallization temperatures.

Thermal analysis techniques are designed to measure the above mentioned transitions both by measurements of heat capacity and mechanical modulus (stiffness). Refer to *Differential Scanning Calorimetry* and *Thermogravirnetric Analysis*. (Source: Cheremisinoff, N.P. *Polymer Characterization: Laboratory Techniques and Analysis*, Noyes Publishers, New Jersey, 1996).

THERMOFORMING

Thermoforming refers to a fabrication process which essentially involves the following three steps:
1. Heating of a plastic sheet to its softening temperature;
2. Forcing the hot, flexible material against the contours of a mold;
3. Cooling the sheet to reatin the mold's shape and detail.

The types of thermoforming processes are vacuum forming, pressure forming, melt phase thermoforming, and matched mold forming. Examples of applications of thermoformed parts are:

Packaging (75-80 % of Total Thermoforming Resin Volume

Barrier Food Containers	Portion (Creamers, Syrup, etc.); Unit Dose Drugs
Convenience and Carry-out Containers, Deli and Dairy Containers	Blister and Bubble Packs (Hardware, Batteries, etc.)
Meat and Poultry Trays, Egg Cartons	Electronics. Tool and Cosmetics Cases and Packages
Form-Fill-Seal (Jelly, Crackers, etc.)	Vending Machine and Other Disposable Cups, Trays, Containers

Other Applications

Automotive Door Innerliners, Utility Shelves	Camper Hardtops and Doors; Motorcycle Windshields, Shrouds, Mudguards

Truck Cab Door Fascia, Instrument Cluster Fascia
Snowmobile Shrouds and Windshields, Golf Cart Shrouds, Seats, Trays
Building Shutters, Skylights, Window Fascia, Storage Modules, Bath and Shower Surrounds
Exterior and Interior Signs
Boat Hulls, Surf Boards

Recreational Vehicle Interior Components, Window Blisters
Tote Bins, Pallets, Trays

Refrigerator Door Liners, Bins

Swimming and Wading Pools
Luggage, Hampers, Carrying Cases, Animal Containers

The most common polymer materials that are converted in thermoforming processes are: ABS, PMMA, Cellulosics, LDPE, HDPE, PET, PP, PS, PVC. The advantages of thermoforming over injection molding are lower equipment costs, ability to make thinner walls, ability to make large surface area parts, shorter possible lead time from conception to production, and less costly model changes due to less expensive tolling. The disadvantages compared to injection molding are first - it is a two-step process (extrusion plus thermoforming); more scrap and regrind is generated; there tends to be more part to part variation; wall thickness is less adjustable and more variable; lower surface gloss; less part complexity. Figures 1 and 2 illustrate some key features of thermoforming. Figure 1 illustrates a sheet fed rotary thermoformer. Figure 2 shows a roll fed continuous thermoformer. Refer also to *Sheet Extrusion*. (Sources: Throne, J. L., *Thermoforming*, Hanser Publishers, 1987; Gruenwald, G, *Thermoforming: A Plastics Processing Guide*, Technomic Publishing Co, 1987).

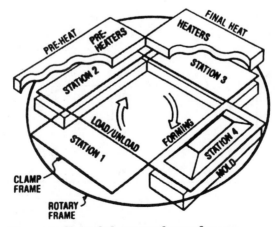

Figure 1. Sheet fed rotary thermoformer.

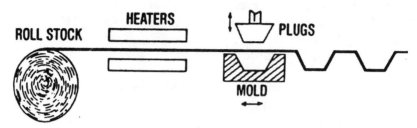

Figure 2. Roll fed continuous thermoformer.

THERMOGRAVIMETRIC ANALYSIS (TGA)

TGA makes a continuous weighing of a small sample (ca 10 mg) in a controlled atmosphere (e.g., air or nitrogen) as the temperature is increased at a programmed linear rate. The thermogram shown in Figure 1 illustrates weight losses due to desorption of gases (e.g., moisture) or decomposition (e.g., HBr loss from halobutyl, CO_2 from calcium carbonate filler). TGA is a very simple technique for quantitatively analyzing for filler content of a polymer compound (e.g., carbon black decomposed in air but not nitrogen). While oil can be readily detected in the thermogram it almost always overlaps with the temperature range of hydrocarbon polymer degradation. The curves cannot be reliably deconvoluted since the actual decomposition range of a polymer in a polymer blend can be affected by the sample morphology. Refer to *Chromatography*. (Source: Cheremisinoff, N.P. *Polymer Characterization: Laboratory Techniques and Analysis*, Noyes Publishers, New Jersey, 1996).

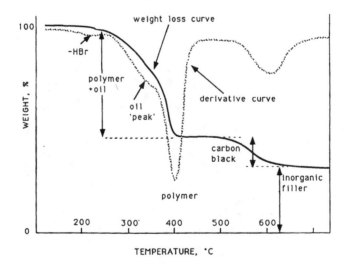

Figure 1. Example of TGA thermogram.

THERMOMECHANICAL ANALYSIS (TMA)

TMA consists of a quartz probe which rests on top of a flat sample (a few mm square) in a temperature controlled chamber. When set up in neutral buoyancy then as the temperature is increased the probe rises in direct response to the expansion of the sample yielding thermal expansion coefficient versus temperature scans. Alternatively, with a penetration probe under dead loading a thermal softening profile is obtained (penetration distance versus temperature). Although this is a simple and versatile experiment, it gives only a semi-quantitative indication of mechanical modulus versus temperature.

THERMOPLASTICS

In the broad classification of plastics there are two generally accepted categories: thermoplastic resins and thermosetting resins. Thermoplastic resins consist of long polymer molecules, each of which may or may not have side chains or groups. The side chains or groups, if present, are not linked to other polymer molecules (i.e., are not cross-linked). Thermoplastic resins, usually obtained as a granular polymer, can therefore be repeatedly melted or solidified by heating or cooling. Heat softens or melts the material so that it can be formed; subsequent cooling then hardens or solidifies the material in the given shape. No chemical change usually takes place during this shaping process.

Examples of common thermoplastics are ABS, PVC, SAN, acetals, acrylics, cellulosics, polyethylenes, polypropylenes, polystyrenes, polycarbonates, polyesters, nylons, and fluoropolymers. Thermoplastics are those polymers that solidify as they are cooled, no longer allowing the long molecules to move freely. When heated, these materials regain the ability to "flow", as the molecules are able to slide past each other with ease. Furthermore, thermoplastic polymers are divided into two classes: amorphous and semi-crystalline polymers.

Amorphous thermoplastics are those with molecules that remain in disorder as they cool, leading to a material with a fairly random molecular structure. An amorphous polymer solidifies, or vitrifies, as it is cooled below its glass transition temperature, Tg. Semicrystalline thermoplastics, on the other hand, solidify with a certain order in their molecular structure. Hence, as they are cooled, they harden when the molecules begin to arrange in a regular order below what is usually referred to as the melting temperature, Tm. The molecules in semi-crystalline polymers that are not transformed into ordered regions remain as small amorphous regions. These amorphous regions within the semi-crystalline domains solidify at the glass transition temperature. Most semi-crystalline polymers have a glass transition temperature at sub-zero temperatures, hence, behaving at room temperature as rubbery or leathery materials. Table 1 presents the most common amorphous and semicrystalline thermoplastics with some of their applications. See *Thermosetting Resins* and *Semi-crystalline Thermoplastics*. (Source: Osswald, T.A. and G. Menges, *Material Science of Polymers for Engineers*, Hanser Publishers, New York, 1996).

Table 1
Common Thermoplastics and Some of their Applications

Polymer	Applications
Amorphous	
Polystyrene	Mass-produced transparent articles, thermoformed packaging, etc. Thermal insulation (foamed), etc
Polymethyl methacrylate	Skylights, airplane windows, lenses, bulletproof windows, stop lights, etc.
Polycarbonate	Helmets, hockey masks, bulletproof windows, blinker lights, head lights, etc.
Unplasticized polyvinyl chloride	Tubes, window frames, siding, rain gutters, bottles, thermoformed packaging, etc.
Plasticized polyvinyl chloride	Shoes, hoses, rotor-molded hollow articles such as balls and other toys, calendered films for raincoats and tablecloths, etc.
Semi-crystalline	
High density polyethylene	Milk and soap bottles, mass production of household goods of higher quality, tubes, paper coating, etc.
Low density polyethylene	Mass production of household goods, grocery bags, etc.
Polypropylene	Goods such as suitcases, tubes, engineering application (fiberglass-reinforced), housings for electric appliances, etc.
Polytetrafluoroethylene	Coating of cooking pans, lubricant-free bearings, etc.
Polyamide	Bearings, gears, bolts, skate wheels, pipes, fishing line, textiles, ropes, etc.

THERMOSETTING RESINS

In thermosetting resins the reactive groups of the molecules form cross-links between the molecules during the fabrication process. The cross-linked or "cured" material cannot be softened by heating. Thermoset materials are usually supplied as a partially polymerized molding compound or as a liquid monomer-polymer mixture. In this uncured condition they can be shaped with or without pressure and polymerized to the cured state with chemicals or heat. Some of the more common thermosets are phenolics, ureas, melamines, epoxies, alkyds, polyesters, silicones, and urethanes. Thermosetting polymers solidify by being chemically cured. Here, the long macromolecules cross-link with each other, during cure, resulting in a network of molecules that cannot slide past each other. The formation of these networks causes the material to lose the ability to "flow" even after reheating. The high density of cross-linking between the molecules makes thermosetting material stiff and brittle. Thermosets also exhibit a glass transition temperature which is sometimes near or above thermal degradation temperatures. Some of the most common thermosets and their applications are reported in Table 1. See *Elastomers*.

Table 1
Common Thermosets and Some of their Applications

Polymer	Applications
Epoxy	Adhesive, automotive leaf springs (with glass fiber), bicycle frames (with carbon fiber), etc.
Melamine	Decorative heat-resistant surfaces for kitchens and furniture, dishes, etc.
Phenolics	Heat-resistant handles for pans, irons and toasters, electric outlets, etc.
Unsaturated polyester	Toaster sides, iron handles, satellite dishes, breaker switch housing (with glass fiber), automotive body panels (with glass fiber), etc.

TIME-DEPENDENT FLUIDS

There is a number of liquids whose flow properties, such as apparent viscosity, change with the time of shearing. In some cases the change is reversible, or at least the viscosity eventually recovers its initial value after a sufficiently long rest period after cessation of shear. In other cases, such as in the mastication of rubber, irreversible changes occur because of changes in the molecular structure. Most formal studies on time-dependent behaviour have been concerned with reversible effects such as thixotropy and rheopexy. Thixotropy is that property of a body by virtue of which the apparent viscosity is temporarily reduced by previous deformation. This means that with thixotropic: material the viscosity depends on the *time* of stirring as compared with a pseudoplastic material which depends on the *rate* of shear. A thixotropic material is often also pseudoplastic but the reverse is not very common. Perhaps the most characteristic feature of thixotropic behavior can be demonstrated using a rotary viscometer. The sample is stiffed for a fixed period of time, for example one minute, at a low shear rate and the torque measured is recorded at some time during this period. The speed is then increased stepwise without stopping the machine and the torque measured at the new shearing rate. This is then repeated over several increased stirring rates. A curved plot will result. Without stopping the viscometer measurements are then made at a series of decreasing stirring rates and the torque measured. In the time-interval between making the measurements when increasing speeds and making measurements at decreasing speeds the viscosity will have dropped and the resulting curve will be below the curve made when making measurements in the sequence of increasing speeds. The loop shown is a characteristic thixotropic: hysteresis loop. The size of the loop will depend on the time for which the material is sheared at each speed, on the number of speeds employed and the maximum shear rate used as well as on the flow characteristics of the material. In consequence, the quantitative measurement of thixotropic behaviour becomes difficult and it is not possible to give a single index of thixotropy. The term anti-thixotropy is used when the apparent viscosity is an increasing function of the duration of flow and when, as with thixotropy, the body recovers its initial state after a long enough rest. Once again a hysteresis loop is produced in the rotary viscometer experiment but is in the opposite sense.

The term *rheopexy* has also been used to describe an increase in the viscosity of a body with time (not necessarily under deformation). It is today more commonly used to denote the solidification of a thixotropic system under the influence of gentle movement. It should, however, be noted that any thixotropic material will increase in viscosity as the intensity of shear is reduced and that the term rheopexy should preferably only be used to refer to an increased *rate* of solidification on gentle movement. There has been a degree of confusion in the terminology of time-independent fluids in the past and even today there are areas of disagreement particularly in the meanings of anti-thixotropy and rheopexy. The term thixotropy has also been used to describe pseudoplastic behaviour and though no longer used by rheologists in this sense it is still employed for the purpose by some polymer technologists. Refer to Figure 1 for a thixotropic flow curve.

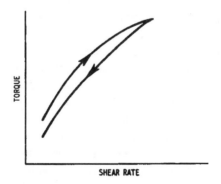

Figure 1. Shows thixotropic behavior.

TITANIUM CATALYST SYSTEMS

Titanium catalysts are a part of the Ziegler supported catalyst systems for ethylene polymerization and copolymerization with propylene. The fixation of the titanium-containing component of a $TiCl_4$-$Al(C_2H_5)_3$ catalytic system on MgO supports increases the reactivity of its active species in the elementary propagation step of polyethylene macrochains, in contrast to alumosilicate-supported and unsupported systems. The Mg containing supports are not only inert "supports" used for the increase in the area of distribution of the titanium component, but are active promoters of Ziegler catalytic systems. The propagating species on the surface of aluminosilicate are more selective on ethylene than those fixed on MgO. At adequate copolymerization conditions, the MgO supported catalyst permits the production of copolymer more enriched propylene than that supported by alumosilicate. A rise in temperature in the reaction zone leads to a marked reduction in the copolymer yield. If the temperature rises from 30 to 70° C, the copolymer yield is reduced practically by one half over the entire range of variation in the comonomer ratio (propylene content 5 to 95 mol%). It is important to note that both the composition and the microstructure of the copolymers depend strongly upon the co-catalyst used. For example, $Mg(C_6H_{13})_2$ (soluble titanium catalyst) results in an alternating-like copolymer in which the inversion of propylene units was negligible. In contrast, $Al(C_2H_5)_3$ gives a block-type copolymer containing long methylene chains. Such a marked difference in the copolymer structure may result from the difference in the reducing power of the co-catalysts, i.e., MgR_2 gives the Ti(III) species, while $Al(C_2H_5)_3$ gives a considerable amount of the Ti(II) species, together with the Ti(III) catalysts. Copolymerization of ethylene with propylene can also be out by using two-catalyst systems. One (Cat-a) can be a solid $MgCl_2$ containing Ti treated with ethyl benzoate together with $Al(C_2H_5)_3$ and the other catalyst (Cat-c) based on a homogeneous mixture of $MgCl_2$ dissolved in 2-ethylhexanol/decane and $TiCl_4$ which is treated with $Al(C_2H_5)_2Cl$. Both catalyst systems exhibit very high activity. The reader can refer to the following reference for a detailed discussion on this catalysts system for the production of EP rubbers: *Elastomer Technology Handbook*, N. P. Cheremisinoff - editor, CRC Press, Boca Raton, Florida, 1993. Also, refer to the subject entries *Vanadium Catalysts* and *Ziegler Polymerization Process*.

TOXICITY BY INGESTION

The term LD_{50} (meaning "lethal dose at the 50th percentile population") signifies that about 50% of the animals given the specified dose by mouth will die. Thus, for a chemical whose LD_{50} is below 50 mg/kg, the toxic dose for 50% of animals weighing 70 kg (150 lb) is 70 x 50 = 3500 mg = 3.5 g, or less than one teaspoonful; it might be as little as a few drops. For a chemical with an LD_{50} of between 5 to 15g/kg, the LD_{50} would be between a pint and a quart for a 150-lb man. All LD_{50} values have been obtained using small laboratory animals such as rodents, cats, and dogs. The substantial risks taken in using these values for estimating human toxicity are the same as those taken when new drugs are administered to humans for the first time.

TOXICITY BY INHALATION (THRESHOLD LIMIT VALUE)

The threshold limit value (TLV) is usually expressed in units of parts per million (ppm) - i.e., the parts of vapor (gas) per million parts of contaminated air by volume at 25° C (77° F) and atmospheric pressure. For a chemical that forms a fine mist or dust, the concentration is given in milligrams per cubic meter (mg/m^3). The TLV is defined as the concentration of the substance in air that can be breathed for five consecutive eight-hour workdays (40-hour work week) by most people without adverse effect. (This definition is given by American Conference of Governmental Industrial Hygienists, "Threshold Limit Values for Substance in Workroom Air, Adopted by ACGIH for 1972").

TP (THERMOPLASTIC ELASTOMER POLYESTER)

Table 1 provides some general properties information on TPs. Refer also to the subject entry *TPE*. In reviewing the general properties of this polymer, note the use of the following legend: A = amorphous - Cr = crystalline - C = clear - E = excellent - G = good - P = poor - O = opaque - T = translucent- R = Rockwell - S = Shore.

Table 1

Average Properties of Thermoplastic Elastomer Poylester

STRUCTURE: A	FABRICATION		
	Bonding	Ultrasonic	Machining
	G	P	G
Specific Density: 1.25	Utilization Temperature (° F)		
Water Absorption Rate (%): 0.3	min	max	
	-94	320	
Elongation (%): 400	Melting Point (° F): 420		
Tensile Strength (Psi): 6000	Transparency: O		
Compression Strength (Psi): -	Uv Resistance: G		
Flexural Strength (Psi): -	CHEMICAL RESISTANCE		
Flexural Modulus (Psi): -	Acids	Alkalis	Solvents
Impact (Izod Ft. Lbs/in): 1.2	G	G	G
Hardness: SD72			

TPE (THERMOPLASTIC ELASTOMERS)

Thermoplastic elastomers (TPEs) continue to grow in commercial importance. These materials combine the functional properties of comparable thermoset elastomers with the fabrication advantages of thermoplastics. As a class, TPEs comprise several types of materials such as elastomeric alloys (EAs), styrenic block copolymers, copolyesters, and thermoplastic polyurethanes.

Thermoplastic elastomers are defined by ASTM D 1566 as "a family of rubber-like materials that, unlike conventional vulcanized rubber, can be processed and recycled like thermoplastic materials". A rubber is defined as "a material that is capable of recovering from large deformations quickly and forcibly and "retracts within 1 min to less than 1.5 times its original length after being stretched at room temperature (18 to 29° C) to twice its length and held for I min before release".

The term thermoplastic elastomer has developed into the accepted generic name for materials as defined in the previous paragraph. This term is generally used as a noun, to distinguish a TPE from a conventional thermoset rubber that has not been vulcanized. Archaic terms for TPEs include "elastoplastic", thermoplastic rubber, thermoplastic vulcanizate,' and impact modified plastic. In addition, there has been a growth of subcategories of TPEs to distinguish between the different types of materials that generally meet the TPE definition. Several examples are "thermoplastic rubber blends", "elastomeric alloys", and "block copolymers". These different terms are used to indicate different morphological structures and different elastomeric performance.

There have been numerous books, book chapters, technical papers, and review articles, patents, trade literature, and symposia on TPEs, most of which were published or held within the last 20 years. Several new organizations with TPEs as a focus have been formed, such as the Thermoplastic Elastomers Special Interest Group (SIG) of the Society of Plastics Engineers and a similar Topical Group within the Rubber Division of the American Chemical Society. As evidenced by this growth in information, the technology of TPEs has grown steadily and, sometimes, by quantum jumps. There is now a wide selection of different TPEs for the materials technologist to consider, depending on specific application needs.

As noted, TPEs are either block copolymers or combinations of a rubber-dispersed phase and a plastic continuous matrix. The attribute contributed by the rubbery phase - such as butadiene or ethylenebutylene in an S-E-S or SEB-S styrenic block copolymer, or the completely vulcanized EPDM rubber particles in a polypropylene (PP)/EPDM EA - is classical elastomeric performance. The elastic properties of a rubber result from long, flexible molecules that are coiled in a random manner. When the molecules are stretched, they uncoil and have a more specific geometry than the coiled molecules. The uncoiled molecules have lower entropy because of the more restricted geometry and, since the natural tendency is an increase in entropy, the entropic driving force is for the molecules to retract, giving elasticity. The soft butadiene or ethylenebutylene segments in the styrenic block copolymer are coiled segments held together by the polystyrene hard block. When acted upon by an external stress, the initial deformation is the uncoiling of the soft rubbery segments, since the energy for displacement is less for the soft segments than for the hard. This behavior gives elastic performance up to the point where the strain is so great that permanent deformation occurs due to loss of bonding between the hard and soft segments, or the hard segment. is deformed beyond its elastic limit, or the temperature is raised to the point where the hard segment softens or even melts. Therefore, under definable conditions of temperature and stress, TPEs behave with classic elastomeric characteristics, just like thermoset rubbers. With this type of performance, TPEs can and are used in many typical thermoset rubber applications, and thus, have one foot in the rubber industry.

The primary morphological difference between TPEs and the thermoset rubbers is the presence of soft rubbery domains bonded to hard plastic domains (with a distinct melting point above which the TPE is molten and suitable for fabrication). The hard plastic domains of a TPE can be formed, destroyed, and reformed repeatedly through the simple process of adding or removing heat energy. Their formation is thus reversible. This capability for the formation of these hard domain cross-links is essentially irreversible. Melting the hard domains by conventional plastic fabrication processes, such as injection molding, blow molding, extrusion, thermoforming, etc., is why TPEs also have their other foot in the plastics industry. While most TPEs do not have typical thermoplastic physical properties, such as high load deflection, they are made into finished articles by typical thermoplastic processing equipment and techniques. It is the nature of TPEs that has allowed thermoplastic processors to expand into the manufacture of rubber parts for merchant production and captive use.

The plastics industry, since John Wesley Hyatt, and the rubber industry, since Charles Goodyear, have both grown into major industries with worldwide sales in the tens of billions of dollars. Although both industries are based on polymer science and technology, there has been little interaction between them. Plastic and rubber materials are processed quite differently. Outside of polyvinyl chloride, few plastics are extensively modified by compounding before fabrication into end-use articles. On the other hand, the rubber polymer is simply the base for a rubber compound developed for specific performance characteristics. The generation of useful rubber articles has created a whole technology based on the compounding and production of specific rubber compounds with a desired set of properties. In addition, conventional thermoset rubbers are vulcanized by chemically crosslinking reactive sites in the base polymer; this operation requires specialized processing equipment for the preparation of rubber parts. Thermoplastic processing is a simpler fabrication process because no chemical modification of the material is required to form the final article. These differences in the rubber and plastics industries have set them apart from each other, to the point where until recently, few companies have done both rubber compounding and/or part manufacture and thermoplastic processing into end-use articles. It is now possible to manufacture rubber parts by using TPEs with plastics processing equipment. Since TPEs behave as rubber up to the temperature-dependent limit of permanent deformation, rubber parts end users have adopted them as rubber. However, since TPEs are processed (fabricated) into rubber parts on conventional thermoplastic processing equipment, the plastics industry has also claimed them. The advent of TPEs has resulted in "rubber-only" companies investing in thermoplastics processing equipment and in plastics processors fabricating rubber parts, thus, spanning the gap between these two major industries.

The birth of TPEs is generally regarded to be the invention and commercialization of thermoplastic polyurethanes by B.F. Goodrich in 1959. Following this development were the introductions of styrenic block copolymers by Shell Chemical Company in the 1960s, copolyesters by E.I. duPont Company, thermoplastic elastomeric olefins (TEOs) by Uniroyal in the 1970s, EAs by Monsanto Chemical Company in 1981, and block copolymers of polyamides by Atochem in 1982. The target area for growth of TPEs is primarily in thermoset rubber replacement. A second area is new elastomeric applications where thermoset rubber would rationally be considered. The third area is in soft thermoplastic replacement where, greater flexibility is needed. Table 1 provides a listing of polymer blends and alloys by supplier.

Table 1

Specialty Polymer Blends and Alloys Along with Trade Names

Product	Supplier	Trade Name	Grade
ABS/nylon	Borg-Warner	ELEMID	RMI
ABS/nylon	Monsanto	TRIAX	All
ABS/PC	Borg-Warner	PROLOY	All
ABS/PC	Borg-Warner	CYCOLOY	All
ABS/PC	Dow	PULSE	All
ABS/PC	Mobay	BAYBLEND	All
ABS/PVC	Borg-Warner	CYCOVIN	All
ABS/PVC	Amoco	MINDEL	A-670, A-650
ABS/PVC	Monsanto	LUSTRAN	865, 860
ABS/PVC	Schulman	POLYMAN	507, 509, 511

Table 1 continued

Product	Supplier	Trade Name	Grade
ABS/PVC	Shuman	-	780
ABS/SMA	Monsanto	CADON	All
Modified Acetal	BASF	ULTRAFORM	N2640X
Modified Acetal	DuPont	DELRIN	ST-100, T-500
Modified Acetal	Hoechst Celanese	DURALOY	1000, 1100, 1200, 21
Modified Acetal	Hoechst Celanese	CELCON	C-400, C-401
Modified ASA	BASF	TERBLEND	SKR2861
Modified ASA	G.E.	GELOY	13320, 1221, 1220
Modified Ionomer	DuPont	SURLYN	HP Series
Modified Nylon	Allied Chemical	CAPRON	8250, 8252, 8350
Modified Nylon	BASF	ULTRAMID	KR4430, B3L, KR4645
Modified Nylon	DuPont	ZYTEL	S Series, 408
Modified Nylon	Emser Industries	GRILON	A28 Series
Modified Nylon	Hoechst Celanese	CELANESE	N-297, N-303
Modified Nylon	INP	THERMOCOMP VF	All
Modified Nylon	Nylon Corp. of America	NYCOA	1417, 2001
Modified Nylon	Thermofil	-	All
Modified Nylon	Wilson Fiberfil	NYAFIL	TN
Modified PBT	BASF	ULTRADUR	KR4070, KR4071
Modified PBT	Comalloy	VOLEX	430, 431
Modified PBT	Comalloy	HILOY	431, 432, 433, 434
Modified PBT	Comalloy	COMTUF	431
Modified PBT	G.E.	VALOX	340

Table 1 continued

Product	Supplier	Trade Name	Grade
Modified PBT	Hoechst Celanese	VANDAR	4424F
Modified PBT	Mobay	POCAN	S1506
Modified PET	Comalloy	VOLEX	460, 461, 462
Modified PET	Comalloy	HILOY	441, 461, 462, 463
Modified PET	Comalloy	COMUIF	461, 462, 464
Modified PET	DuPont	RYNITE SST	All
Nylon/PP	Dexter	DEXLON	All
PBT/PC	Comalloy	-	-
PBT/PET	Comalloy	HILOY	432
PBT/PET	G.E.	VALOX	860, 855, 815, 830
PBT/PET	Hoechst Celanese	CELANEX	5330
PC/Nylon	Dexter	DEXCRAB	All
PC/PE	G.E.	LEXAN	191
PC/PE	Mobay	MAKROLON	T-7855-T-7955
PC/Polyester	G.E.	VALOX	508m, 533
PC/Polyester	G.E.	XENOY	All
PC/Polyester	Mobay	MARKOBLEND	UT
PC/Polyester	Thermofil	R2	All
PC/SMA	Arco Chemical	ARLOY	All
PC/TPU	Mobay	TEXIN	902, 3203
PET/PC	Comalloy	-	-
PET/PSF	Amoco	MINDEX	B-390, B-322
PPE/PS	G.E.	NORYL	All
PPO/Nylon	BASF	ULTRANYL	All
PPO/Nylon	G.E.	NORYL GTX	All
PPO/Nylon	BASF	ULTRANYL	All

Table 1 continued

Product	Supplier	Trade Name	Grade
PPO/PBT	G.E.	NORYL GTX	All
PPO/PS	BASF	LURANYL	All
PPO/PS	Borg-Warner	PREVEX	All
PP/EPDM	Dexter	ONTEX	All
PP/EPDM	Ferro	FERROFLEX	All
PP/EPDM	Schulman	POLYTROPE	All
PP/EPDM	Teknor Apex	TELCAR	All
PP/Nylon	Dexter	DEXPRO	All
PVC/Acrylic	Rohm & Haas	KYDEX	All
PVC/Acrylic	Polycast Technology	-	All
PVC/Urethane	Dexter	VYTHENE	All
SAN/EPDM	Dow	ROVEL	701, 705, 401, 501
SMA/HIPS	Arco Chemical	DYLARK	238

TPO (THERMOPLASTIC ELASTOMER POLYOLEFIN)

Table 1 provides average properties data for TPOs. In reviewing the general properties of this polymer, note the use of the following legend: A = amorphous - Cr = crystalline - C = clear - E = excellent - G = good - P = poor - O = opaque - T = translucent- R = Rockwell - S = Shore.

Table 1. Average Properties of TPO

STRUCTURE: A	FABRICATION		
	Bonding	Ultrasonic	Machining
	P	P	P
Specific Density: 0.89	Utilization Temperature (° F)		
Water Absorption Rate (%): 0.03	min: -60		max: 248
Elongation (%): 500	Melting Point (° F): 248		
Tensile Strength (Psi): 4600	Coefficient of Expansion: 0.000055		

Table 1 continued

Flexural Strength (Psi): 1400	Arc Resistance: 180
	Dielectric Strength (kV/mm): 24
Flexural Modulus (Psi):	Transparency: O
Impact (Izod Ft. Lbs/in): NB	UV Resistance: P
Hardness: SD50	**CHEMICAL RESISTANCE**

Acids	Alkalis	Solvents
P	G	P

TPU (THERMOPLASTIC POLYURETHANE)

Table 1 provides average properties of TPUs. In reviewing the general properties of this polymer, note the use of the following legend: A = amorphous - Cr = crystalline - C = clear - E = excellent - G = good - P = poor - O = opaque - T = translucent- R = Rockwell - S = Shore.

Table 1. Average Properties of TPU

STRUCTURE: A	FABRICATION		
	Bonding	Ultrasonic	Machining
	G	P	P
Specific Density: 1.2	Utilization Temperature (° F)		
Water Absorption Rate (%): 0.8	min: -110 max: 239		
Elongation (%): 500	Melting Point (° F): 284		
Tensile Strength (Psi): 5800	Coefficient of Expansion: ND		
Compression Strength (Psi):	Arc Resistance: 120		
Flexural Strength (Psi):	Dielectric Strength (kV/mm): 20		
Flexural Modulus (Psi):	Transparency: T		
Impact (Izod ft. lbs/in): NB	UV Resistance: G		
HARDNESS: R60	**CHEMICAL RESISTANCE**		
	Acids	Alkalis	Solvents
	G	G	G

TRADE NAMES

The following is a list of trade names for commercially available polymers. All of these suppliers can be accessed along with their product information on the Internet.

Trade Name	Product	Company
Aclar	CTFE	Allied Signal
Acrylite	Acrylic Resin and Sheet	Cyro
Acrysteel	Acrylic Sheet	Aristech
Affinity	Polyolefin Plastomers	Dow
Aim	Advanced Styrenic Resin	Dow
Alathon	HDPE	Occidental
Alcryn	Halogenated polyolefin	Dupont
Amodel	Polyphthalamide	Amoco
APEC	High Heat PC	Bayer
Arcel	Expandable PE Copolymer	Arco
Aristech	Acrylic Resin	Aristech
Arloy	SMA/PC alloy	Arco
Arpak	Expanded PE Beads	Arco
Arpro	Expanded PE Beads	Arco
Astrel	Polyarylsulfone	3M
Aurum	Polyimide Resin	Mitsui Toatsu
Avimid K	Polyimide	DuPont
Barex	Acrylonitrile CoPolymer	BP Chemical
Bayblend	ABS/Polycarbonate	Bayer
Beetle	Urea Molding Compound	A.C. Molding

Trade Name	**Product**	**Company**
Cadon	SMA and SMA/ABS 2	Bayer
Calibre	PC	Dow
Capron	Nylon 6	Allied Signal
Celanex	PBT, PBT/PET alloy	Hoechst Celanese
Celazole	Polybenzimidazole	Hoechst Celanese
Celcon	Acetal and acetal elastomer	Hoechst Celanese
Centrex	AES/ASA	Bayer
Crystalor	Polymethylpentene	Phillips
Cycolac	ABS	G.E.
Cycoloy	ABS/PC alloy	G.E.
Cyrolite	Acrylic Resin	Cyro
Delrin	Acetal, acetal/elastomer alloy	Dupont
Delta	ABS	GPC (Taiwan)
Denka	ABS	Showa (Japan)
Desmopan	TPU	Bayer
Diakon	Acrylic Resin	ICI
Dimension	PPE/Nylon Alloy	Allied Signal
Dowlex	LDPE/LLDPE	Dow
Duraloy	Impact Modified Polyester-	Celanese
Durethan	Nylon 6	Bayer
Durel	PAR	Occidental
Dylark	SMA and SMA/PS alloy	Arco

Trade Name	**Product**	**Company**
Dylene	PS	Arco
Eastar	Polyester-PCTG	Eastman Chemical
Eastpac	Polyester-PET	Eastman Chemical
Ecdel	Polyetherester (TPE)	Eastman
Ecolan	Starch Modified PU	Porvair
Ektar	Polyester-PETG	Eastman
	Polyester-PCTG	
	Polyester-PET	
	Polyester-PCT	
	Polypropylene	
	PC/Polyester Alloy	
Engage	Polyolefin Elastomers	Dow Elastomers
Enuran	PBT Alloy	G.E.
Estamid	Polyamide CTPE	Dow
Estane	Urethane (TPE)	B.F. Goodrich
Eval	EVOH	Eval
Fiberloc	PVC Glass Reinforced	Geon
Flexomer Carbide	LLDPE/EVA	Union
Fluon	TFE	ICI
Foraflon	PVDF	Atochem
Forar	HDPE	Amoco
Fortiflex	HDPE, MDPE	Solvay
Fortilene	Polypropylene	Solvay
Fortron	Polyphenylene Sulfide	Hoechst Celanese
Fostarene	Crystal PS	Hoechst

Trade Name	Product	Company
Fostalite	Light stable PS	Hoechst
Fosta Tuf-Flex	Impact PS	Hoechst
Geloy	ASA and ASA/PVC alloys	G.E.
Geolast	PP/Nitrile (TPE)	Advanced Elastomer Systems
Geon	PVC	Geon
Grilamid	Nylon 12	EMS American Grilon
GTX	PPE/Nylon	G.E.
Halar	ECTFE	Ausimont
Hostalen GUR	UHMWPE	Hoechst Celanese
Hostalen HMW	HDPE	Hoechst Celanese
Hytrel	Polyetherester (TPE)	DuPont
Isoplast	Rigid PU	Dow
Ixef	Polyarylamide	Solvay
K-Resin	TPE	Phillips Chemical
Kadel	Polyketone	Amoco
Kamax	Acrylic	Rohn & Haas
Kapton	Polyimide Film	DuPont
Kel-F	CTFE	3M
Kevlar	Aramid Fiber	DuPont
Korad	Acrylic Film	PEP
Krayton	TPE	Shell
Kynar	PVDF	Atochem
Lenzing	Polyimide	Lenzing
Lexan	PC	G.E.

Trade Name	Product	Company
Lucite	Acrylic	DuPont
Lupiace	PPE	Mitsubishi
Lupital	Acetal Resin	Mitsubishi
Luran	SAN	BASF
Lustran	ABS, SAN Copolymer	Bayer
Lustrex	PS	Novacor
Magnum	ABS	Dow
Makroblend	PC Blends	Bayer
Makrolon	PC	Bayer
Maranyl	PA 66	LNP
Marlex	HDPE	Phillips
Merlon	PC	Bayer
Mindel	Polysulfone Blends	Amoco
Minlon	Mineral reinforced PA66	DuPont
Mylar	PET film	DuPont
NAS 10	Acrylic Styrene Copolymer	Novacor
Noryl GTX	PPO/PS alloy	G.E.
Novodur	ABS	Bayer
Oleplate	PP	Amoco
Olevac	PP	Amoco
Ontex	TPE	Research Polymers
Paxon	HDPE	Paxon Polymers
Pellethane	TPU	Dow
Petrothene	PE	Quantum

Trade Name	Product	Company
Plexiglas	Acrylic Sheet	AtoHaas
Polyman	ABS/PVC alloy	Schulman
Polystrol	PS	BASF
Prevail	ABS/TPU	Dow
Prevex	PPE/PS alloy	G.E.
Pro-Fax	PP	Himont
Pulse	PC/ABS alloy	Dow
Radel	Polyethersulfone	Amoco
Rilsan	PA 11, 12	Atochem
Royalite	ABS and ABS alloys	Uniroyal
Rynite	PET	DuPont
Ryton	PPS	Phillips
Saran	PVDC	Dow
Selar	PA/PE alloy	DuPont
SP	EMAC	Chevron
Stereon	Styrene/butadiene copolymer(TPE)	Firestone
Styron	PS	Dow
Supec	Polyphenylene Sulfide	G.E.
Surlyn	Ionomer	DuPont
Teflon	FEP,PFA,TFE film	DuPont
Tefzel	ETFE	DuPont
TempRite	CPVC	B.F. Goodrich
Tenite	Cellulosic Compounds	Eastman Chemical
Terblend	ASA/PC alloy	BASF

Trade Name	Product	Company
Terluran	ABS	BASF
Texin	TPU	Bayer
Torlon	Polyamideimide	Amoco
Toyolac	ABS	Toray
TPX	Polymethylpentene	Mitsui
Triax	ABS/PA	Bayer
Tyril	SAN Copolymer	Dow
Udel	Polysulfone	Amoco
Ultem	Polyetherimide	G.E
Ultradur	Polyester-PBT	BASF
Ultraform	Acetal	BASF
Ultramid	PA	BASF
Ultrason	PES	BASF
Uvex	CAB	Eastman
Valox	PBT	G.E
Vectra	Liquid Crystal Polymer	Hoechst Celanese
Vespel	PA	DuPont
Vydyne	PA	Monsanto
Xenoy	PC/Polyester alloy	G.E.
XT	Acrylic	Cyro
Xydar	Liquid Crystal Polymer	Amoco
Zylar	Acrylic	Novacor
Zytel	PA	DuPont

TRANSFER MOLDING

In transfer molding, the thermosetting molding powder is placed in a chamber or pot outside the molding cavity and subjected to heat and pressure to liquefy it. When liquid enough to start flowing, the material is forced by the pressure into the molding cavity, either by a direct sprue or through a system of runners and gates. The material sets hard to the cavity shape after a certain time (cure time) has elapsed. When the mold is disassembled, the molded part is pushed out of the mold by ejector pins, which operate automatically. Figure 1 illustrates the molding cycle of pot-type transfer molding.

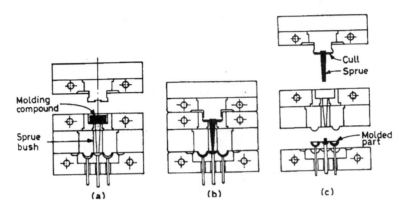

Figure 1. Molding cycle of a pot-type transfer mold.

Moving from left to right in this figure, the molding compound is placed in the transfer pot. In the middle sketch, the molding compound is forced under pressure when hot through an orifice and into a closed mold. When the mold opens (right-most figure), the sprue remains with the cull in the pot, and the molded part is lifted out of the cavity by ejector pins. (Source: Chanda, M. and S. K. Roy, Plastics Technology Handbook, Marcel Dekker, Inc., New York, NY, 1987). *See also Plunger-TypeTransfer Molding.*

TRANSMISSION ELECTRON MICROSCOPY

TEM (refer to Figure 1) is used whenever a more in-depth study (when domain sizes are less than 1 micron or so) is required on polymer phase morphologies such as dynamically vulcanized alloys and Nylon/EP filler location as in carbon black in rubber compounds and also in the morphology of block copolymers. Thin sections are required and take anywhere from one hour to one day per sample depending on the nature of the sample. They must be 100 nm in thickness and are prepared usually by microtoming with a diamond knife at near liquid nitrogen temperatures (-150° C). The same contrasting media for SEM apply to TEM. In addition, PIB backbone polymers scission and evaporate in the TEM which helps locate these polymers domains in blends. Refer to *Scanning Electron Microscopy*. (Source: Cheremisinoff, N.P., *Polymer Characterization: Laboratory Techniques and Analysis*, Noyes Publishers, New Jersey, 1996).

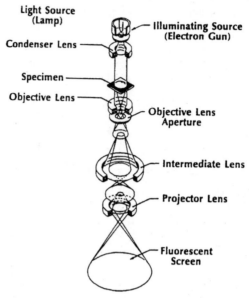

Figure 1. Schematic of TEM.

U

ULTRASONIC PARTS ASSEMBLY

Ultrasonic assembly techniques are a fast, clean and efficient method of assembling or processing rigid thermoplastic parts or films, as well as synthetic fabrics. There are a variety of methods used to join plastics and plastic to metal parts or other nonplastic materials, thereby replacing the use of solvents, adhesives, mechanical fasteners, or other consumables. In this regard, the techniques may be viewed as a pollution prevention technology. A typical assembly converts 50/60 Hz current to 20 to 40 kHz electrical energy through a solid-state power supply. This high frequency electrical energy is sent to a converter, a component that changes electrical energy into mechanical vibratory energy at ultrasonic frequencies. The vibratory energy is then transmitted through an amplitude-modifying device to a horn, which is an acoustic tool that transfers the vibratory energy directly to the assembly parts. The vibrations are transmitted through the workpiece to the joint area where the vibratory energy is converted to heat through friction that melts the plastic. When this molten state is reached at the part interface, the vibration is stopped; pressure is maintained briefly on the parts while the molten plastic solidifies. This action creates a strong molecular bond between the parts. Cycle times are usually less than one second, and the weld that is obtained approaches that of the parent material.

UNDERWRITERS LABORATORIES (UL)

UL is a not-for-profit organization whose goods are to establish, maintain, and operate laboratories for the investigation of materials, devices, products equipment, constructions, methods, and systems with respect to hazards affecting life and property. There are five testing facilities in the U.S. and over 200 inspection centers. More than 700 engineers and 500 inspectors conduct tests and follow-up investigations to insure that potential hazards are evaluated and proper safeguards provided.

UL has six basic services it offers to manufacturers, inspection authorities, or government officials. These are product listing service, classification, service, component recognition service, certificate service, inspection service, and fact finding and research. UL's Electrical Department is in charge of evaluating individual plastics and other products using plastics as components. The Electrical Department evaluates consumer products such as TV sets, power tools, appliances, and industrial and commercial electrical equipment and components.

In order for a plastic material to be recognized by UL it must pass a variety of UL tests including the UL 94 flammability test and the UL 746 series, and short- and long-term property evaluation tests. When a plastic material is granted Recognized Component Status, a yellow card is issued. The card contains precise identification of the material including supplier, product designation, color, and its UL 94 flammability classification at one or more thickness, Also included are many of the property values such as temperature index, hot wire ignition, high-current arc ignition, and arc resistance.

These data also appear in the recognized component directory. UL publishes the names of the companies who have demonstrated the ability to provide a product conforming to the established requirements, upon successful completion of the investigation and after agreement of the terms and conditions of the listing and follow-up service. Listing signifies that production samples of the product have been found to comply with the requirements and that the manufacturer is authorized to use the UL's listing mark on the listed products which comply with the requirements. UL's consumer advisory council was formed to advise UL in establishing levels of safety for consumer products to provide UL with additional user field experience and failure information in the field of product safety and to aid in educating the general public in the limitations and safe use of specific consumer products.

UNITS OF MEASUREMENT

The following is a summary of most of the units of measurement to be found in use around the world today, together with the appropriate conversion factors needed to change them into a "standard" unit of the SI. The units may be found either by looking under the category in which they are used [such as length, mass, density, energy etc.], or else by picking one unit from an alphabetically ordered list of units. There is an outline of the SI; a list of its basic defining standards and also some of its derived units; then another list of all the SI prefixes and some notes on conventions of usage. Figure 1 provides a summary of engineering parameters that are commonly used. Table 1 is a summary table of conversion factors most often required. Note that **x** means "multiply by", **/** means "divide by", **#** means it is an exact value. All other values are given to an appropriate degree of accuracy.

Length	Area	Volume	Mass	Temperature
Density	Pressure & Stress	Speed	Fuel Consumption	Power
Energy (Work)	Specific Energy by Mass	Specific Energy by Volume	Force	Torque
Flow Rate by Mass	Flow Rate by Volume	Spread Rate by Mass (inc. Rainfall)	Spread Rate by Volume (inc. Rainfall)	

Figure 1. Common engineering parameters.

Table 1
Conversion Factors

To change	Into	Do this	To change	Into	Do this
acres	hectares	x 0.4047	kilograms	ounces	x 35.3
acres sq.	kilometers	/ 247	kilograms	pounds	x 2.2046
acres sq.	meters	x 4047	kilograms	tonnes	/ 1000 #
acres sq.	miles	/ 640 #	kilograms	tons (UK/long)	/ 1016

Table 1 continued

To change	Into	Do this	To change	Into	To change
barrels (oil)	cu.meters	/ 6.29	kilograms	tons (US/short)	/ 907
barrels (oil)	gallons (UK)	x 34.97	kilometers	meters	x 1000 #
barrels (oil)	gallons (US)	x 42 #	kilometers	miles	x 0.6214
barrels (oil)	literss	x 159	liters	cu.inches	x 61.02
centimeters	feet	/ 30.48 #	liters	gallons (UK)	x 0.2200
centimeters	inches	/ 2.54 #	liters	gallons (US)	x 0.2642
centimeters	meters	/ 100 #	liters	pints (UK)	x 1.760
centimeters	millimeters	x 10 #	liters	pints (US liquid)	x 2.113
cubic cm	cubic inches	x 0.06102	meters	yards	/ 0.9144 #
cubic cm	liters	/ 1000 #	meters	centimeters	x 100 #
cubic cm	milliliters	x 1 #	miles	kilometers	x 1.609
cubic feet	cubic inches	x 1728 #	millimeters	inches	/ 25.4 #
cubic feet	cubic meters	x 0.0283	ounces	grams	x 28.35
cubic feet	cubic yards	/ 27 #	pints (UK)	liters	x 0.5683
cubic feet	gallons (UK)	x 6.229	pints (UK)	pints (US liquid)	x 1.201
cubic feet	gallons (US)	x 7.481	pints (US liquid)	liters	x 0.4732
cubic feet	liters	x 28.32	pints (US liquid)	pints (UK)	x 0.8327
cubic inches	cubic cm	x 16.39	pounds	kilograms	x 0.4536

Table 1 continued

To change	Into	Do this	To change	Into	Do this
cubic inches	liters	x 0.01639	pounds	ounces	x 16 #
cubic meters	cubic feet	x 35.31	square cm	sq. inches	x 0.1550
feet	centimeters	x 30.48 #	square feet	sq. inches	x 144 #
feet	meters	x 0.3048 #	square feet	sq. meters	x 0.0929
feet	yards	/ 3 #	square inches	square cm	x 6.4516 #
fl.ounces (UK)	fl.ounces (US)	x 0.961	square inches	square feet	/ 144 #
fl.ounces (UK)	milliliters	x 28.41	square km	acres	x 247
fl.ounces (US)	fl.ounces (UK)	x 1.041	square km	hectares	x 100 #
fl.ounces (US)	milliliters	x 29.57	square km	square miles	x 0.3861
gallons	pints	x 8 #	square meters	acres	/ 4047
gallons (UK)	cubic feet	x 0.1605	square meters	hectares	/ 10 000 #
gallons (UK)	gallons (US)	x 1.2009	square meters	square feet	x10.76
gallons (UK)	liters	x 4.54609 #	square meters	square yards	x 1.196
gallons (US)	cubic feet	x 0.1337	square miles	acres	x 640 #
gallons (US)	gallons (UK)	x 0.8327	square miles	hectares	x 259
gallons (US)	liters	x 3.785	square miles	square km	x 2.590
grams	kilograms	/ 1000 #	sq. yards	sq. meters	x1.196

Table 1 continued

To change	Into	Do this	To change	Into	Do this
grams	ounces	/ 28.35	tonnes	kilograms	x 1000 #
hectares	acres	x 2.471	tonnes	tons (UK/long)	x 0.9842
hectares	square km	/ 100 #	tonnes	tons (US/short)	x 1.1023
hectares	square meters	x 10000 #	tons (UK/long)	kilograms	x 1016
hectares	square miles	/ 259	tons (UK/long)	tonnes	x 1.016
hectares	square yards	x 11 960	tons (US/short)	kilograms	x 907.2
inches	centimeters	x 2.54 #	tons (US/short)	tonnes	x 0.9072
inches	feet	/ 12 #	yards	meters	x 0.9144 #

The System International [SI]

Le Systeme international d'Unites officially came into being in October 1960 and has been adopted by nearly all countries, though the amount of actual usage varies considerably. It is based upon 7 principal units, 1 in each of 7 different categories which are given below. Definitions of these basic units are given. Each of these units may take a prefix. From these basic units many other units are derived and named.

Category	Name	Abbreviation
Length	meter	m
Mass	kilogram	kg
Time	second	s
Electric current	ampere	A
Temperature	kelvin	K
Amount of substance	mole	mol
Luminous intensity	candela	cd

Definitions of the Seven Basic SI Units

meter [m]

The meter is the basic unit of length. It is the distance light travels, in a vacuum, in 1/299792458th of a second.

kilogram [kg]

The kilogram is the basic unit of mass. It is the mass of an international prototype in the form of a platinum-iridium cylinder kept at Sevres in France. It is now the only basic unit still defined in terms of a material object, and also the only one with a prefix [kilo] already in place.

second [s]

The second is the basic unit of time. It is the length of time taken for 9192631770 periods of vibration of the caesium-133 atom to occur.

ampere [A]

The ampere is the basic unit of electric current. It is that current which produces a specified force between two parallel wires which are 1 meter apart in a vacuum. It is named after the French physicist Andre Ampere (1775-1836).

kelvin [K]

The kelvin is the basic unit of temperature. It is 1/273.16th of the thermodynamic temperature of the triple point of water. It is named after the Scottish mathematician and physicist William Thomson 1st Lord Kelvin (1824-1907).

mole [mol]

The mole is the basic unit of substance. It is the amount of substance that contains as many elementary units as there are atoms in 0.012 kg of carbon-12.

candela [cd]

The candela is the basic unit of luminous intensity. It is the intensity of a source of light of a specified frequency, which gives a specified amount of power in a given direction.

Derived Units of the SI

From the 7 basic units of the SI many other units are derived for a variety of purposes. Only some of them are explained here. The units printed in bold are either basic units or else, in some cases, are themselves derived.

farad [F]

The farad is the SI unit of the capacitance of an electrical system, that is, its capacity to store electricity. It is a rather large unit as defined and is more often used as a microfarad. It is named after the English chemist and physicist Michael Faraday (1791-1867).

hertz [Hz]

The hertz is the SI unit of the frequency of a periodic phenomenon. One hertz indicates that 1 cycle of the phenomenon occurs every **second**. For most work much higher frequencies are needed such as the kilohertz [kHz] and megahertz

[MHz]. It is named after the German physicist Heinrich Rudolph Hertz (1857-94).

joule [J]

The joule is the SI unit of work or energy. One joule is the amount of work done when an applied force of 1 newton moves through a distance of 1 **meter** in the direction of the force. It is named after the English physicist James Prescott Joule (1818-89).

newton [N]

The newton is the SI unit of force. One newton is the force required to give a mass of 1 kilogram an acceleration of 1 **meter** per **second**. It is named after the English mathematician and physicist Sir Isaac Newton (1642-1727).

ohm [Ω]

The ohm is the SI unit of resistance of an electrical conductor. Its symbol is the capital Greek letter 'omega'. It is named after the German physicist Georg Simon Ohm (1789-1854).

pascal [Pa]

The pascal is the SI unit of pressure. One pascal is the pressure generated by a force of 1 newton acting on an area of 1 square **meter**. It is a rather small unit as defined and is more often used as a kilopascal [kPa]. It is named after the French mathematician, physicist and philosopher Blaise Pascal (1623-62).

volt [V]

The volt is the SI unit of electric potential. One volt is the difference of potential between two points of an electical conductor when a current of 1 **ampere** flowing between those points dissipates a power of 1 **watt**. It is named after the Italian physicist Count Alessandro Giuseppe Anastasio Volta (1745-1827).

watt [W]

The watt is used to measure power or the rate of doing work. One watt is a power of 1 **joule** per second. It is named after the Scottish engineer James Watt (1736-1819).

Note that prefixes may be used in conjunction with any of the above units.

The Prefixes of the SI

The SI allows the sizes of units to be made bigger or smaller by the use of appropriate prefixes. For example, the electrical unit of a watt is not a big unit even in terms of ordinary household use, so it is generally used in terms of 1000 watts at a time.

The prefix for 1000 is kilo so we use kilowatts [kW] as our unit of measurement. For makers of electricity, or bigger users such as industry, it is common to use megawatts [MW] or even gigawatts [GW].

The full range of prefixes with their [symbols or abbreviations] and their multiplying factors which are also given in other forms is:

yotta [Y] 1 000 000 000 000 000 000 000 000 = 10^{24}
zetta [Z] 1 000 000 000 000 000 000 000 = 10^{21}
exa   [E] 1 000 000 000 000 000 000 = 10^{18}
peta [P] 1 000 000 000 000 000 = 10^{15}
tera [T] 1 000 000 000 000 = 10^{12}
giga [G] 1 000 000 000   (a thousand millions = a billion)
mega [M] 1 000 000   (a million)
kilo [k] 1 000   (a thousand)
hecto [h] 100
deca [da]10
          1
deci [d] 0.1
centi [c] 0.01
milli [m] 0.001   (a thousandth)
micro [μ] 0.000 001   (a millionth)
nano [n] 0.000 000 001   (a thousand millionth)
pico [p] 0.000 000 000 001 = 10^{-12}
femto [f] 0.000 000 000 000 001 = 10^{-15}
atto [a] 0.000 000 000 000 000 001 = 10^{-18}
zepto [z] 0.000 000 000 000 000 000 001 = 10^{-21}
y o c t o [y] 0 . 0 0 0   0 0 0   0 0 0   0 0 0   0 0 0   0 0 0   0 0 0   0 0 1
          = 10^{-24}

[μ] the symbol used for micro is the Greek letter known as 'mu'.

Nearly all of the SI prefixes are multiples or sub-multiples of 1000. However, these are inconvenient for many purposes and so hecto, deca, deci, and centi are also used.

deca also appears as deka [da] or [dk] in the USA and Contintental Europe.

Conventions of Usage in the SI

There are various rules laid down for the use of the SI and its units as well as some observations to be made that will help in its correct use.

Any unit may take only ONE prefix. For example "millimillimeter" is incorrect and should be written as "micrometer."

Most prefixes which make a unit bigger are written in capital letters (M, G, T, etc.), but when they make a unit smaller then lower case (m, n, p, etc.) is used. Exceptions to this are the kilo [k] to avoid any possible confusion with kelvin [K]; hecto [h]; and deca [da] or [dk].

A unit which is named after a person is written all in lower case (newton, volt, pascal, etc.) when named in full, but starting with a capital letter (N, V, Pa, etc.) when abbreviated. An exception to this rule is the liter which, if written as a lower case "l" could be mistaken for a "1" (one) and so a capital "L" is allowed as an alternative. It is intended that a single letter will be decided upon some time in the future when it becomes clear which letter is being favored most in use.

Units written in abbreviated form are NEVER pluralised. So"m" could always be either "meter" or "meters". "ms" could represent "meter second" (whatever that is) or, more correctly, "millisecond".

An abbreviation (such as J, N, g, Pa, etc.) is NEVER followed by a full-stop unless it is the end of a sentence.

To make numbers easier to read they may be divided into groups of 3 separated by spaces (or half-spaces) but NOT commas.

The SI preferred way of showing a decimal fraction is to use a comma (123,456) to separate the whole number from its fractional part. The practice of using a point, as is common in English-speaking countries, is acceptable providing only that the point is placed ON the line of the bottom edge of the numbers (123.456).

It will be noted that many units are eponymous, that is they are named after persons. This is always someone who was prominent in the early work done within the field in which the unit is used.

Metric System of Measurements

Length

10 millimeters = 1 centimeter
10 centimeters = 1 decimeter
10 decimeters = 1 meter
10 meters = 1 decameter
10 decameters = 1 hectometer
10 hectometers = 1 kilometer
1000 meters = 1 kilometer

Volume

1000 cu. mm = 1 cu. cm
1000 cu. cm = 1 cu. decimeter
1000 cu. dm = 1 cu. meter
1 million cu. cm = 1 cu. meter

Mass

1000 grams = 1 kilogram
1000 kilograms = 1 tonne

Area

100 sq. mm = 1 sq. cm
10 000 sq. cm = 1 sq. meter
100 sq. meters = 1 are
100 ares = 1 hectare
10 000 sq. meters = 1 hectare
100 hectares = 1 sq. kilometer
1 000 000 sq. meters = 1 sq. kilometer

Capacity

10 milliliters = 1 centiliter
10 centiliter = 1 deciliter
10 deciliters = 1 liter
1000 liters = 1 cu. meter

The distinction between "Volume" and "Capacity" is artificial and kept here only for historic reasons. A millitre is a cubic centimeter and a cubic decimeter is a liter.

The U K (Imperial) System of Measurements

Length

12 inches = 1 foot
3 feet = 1 yard
22 yards = 1 chain
10 chains = 1 furlong
8 furlongs = 1 mile
5280 feet = 1 mile
1760 yards = 1 mile

Area

144 sq. inches = 1 square foot
9 sq. feet = 1 square yard
4840 sq. yards = 1 acre
640 acres = 1 square mile

Volume

1728 cu. inches = 1 cubic foot
27 cu. feet = 1 cubic yard

Capacity

20 fluid ounces = 1 pint
4 gills = 1 pint
2 pints = 1 quart
4 quarts = 1 gallon (8 pints)

Mass (Avoirdupois)

437.5 grains = 1 ounce
16 ounces = 1 pound (7000 grains)
14 pounds = 1 stone
8 stones = 1 hundredweight [cwt]
20cwt = 1 ton (2240 pounds)

Troy Weights

24 grains = 1 pennyweight
20 pennyweights = 1 ounce (480 grains)
12 ounces = 1 pound (5760 grains)

Apothecaries' Measures

20minims = 1 fl.scruple
3 fl.scuples = 1 fl.drachm
8 fl.drachms = 1 fl.ounce
20 fl.ounces = 1 pint

Apothecaries' Weights

20 grains = 1 scruple
3 scruples = 1 drachm
8 drachms = 1 ounce (480 grains)
12 ounces = 1 pound (5760 grains)

The old Imperial (now UK) system was originally defined by three standard measures - the yard, the pound and the gallon which were held in London. They are now defined by reference to the SI measures of the meter, the kilogram and the liter. These equivalent measures are exact.

1 yard = 0.9144 meters - same as US
1 pound = 0.453 592 37 kilograms - same as US
1 gallon = 4.546 09 liters

Note particularly that the UK gallon is a different size to the US gallon so that NO liquid measures of the same name are the same size in the UK and US systems.

Also that the ton (UK) is 2240 pounds while a ton(US) is 2000 pounds. These are also referred to as a long ton and short ton, respectively.

The U.S. System of Measurements

Most of the US system of measurements is the same as that for the UK. The biggest differences to be noted are in Capacity which has both liquid and dry measures as well as being based on a different standard - the US liquid gallon is smaller than the UK gallon. There is also a measurement known at the US survey foot. It is gradually being phased out as the maps and land plans are re-drawn under metrication. (The changeover is being made by putting 39.37 US survey feet = 12 meters)

Length

12 inches = 1 foot
3 feet = 1 yard
220 yards = 1 furlong
8 furlongs = 1 mile
5280 feet = 1 mile
1760 yards = 1 mile

Area

144 sq. inches = 1 square foot
9 sq. feet = 1 square yard
4840 sq. yard = 1 acre
640 acres = 1 square mile
1 sq.mile = 1 section
36 sections = 1 township

Volume

1728 cu. inches = 1 cubic foot
27 cu. feet = 1 cubic yard

Capacity (Dry)

16 fluid ounces = 1 pint
2 pints = 1 quart
8 quarts = 1 peck
4 pecks = 1 bushel

Capacity (Liquid)

4 gills = 1 pint
2 pints = 1 quart
4 quarts = 1 gallon (8 pints)

Mass

437.5 grains = 1 ounce
16 ounces = 1 pound (7000 grains)
14 pounds = 1 stone

100 pounds = 1 hundred
20 cwt = 1 ton (2000 pounds)

Troy Weights
24 grains = 1 pennyweight
20 pennyweights = ounce(480 grains)
weight [cwt]12 ounces = 1 pound (5760 grains)

Apothecaries' Measures
60 minims = 1 fl.dram
8 fl.drams = 1 fl.ounce
16 fl.ounces = pint
12 ounces = 1 pound (5760 grains)

Apothecaries' Weights
20 grains = 1 scruple
3 scruples = 1 dram
8 drams = 1 ounce (480 grains)

As with the UK system these measures were originally defined by physical standard measures - the yard, the pound, the gallon and the bushel. They are now all defined by reference to the SI measures of the meter, the kilogram and the litre. These equivalent measures are exact.

1 yard = 0.9144 meters - same as UK
1 pound = 0.453 592 37 kilograms - same as UK
1 gallon (liquid) = 3.785 411 784 liters
1 bushel = 35.239 070 166 88 liters

Note particularly that the US gallon is a different size to the UK gallon so that NO liquid measures of the same name are the same size in the US and UK systems.

Also that the ton (US) is 2000 pounds while a ton (UK) is 2240 pounds. These are also referred to as a short ton and long ton respectively.

Note than in matters concerned with land measurements, for the most accurate work, it is necessary to establish whether the US survey measures are being used or not.

UN HAZARD CLASSES AND DIVISIONS

The hazard class of a material is indicated either by its class (or division) number, or its class name. For a placard corresponding to the primary hazard class of a material, the hazard class or division number must be displayed in the lower corner of the placard. The UN (United Nations) hazard classes are as follows:

Class 1   *Explosives*
   Division 1.1   Explosives with a mass explosion hazard
   Division 1.2   Explosives with a projection hazard
   Division 1.3   Explosives with predominantly a fire hazard
   Division 1.4   Explosives with no significant blast hazard

Division 1.5   Very insensitive explosives; blasting agents

Division 1.6   Extremely insensitive detonating   substances

Class 2   *Gases*

Division 2.1   Flammable gas

Division 2.2   Non-flammable, non-poisonous compressed gas

Division 2.3   Gas poisonous by inhalation

Division 2.4   Corrosive gas

Class 3   *Flammable liquid and Combustible liquid*

Class 4   *Flammable Solid; Spontaneously combustible material; and Dangerous when wet material*

Class 5   *Oxidizers and Organic Peroxides*

Division 5.1   Oxidizer

Division 5.2   Organic peroxide

Class 6   *Poisonous material and infectious substance*

Division 6.1   Poisonous materials

Division 6.2   Infectious substance

Class 7   *Radioactive material*

Class 8   *Corrosive material*

Class 9   *Miscellaneous hazardous material*

U-TUBE HEAT EXCHANGER

In the U-tube exchanger, a bundle of nested tubes, each bent in a series of concentrically tighter U-shapes, is attached to a single tubesheet, as illustrated in Figure 1. Each tube is free to move relative to the shell, and relative to one another, so the design is ideal for situations that accommodate large differential temperatures between the shellside and the tubeside fluids during service. Such flexibility makes the U-tube exchanger ideal for applications that are prone to thermal shock or intermittent service. As with other removable-bundle exchangers, the inside of the shell, and to the outside of the tubes.

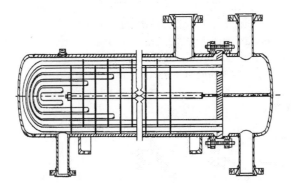

Figure 1. U-tube design.

However, unlike the straight-tube exchanger, whose tube internals can be mechanically cleaned, there is no way to physically access the U-bend region inside each tube, so chemical methods are required for tubeside maintenance. As a rule of thumb, non-fouling fluids should be routed through the tubes, while fouling fluids should be reserved for shellside duty. This inexpensive exchanger allows for multi-tube pass arrangements. However, because the U-tube cannot be made single pass on the tubeside, true countercurrent flow is not possible. Common TEMA designations are BEU and AEU, and typical applications include oil cooling, chemical condensing, and steam heating applications.

UREAS

The ureas are a group of chemical compounds that are generally compounded with wood flour as their chief reinforcement, providing a free-flowing granular compound that can be preformed, preheated, or preplasticated. These compounds can be readily molded with both compression and transfer molding methods, however they require special formulations when they are processed with screw injection equipment due to the need to extend their flow life at the higher barrel temperatures.

V

VACUUM FORMING

This term refers to a thermoforming process in which vacuum is used to provide atmospheric pressure to force material against the contours of a mold. PVC and PS blister packs are product examples. See *Thermoforming*.

VANADIUM CATALYST

Perchloroacetate and perchlorocrotonate improve catalytic activity for copolymerization of ethylene with propylene. There are numerous investigations on the preparation of various types of vanadium complexes with B-diketones, 0,8-triketones, ketophenols, and copolymerization of ethylene with propylene by the vanadium catalyst and activating agents such as butyl-perchlorocrotonate or ethyltrichloroacetate. The copolymerization is carried out in a liquid monomer system and in a heptane solution system. Attempted correlations of catalytic activity with the coordination geometry, coordination number of vanadium, and steric and electric effect of the ligands did not give regular trends. Neither improvement on the features of the copolymer produced nor on the catalyst activity have been reported over the commonly known vanadium catalysts. There are a large number of activating agents that can be used, such as butyl perchlorocrotonate, ethyl trichloroacetate,   trichloromethylbenzene,   and   1,1,2,2-tetrachloro-1,2-

diphenylethane. Productivity improvements are always observed. The results are interpreted on the basis of the enhanced oxidation of the vanadium(II) species by radical intermediates resulting from the reaction between tin (IV) hydrides and the chlorinated organic activator. Extremely short-lived radical species involved in the hydrostannolysis reaction of the chlorinated activator, probably plays a central role as demonstrated by the enhancement of the reaction rate constant when the oxidation of V(II) to V(III) by aralkyl halides is carried out in the presence of tributyltin hydride. The reader can refer to the following reference for a detailed discussion on this catalysts system for the production of EP rubbers: *Elastomer Technology Handbook*, N. P. Cheremisinoff - editor, CRC Press, Boca Raton, Florida, 1993.

VAPOR (GAS) SPECIFIC GRAVITY

The value is the ratio of the weight of vapor to the weight of an equal volume of dry air at the same conditions of temperature and pressure. Buoyant vapors have a vapor specific gravity less than one. The value may be approximated by the ratio M/29, where M is the molecular weight of the chemical. In some cases the vapor may be at a temperature different from that of the surrounding air. For example, the vapor from a container of boiling methane at -172°F sinks in warm air, even though the vapor specific gravity of methane at 60°F is about 0.6.

VARNISH

Solution of gum or of natural or synthetic resins that dries to a thin, hard, usually glossy film; it may be transparent, translucent, or tinted. Oil varnishes are made from hard gum or resin dissolved in oil. Spirit varnishes, e.g., shellac, are usually made of soft gums or resins dissolved in a volatile solvent. Enamel is varnish with added pigments. Lacquer may be either a synthetic or natural varnish. As a decorative or protective coating, varnish has been used for oil paintings, for string instruments, and, by the ancient Egyptians, for mummy cases.

VISCOELASTIC BEHAVIOR

A polymer at a specific temperature and molecular weight may behave as a liquid or a solid depending on the speed (time scale) at which its molecules are deformed. This behavior, which ranges between liquid and solid, is generally referred to as the viscoelastic behavior or material response.

Linear viscoelasticity is valid for polymer systems that are undergoing *small deformations*.

Non-linear viscoelasticity describes large deformations such as those encountered in flowing polymer melt.

In linear viscoelasticity the *stress relaxation test* is often used, along with the *time-temperature superposition principle* and the *Boltzmann superposition principle*, to explain the behavior of polymeric materials during deformation. In a stress relaxation test, a polymer test specimen is deformed a fixed amount and the stress required to hold that amount of deformation is recorded over time. This test is very cumbersome to perform, so the design engineer and the material scientist have tended to ignore it. In fact, the standard relaxation test ASTM D2991 was recently dropped by ASTM. Rheologists and scientists, however, have been consistently using the stress relaxation test to interpret the viscoelastic behavior of polymers. Stress relaxation is time and temperature dependent, especially around the glass transition temperature. It is well known that high temperatures lead to small molecular relaxation times and low temperatures lead to materials with large relaxation times.

When changing temperature, the shape of creep or relaxation test results remain the same except that they are horizontally shifted to the left or right, which represent lower or higher response times, respectively. The time-temperature equivalence seen in stress relaxation test results can be used to reduce data at various temperatures to one general master curve for a reference temperature, T_{ref}.

To generate a master curve at the reference temperature, the curves shown in the left of Figure 1 must be shifted horizontally, maintaining the reference curve stationary. Density changes are usually small and can be neglected, eliminating the need to perform tedious corrections. The master curve for the data in this plot is shown on the right side of the figure. Each curve was shifted horizontally until the ends of all the curves became superimposed. (Source: Osswald, T.A. and G. Menges, *Material Science of Polymers for Engineers*, Hanser Publishers, New York, 1996).

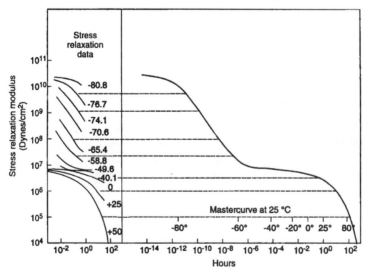

Figure 1. Relaxation curves and corresponding master curve at one temperature.

VISCOSITY INDEX (V.I.) IMPROVER

Lubricant additive, usually a high molecular-weight polymer, that reduces the tendency of an oil to change viscosity with temperature. Multi-grade oils, which provide effective lubrication over a broad temperature range, usually contain V.I. improvers.

VISCOSITY-TEMPERATURE RELATIONSHIP

Refers to the manner in which the viscosity of a given fluid varies inversely with temperature. Because of the mathematical relationship that exists between these two variables, it is possible to predict graphically the viscosity of a fluid at any temperature within a limited range if the viscosities at two other temperatures are known. Viscosity-temperature points of a fluid are located on the chart and a straight line drawn through them, other viscosity-temperature values of the fluid will fall on this.

VOLATILITY

Expression of evaporation tendency. The more volatile a liquid, the lower its boiling point and the greater its flammability. The volatility of solvents can be precisely determined by tests for evaporation rate; also, it can be estimated by tests for flash point and vapor pressure, and by distillation tests.

VULCANIZATION

Vulcanization is the process to chemically convert a polymer compound (rubber) to an elastic and final end-use product (i.e., in its final shape). The process is dependent on time, temperature and the processing method. Typical equipment used include a press, autoclave, oven, fluid beds, microwave, combinations of techniques.

The methods of vulcanization may be batch or continuous in nature. In continuous vulcanization modes for high volume productions, rubber article production is normally accomplished by extrusion, whereby the part is formed with a die, and the remainder of the process can be individually or a combination of curing processes such as the use of a shear head, a UHF line, hot air (oven drying method), fluid bed (using ballotini), or liquid curing medium (LCM). The final stages involve passing the extruded article through a cooling bath and parts finishing.

The following provides a generalized comparison of the continuous vulcanization modes.

	LCM	Hot Air	Fluid Bed	Shear Head	UHF
Stages	1 LCM Bath	3 Hot Air Ovens in Series	1 Fluid Bed	Shear Head Plus 1 Hot Air Oven	UHF Plus 2 Hot Air Ovens in Series
Total Length, m	35	35 x 3	30	20	35
Residence Time Index	100	400	120	150	180
Energy Consumption Index	100	140	105	125	133
Relative Investment Costs	100	140	105	125	133

The following is a generalized comparison of the advantages and the limitations of each of the continuous vulcanization modes. See also *Cure Time*.

Vulcanization Mode	Advantages	Limitations
LCM	Good Heat Transfer	Safety; Collapse Resistance of tubing and weatherstrips a problem; Salt Losses; Pollution
Hot Air	Easy Process	Poor Heat Transfer; No peroxide cures can be used with rubbers
Fluid Bed	Good Heat Transfer	Glass balls (they fracture - need replacement); No peroxide cures can be used with rubbers
Shear Head	Short Line; Low Energy Demand	Limited Profiles Range; No peroxide cures can be used with rubbers
UHF	Flexibility; Low Energy Demands, Clean process	No peroxide cures can be used with rubbers

VULCANIZING AGENTS

Vulcanization is a chemical reaction between rubber and a functional group, usually brought about by heat. The result is a product that is stronger, more elastic, more resilient, and less sensitive to temperature changes and the action of solvents than the original polymer. The first accelerators used were inorganic basic materials. They included basic lead carbonate, lime, magnesia, and litharge. Inorganic accelerators remained the most important type for many years. Today they have largely been superseded by organic accelerators because the latter are more powerful and produce vulcanizates of much greater strength and higher quality and durability.

Organic accelerators are most effective when combined with zinc oxide. The chemicals listed in below are organized alphabetically by chemical family name. Table 1 provides a list of these chemicals along with trade names and manufacturer's names.

Table 1

Trade Names of Vulcanizing Agents

Chemical Family	Chemicals	Trade names	Manufacturer
Alkaline Earth Oxides	Magnesium Oxide and Zinc Oxide	Liquispers MBZ	Basic Chemical Division, Basic Incorporated
Aliphatic Amines	Hexamethylene Diamine	Numerous	Numerous
Metal Oxides	Proprietary	Maglite D and K	Merck & Co., Inc.
Organic Peroxide	t-Butyl Perbenzoate, Dicurnyl Peroxide, Calcium Carbonate, Silica, t-Butyl Peroxymaleic Acid, t-Butylperoxy Valerate	t-Butyl Perbenzoate, Di-Cup R, Di-Cup 40C, Di- Cup 40KE, Luperco 130-XL, Luperco 10 1 -XL, Luperox PMA, Luperco 230-XL	Pennwalt Corp., Hercules, Inc.
Nonmetals and Sulfenamides	Sulfur, N-Oxydiediylenebenz othiazole-2-Sulfenamide, Insoluble Sulfur Oil treated	Amax, Crystex	R. T. Vanderbilt Co., Akzo Corp.

Table 1 continued

Chemical Family	Chemicals	Trade Names	Manufacturer
Miscellaneous	Dibenzoyl p-Quinonedioxime, Hcxamethylenediamine Carbamate, n,n'-Dicinnamylidene-1,6-Hexanediamine, Lead Oxides	Dibenzo GMF, Diak No. 1, Diak No. 3, Poly-Dispersion	Uniroyal Chemical, Inc., E. I. Du Pont de Nemours & Co., Wyrough and Loser, Inc., American Cyanamid Co.

The following provides key properties information on some of the major agents. Also, refer to the subject entry *Peroxides*.

Magnesium Oxide: *Synonyms*: Calcined Brucite; Calcined Magnesia; Calcined Magnesite; Gramnag; Magcal; Maglite; Magnesia; Magnesia USTA; Magnezu Tlenek (Polish); Magox; Pericalse; Seawater Magnesia; *CAS*: 1309-48-4; *Formula*: MgO; *Molecular Weight*: 40.31; *Chemical Class*: Metal oxide; *Physical Description*: White fume; *Specific Gravity*: 2.58.

Zinc Oxide: *Synonyms*: Flowers of Zinc, Zinc White, Chinese White; *CAS*: 1314-132; *Formula*: OZn; *Molecular Weight*: 81.37; *Chemical Class*: Metal oxide; *Physical Description*: Odorless, white or yellow powder; *Vapor Gravity*: 5.47.

Hexamethylene Diamine: *Synonyms*: 1,6-Hexanediamine; 1,6-Diaminohexane; Hexamethylenediamine; 1,6-Hexamethylenediamine; *CAS*: 124-09-4; *Formula*: $C_6H_{16}N_2$; *Molecular Weight*: 116.24; *Chemical Class*: Aliphatic amine; *Physical Description*: Colorless solid or watery liquid with a weak ammonia odor; *Boiling Point*: 204.8°C; *Melting Point*: 42°C; *Flash Point*: 80.8°C; *Specific Gravity*: 0.799 @ 60° C; *Water Solubility*: Freely Soluble.

Tert-butyl Peroxybenzoate: *Synonyms*: t-Butyl Perbenzoate; Tert-Butylperbenzoan (Czech); t-Butyl Peroxy Benzoate; Perbenzoate De Butyle Tertiaire (French); *CAS*: 614-45-9; *Formula*: $C_{11}H_{14}O_3$; *Molecular Weight*: 194.23; *Chemical Class*: Organic peroxide; *Physical Description*: Colorless to slightly yellow liquid; *Boiling Point*: 76° C; *Melting Point*: 8° C; *Flash Point*: 87.7° C; *Specific Gravity*: 1.0; *Water Solubility*: Insoluble.

Organic Peroxide Mixture: *Synonyms*: Di-Cup R Organic Peroxide; *Chemical Family*: Organic Peroxide; *CAS*: 80-43-3; *Ingredients And Composition Data*: 99 % Dicumyl Peroxide [peroxide, bis (alpha alpha-dimethylbenzyl)]; *Specific Gravity*: 0.9; *Vapor Pressure (mm Hg)*: 0.004 mm Hg at 500° C; *Physical Description*: Yellow, acrid color. Material is granular solid at room temperature; *Flash Point*: 127° C. In addition to this example, there are a number of other commercially available organic peroxide mixtures. These include Di-Cup Organic Peroxide (a mixture of Dicumyl Peroxide, Calcium Carbonate and Cumene), Di-Cup 40 kE Organic Peroxide (a mixture of Dicumyl Peroxide, Silane modified clay, and Cumene). The reader can refer to the following reference for detailed descriptions and chemical properties: Cheremisinoff, N. P., *Hazardous Chemicals in the Polymer Industry*, Marcel Dekker Inc., New York, 1994.

Dicumyl Peroxide: *Synonyms*: Active Dicumyl Peroxide; Bis(alpha,alpha-Dimethylbenzyl) - Peroxide; Cumene Peroxide; Curnyl Peroxide; Dicurnyl Peroxide; Dicurnyl Peroxide (DOT); Di-alpha-Cumyl Peroxide; DI-CUP; Diisopropylbenzene Peroxide; Isopropylbenzene Peroxide; *CAS*: 80-43-3; *Formula*: $C_{18}H_{22}O_2$; *Molecular Weight*: 270.40; *Chemical Class*: Organic peroxide; *Melting Point*: 41 °C; *Flash Point*; > 10 °C.

2,5-Dimethyl-2,5-Di(Tert-Butylperoxy)Hexyne-3: *Synonyms*: 2,5-Dimethyl-2,5-Di(t-Butylperoxy)Hexyne-3; 3-Hexyne, 2,5-Dimethyl-2,5-Di (T-Butylperoxy); *CAS*: 1068-27-5; *Formula*: $C_{16}H_{30}O_4$; *Molecular Weight*: 286.46; *Chemical Class*: Organic peroxide; *Boiling Point*: 67° C at 2 mm Hg; *Flash Point*: 85° C; *Specific Gravity*: 0.881.

2,5-Dimethyl-2,5-Di(Tert-Butylperoxy)Hexane: *CAS*: 78-63-7; *Formula*: $C_{16}H_{34}O_4$; *Molecular Weight*: 290.50; *Physical Description*: Colorless to light-yellow liquid; *Boiling Point*: 250 °C; *Melting Point*: 8 °C; *Specific Gravity*: 0.85; *Water Solubility*: Insoluble.

Tert-Butyl Peroxymaleic Acid: *Synonyms*: Luperox PMA; *Chemical Family*: Organic peroxide; *Melting Point*: 114 to 116° C.

Sulfur: *Synonyms*: Bensulfoid; Brimstone; Collokit; Colloidal-S; Colloidal Sulfur; Coisul; Corosul D; Corosul S; Cosan; Cosan 80; Crystex; Flour Sulfur; Flowers of Sulfur; Ground Vocle Sulfur; Hexasul; Kolofog; Kolospray; Kumulus; Magnetic 70; Magnetic 90; Magnetic 95; Microflotox; Precipitated Sulfur; Sofril; Sperlox-S; Spersul; Spersul Thiovit; Sublimed Sulfur; Sulfidal; Sulforon; Sulfur, Solid; Sulkol; Super Cosan; Sulphur; Sulfur (DOT); Sulsol; Technetium Tc 99M Sulfur Colloid; Tesuloid; Thiolux; Thiovit; *CAS*: 7704-34-9; *Formula*: S; *Molecular Weight*: 32.06; *Chemical Class*: Nonmetal; *Melting Point*: 119° C; *Specific Gravity*: 2.070.

W

WAX (PETROLEUM)

Any of a range of relatively high-molecular-weight hydrocarbons (approximately C_{16} to C_{50}) solid at room temperature, derived from the higher-boiling petroleum fractions. There are three basic categories of petroleum-derived wax: paraffin (crystalline), microcrystalline and petrolatum. Paraffin waxes are produced from the lighter lube oil distillates, generally by chilling the oil and filtering the crystallized wax; they have a distinctive crystalline structure, are pale yellow to white (or colorless), and have a melting point range between 48° C and 71° C. Fully refined paraffin waxes are dry, hard, and capable of imparting good gloss. Microcrystalline waxes are produced from heavier lube distillates and residua usually by a combination of solvent dilution and chilling. They differ from paraffin waxes in having poorly defined crystalline structure, darker color, higher viscosity, and higher melting points--ranging from 63° C to 93° C.

The microcrystalline grades also vary much more widely than paraffins in their physical characteristics: some are ductile and others are brittle or crumble easily. Both paraffin and microcrystalline waxes have wide uses in food packaging, paper coating, textile moisture proofing, candle-making, and cosmetics. Petrolatum is derived from heavy residual lube stock by propane dilution and filtering or centrifuging. It is microcrystalline in character and semi-solid at room temperature.

The best known type of petrolatum is the "petroleum jelly" used in ointments. There are also heavier grades for industrial applications, such as corrosion preventives, carbon paper, and butcher's wrap. Traditionally, the terms *slack wax,* and *refined wax* were used to indicate limitations on oil content. Today, these classifications are less exact in their meanings, especially in the distinction between slack wax and scale wax.

WAX APPEARANCE POINT (WAP)

Temperature at which wax begins to precipitate out of a distillate fuel, when the fuel is cooled under conditions prescribed by test method ASTM D 3117. WAP is an indicator of the ability of a distillate fuel, such as diesel-fuel, to flow at cold operating temperatures. It is very similar to cloud point.

WELDING

Some plastics can be joined by welding, in the same manner as metals--PVC and polyethylene tanks and ductwork being prime examples. More commonly, surfaces are joined by being brought into contact with one another and heated by conduction or by dielectric heating. Heat sealing of bags made from tubes of blow-extruded polyolefins such as polyethylene and polypropylene usually requires contact with a hot sealing bar. PVC has a high enough dielectric loss that heat can be generated throughout the material by exposure to a high-frequency, high-voltage electric field.

WHITE OIL

Highly refined straight mineral oil, essentially colorless, odorless, and tasteless. White oils have a high degree of chemical stability. The highest purity white oils are free of unsaturated components and meet the standards of the United States Pharmacopeia (USP) for food, medicinal, and cosmetic applications. White oils not intended for medicinal use are known as technical white oils and have many industrial applications--including textile, chemical, and plastics manufacture--where their good color, non-staining properties and chemical inertness are highly desirable.

X

X-RAY PHOTOELECTRON SPECTROSCOPY

X-ray photoelectron spectroscopy (XPS) is also commonly known as electron spectroscopy. This technique is used for chemical analysis. XPS is a versatile surface analytical technique with wide-ranging applicability. The sample to be analyzed can be a solid, a liquid, or a gas. Although polymers such as fluoropolymers and poly(vinyl chloride) have been shown to suffer X-ray damage after prolonged exposure, the technique is essentially nondestructive. It can provide useful chemical-state information, and interpretation of spectra is relatively straightforward. XPS has been used extensively in the study of polymer surfaces, including surface modification, polymer chain mobility, contamination, degradation, adhesion failure, chemical reaction, and biocompatibility. In polymer analyses, XPS has one noticeable limitation: lack of molecular specificity. For example, polyethylene and polypropylene cannot be differentiated by core-level XPS. Secondary ion mass spectrometry (SIMS), which has a very high degree of molecular specificity, is a powerful complementary technique. In the early development of XPS, one of its drawbacks was the need for a relatively large sampling area (a few millimeters by a few millimeters). With the development of small-area XPS systems, the analysis area can be as small as 20 mm in diameter. During an XPS experiment, a sample surface is irradiated with X-rays. An X-ray photon interacts with an inner-shell electron of an atom. The interaction causes a complete transfer of the photon energy to the electron, which then has enough energy to leave the atom and escape from the surface of the sample. This electron is referred to as a photoelectron.

XYLENE

Aromatic hydrocarbon, C_8H_{10}, with three isomers plus ethylbenzene. It is used as a solvent in the manufacture of synthetic rubber products, printing inks for textiles, coatings for paper, and adhesives, and serves as a raw material in the chemical industry.

Y

YIELD

Yield is defined conceptually as the onset of plastic deformation in a material under an applied load. A plastic deformation is one that remains after the load is removed and is also called a permanent or nonrecoverable deformation. By contrast, at small enough loads, deformation is elastic and is recovered after the load is removed (i.e., the specimen returns to its original length). Yield thus represents the transition from elastic to plastic deformation. As an example, consider a material under an applied tensile load. The length of the specimen will increase, as measured by the elongation or increase in length of the specimen, when the load is applied. Figure 1 is a schematic drawing of a typical tensile load-elongation curve. As the elongation increases, the load at first increases linearly but then increases more slowly and eventually passes through a maximum, where the elongation increases without any increase in load. The peak in the load-elongation curve is the point at which plastic flow becomes dominant and is commonly defined as the *yield point*. Not all load-elongation curves look like Figure 1, which is typical of a material that clearly exhibits plastic deformation prior to rupture. Other materials fracture before reaching a maximum, as shown in Figure 2. Even though no maximum is reached, there may be plastic deformation and it is usually assumed that yield occurs at some arbitrarily chosen value of strain e (elongation divided by original length). For polymers, the strain at which plastic deformation begins is often taken to be 2%, whereas in metals a value of 0.2% is common, and in ceramics 0.05% is used. One then draws a line parallel to the initial linear portion of the load-elongation curve but offset by the desired amount. The point where this line intersects the load-elongation curve is called the *offset yield point,* as seen in Figure 2. Use of an offset yield point is common in metals and ceramics where plastic deformation follows immediately after linear elastic deformation. Polymers, however, are more complex and generally exhibit considerable nonlinear elasticity before plastic deformation. Nonlinear elasticity also produces a deviation from the initial linear behavior but is fully recoverable. Since it is not obvious whether the deviation from linearity is caused by plasticity or nonlinearity, yield in polymers is preferably defined as the maximum in the load-elongation curve, if there is a maximum, and as the offset yield point only if there is no maximum. In some cases, the load rises linearly to fracture with no plastic deformation. The material is then said to be brittle, in distinction to the foregoing behavior where plastic deformation occurs and the material is said to be ductile. A given material will exhibit ductile or brittle behavior depending on the test conditions. A material that is brittle at low temperature will become ductile at high temperature. The brittle to ductile transition temperature is then an important material characteristic. For polyethylene, the temperature is about 125° K, depending somewhat on the specific polyethylene and the test conditions. A brittle to ductile transition is also observed in metals and ceramics. (Source: *Handbook of Polymer Science and Technology, Volume 2: Performance Properties of Plastics and Elastomers*, N. P. Cheremisinoff, Macrel Dekker Publishers, New York, 1989).

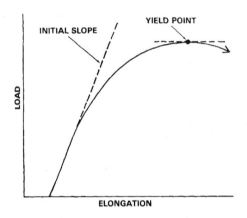

Figure 1. Typical load-elongation curve.

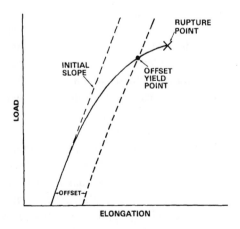

Figure 2. Analysis of load-elongation.

Z

ZIEGLER NATTA CATALYSTS

These are an important class of mixtures of chemical compounds remarkable for their ability to effect the polymerization of olefins (hydrocarbons containing a double carbon-carbon bond) to polymers of high molecular weights and highly ordered (stereoregular) structures. These catalysts were originated in the 1950s by the German chemist Karl Ziegler for the polymerization of ethylene at atmospheric pressure. Ziegler employed a catalyst consisting of a mixture of titanium tetrachloride and an alkyl derivative of aluminum. Giulio Natta, an Italian chemist, extended the method to other olefins and developed further variations of the Ziegler catalyst based on his findings on the mechanism of the polymerization reaction. The Ziegler-Natta catalysts include many mixtures of halides of transition metals, especially titanium, chromium, vanadium, and zirconium, with organic derivatives of nontransition metals, particularly alkyl aluminum compounds.

ZIEGLER POLYMERIZATION PROCESS

This type of polymerization is sometimes referred to as coordination polymerization since the mechanism involves a catalyst-monomer coordination complex or some other directing force that controls the way in which the monomer approaches the growing chain. The coordination catalysts are generally formed by the interaction of the alkyls of group I-III metals with halides and other derivatives of transition metals in groups IV-VIII of the periodic table. In a typical process the catalyst is prepared from titanium tetrachloride and aluminum triethyl or some related material. In a typical process ethylene is fed under low pressure into the reactor which contains liquid hydrocarbon to act as diluent. The catalyst complex may be prepared first and fed into the vessel or may be prepared in situ by feeding the components directly into the main reactor. Reaction is carried out at some temperature below 100° C (typically 70° C) in the absence of oxygen and water, both of which reduce the effectiveness of the catalyst through poisoning. The catalyst remains suspended and the polymer, as it is formed, becomes precipitated from the solution. A slurry is formed which progressively thickens as the reaction proceeds. Before the slurry viscosity becomes high enough to interfere seriously with removing the heat of reaction, the reactants are charged into a catalyst decomposition tank. Here the catalyst is destroyed by the action of ethanol, water or caustic alkali. In order to reduce the amount of metallic catalyst fragments to the lowest possible amount, the process of catalyst decomposition and subsequent purification steps become critical.

INDEX